COMMUNICATION ARCHITECTURES FOR SYSTEMS-ON-CHIP

Embedded Systems

Series Editor

Richard Zurawski

SA Corporation, San Francisco, California, USA

COMMUNICATION ARCHITECTURES FOR SYSTEMS-ON-CHIP

Edited by
José L. Ayala

CRC Press
Taylor & Francis Group
Boca Raton   London   New York

CRC Press is an imprint of the
Taylor & Francis Group, an **informa** business

CRC Press
Taylor & Francis Group
6000 Broken Sound Parkway NW, Suite 300
Boca Raton, FL 33487-2742

First issued in paperback 2017

© 2011 by Taylor and Francis Group, LLC
CRC Press is an imprint of Taylor & Francis Group, an Informa business

No claim to original U.S. Government works

ISBN 13: 978-1-138-11794-5 (pbk)
ISBN 13: 978-1-4398-4170-9 (hbk)

Visit the Taylor & Francis Web site at
http://www.taylorandfrancis.com

and the CRC Press Web site at
http://www.crcpress.com

Dedicado a quien me enseña cada día,
a quien me hace pensar,
a quien me sonríe y me apoya,
a quien me estimula y causa admiración.
Dedicado a ti.

Contents

List of Figures

List of Tables

Preface

The purpose of the book *Communication Architectures for Systems-on-Chip* is to provide a reference for the broad range of professionals, researchers, and students interested in the design of systems-on-chip and, in particular, with a clear emphasis on the technologies and mechanisms used to perform the data communication in these devices.

The book covers from what could be considered "traditional" communication architectures (buses) and novel architectures (networks-on-chip), to recent technologies (nanoelectronics, optics, radiofrequency, etc.) and design issues (security) found in system-on-chip coomunications.

This book is structured in eight chapters that provide a comprehensive knowledge to this area. It has contributors from several countries (Switzerland, Spain, Italy, United States, France) that come from industry and academia. These authors are directly involved in the creation and management of the ideas presented in this book, and both the research and implementation parts are covered. Therefore, all those interested in more practical or theoretical issues will find something interesting for them.

The book can be used as a reference for university (post)graduate courses and PhD students, as well for an advanced course on Systems-on-Chip. It is an excellent companion for those who seek to learn more about this technology and find themselves involved in their design from a professional or academic level.

Acknowledgments

My gratitude goes to Richard Zurawski who proposed me to get involved in this project and encouraged me to work on such an exciting challenge. All those who helped with identifying the contributing authors for the chapters on "technology," in particular to Professor Yuan Xie. They made my life much easier. I would like to express my gratitude to all the contributing authors for their tremendous cooperation and outstanding work. I would also like to thank my publisher Nora Knopka and other Taylor & Francis staff involved in the publication of this book, particularly Jennifer Ahringer. And thank you to all those so close to me who have suffered the countless hours that I spent working on this book.

Author

José L. Ayala got a MS in telecommunications engineering from Politecnica University of Madrid in 2001, and a MS in physics from the Open University of Spain in 2002. After that, he pursued his PhD in electronic engineering at the Department of Electrical Engineering of the Politecnica University of Madrid, getting in 2005 the "cum laude" qualification and the award for the "best PhD dissertation."

From that time, he has been actively working on research in the area of embedded systems design, with a special focus on the optimization of these systems for power and energy consumption, as well as thermal dissipation. In particular, he has developed hardware customizations for the power reduction and thermal optimization of embedded processors; he has provided compiler transformations for the reduction of thermal-driven reliably factors in the register file of complex microprocessor architectures; and he has proposed run-time OS-managed policies for the liquid-based cooling of multiprocessor systems.

Professor Ayala has collaborated and works with several professors of international universities: Professor Alexander Veidenbaum from the University of California Irvine, Professor Luca Benini from the University of Bologna, Professor Giovanni de Micheli from EPFL, Professor Francky Katthoor from IMEC, and so on. Also, he has spent several years in these institutions to gain expertise and work on the common research fields.

He has published numerous journal and conference papers that support his research. He also counts with several book chapters and international patents, and he has given various tutorials and invited talks in high-quality conferences. Professor Ayala is currently an associate professor at the Department of Computer Architecture in the Complutense University of Madrid, and a permanent visiting professor at EPFL Switzerland. He is a member of CEDA (Council for Electronic Design Automation), IEEE, IFIP (International Federation for Information Processing), TPC, and Organizing Committee of several international conferences (VLSI-SoC, GLSVLSI, etc.).

1

Introduction

José L. Ayala

Complutense University of Madrid, Spain

CONTENTS

1.1 Today's Market for Systems-on-Chip

For more than 20 years, integrated electronics has been the major new technological force shaping our everyday lives. Today's trend is shifting from personal computers to personal communication and computing, where system knowledge and expertise is now being encapsulated into single-chip solutions incorporating both hardware and software. This revolution is enabled and fueled by deep submicron Complementary Metal-Oxide-Semiconductor (CMOS) technologies, through which gigascale integration will be possible in the very near future.

The broad range of use of these electronic systems has allowed their integration in multiple application domains. We can find examples in small integrated electronic applications, like electronic tags and smartcards. Another field of application is the area of microcontrollers, where system-on-a-chip (SoC) can be found in common consumer-oriented devices, such as washing machines, microwave ovens, etc. Also, many computer peripherals (keyboards, hard drive controllers, etc.) rely on system-on-a-chip to perform their control operations. Automotive has also taken advantage of this technology, using these devices in air bags, Anti-lock Braking System (ABS), Electricity Supply Board (ESB), or engine control. However, those sectors that have benefited the most from the development of the embedded systems are communications and multimedia. The former includes mobile phones, network routers, modems, and software radio; while the latter includes cable and satellite TV decoders, HDTV, DVD players, and video games. According to In-Stat/MDR, the market for smart appliances in digital home experienced a 70% compound annual growth rate from 2001 to 2006 [19]. Moving forward, Gartner Market Report predicted that $500 millions market for SoC in 2005 will grow over 80% by 2010 [10]. The annual growth rate is about 2x faster than general-purpose microprocessors [3].

The well-organized integrated approach of multiple disciplines, such as device modeling, system design, and development of applications has motivated the success of system-on-a-chip (SoC) design. The continuous advances in integration technologies have also contributed by the rapid development of Very Large Scale Integration Circuits (VLSI) from the device perspective, due to the integration of billions of transistors on a single chip (see Figure 1.1) [12, 13, 17]. Therefore, a modern SoC can have billions of transistors, supporting a wide range of complex functions.

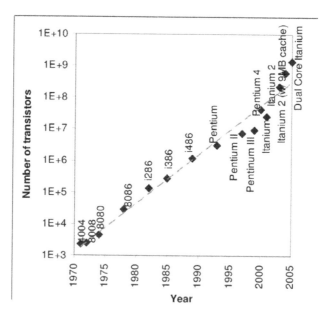

FIGURE 1.1
Transistors per die [1].

Moreover, 3D integration technologies increase the density of transistors per volume unit and target much more complex system architectures. The reasons for the development of 3D SoCs can be understood by the definition of much more complex applications that require the implementation of dense logic that cannot be fit in a traditional chip. Also, the current requirements of power consumption, memory bandwidth/latency, thermal management and testing are supposed to be managed in an easier way when the three dimensions are met. Other design constraints like verification and testing, design for reliability, or mitigation of the technology variations have now been incorporated into modern system design.

These modern design constraints are depicted in Figure 1.2, where these factors are classified according to their current state of development.

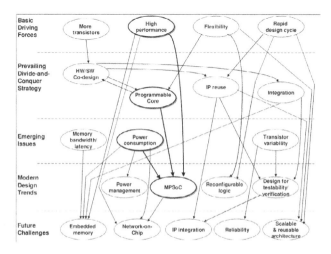

FIGURE 1.2
Design constraints in modern system-on-chip design [1].

The design of SoCs can be characterized by the combination of technology intensive techniques, to interface and interact with the physical world, and software intensive mechanisms, to create the desired functionality and integration. A lot of skills are needed to create these systems where all the emerging qualities of the system match with the needs of the user of the system. Along the following sections, we will analyze the ideas that define the system-on-chip design and the limitations that current designers find when using these methodologies.

1.2   Basics of the System-on-Chip Design

Systems design is the process of deriving, from requirements, a model from which a system can be generated more or less automatically. A model is an abstract representation of a system. These concepts can be applied to the software design, that consists of deriving a program that can be compiled; and to the hardware design, that consists of deriving a hardware description from which a circuit can be synthesized. In both domains, the design process usually mixes bottom-up and top-down activities: the reuse and adaptation of existing component models; and the successive refinement of architectural models in order to meet the given requirements.

Hardware systems are designed as the composition of interconnected, inherently parallel components. The individual components are represented by analytical models (equations), which specify their transfer functions. These models are deterministic (or probabilistic), and their composition is defined by specifying how data flows across multiple components. On the contrary, *software systems* are designed from sequential components, such as objects and threads, whose structure often changes dynamically (components are created, deleted, and may migrate). The components are represented by computational models (programs), whose semantics are defined operationally by an abstract execution engine (also called a *virtual machine*). Abstract machines may be nondeterministic, and their composition is defined by specifying how control flows across multiple components; for instance, the atomic actions of independent processes may be interleaved, possibly constrained by a fixed set of synchronization primitives.

Thus, the basic operation for constructing hardware models is the composition of transfer functions; the basic operation for constructing software models is the definition of an automata or finite state machine.

1.2.1 Main Characteristics of the Design Flow

The design flow of systems-on-chip can be characterized by several factors. These define the challenges and limitations that have to be fulfilled in the future definitions of the design flow.

1.2.1.1 Interoperability

Today's systems are required to be more and more directly connected. In the past, humans linked the different systems together, applying intelligence and adaptations where needed. In the electronic age the adaptation must take place in the systems themselves.

However, connectivity is not sufficient to justify the integration cost of complex systems; only if systems interoperate in an efficient way, is true added value created.

An additional challenge for interoperability are the many dimensions in which interoperability is required: applications, languages, vendors, networks, standards, and releases.

1.2.2 Reliability

The amount of software (and technology) in products is increasing exponentially. Many other products show the same growth in the amount of integrating software parts: mobile phones, portable multimedia devices, etc.

This software parts are far from errorless. Studies of the density of errors in actual code show that 1000 lines of code (LOC) typically contain 3 errors. In case of very good software processes and mature organizations this figure is 1 to 2 errors per kloc; in poor software designs, it can be much worse. Incremental increase of the code size will increase the number of hidden errors in a system also exponentially, as shown in Figure 1.3.

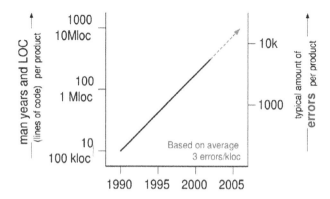

FIGURE 1.3
Reliability as opposed to software complexity [14].

1.2.3 Power Consumption

Many technological improvements show an exponential increase: circuit density, storage capacity, processing power, network bandwidth. An exception is the energy density in batteries, which has been improved lately, but much less dramatically than the increase on power consumption experienced by the electronic components. The power usage of processing per Watt has improved significantly, but unfortunately the processing needs have increased also rapidly, as well as the number of integrated transistors.

Figure 1.4 shows these opposing forces: the need for less power consumption and for more performance. There are multiple reasons to strive for less power consumption:

- less heat dissipation, easier transport of waste heat

- increased standby time

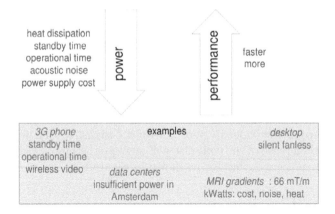

FIGURE 1.4
Power consumption and its relation with performance [14].

- increased operational time

- decreased acoustic noise of cooling

- decreased power supply cost

Power consumption plays a role in all domains, for example in Global System for Mobile Communications (GSM) phones (standby time, operational time, wireless video), data centers (some cities cannot accommodate more data centers, due to the availability of electrical power), Magnetic Resonance Imaging (MRI) gradients (faster switching of higher gradient fields, amplifier costs, acoustics, and cooling are significant technical challenges), and desktop PCs (where a current trend is to silent fanless desktops).

1.2.4 Security

Several stakeholders have significant different security interests. Figure 1.5 shows three categories with different interests and security solutions:

1. Government and companies, which implement restrictive rules, which can be rather privacy intrusive

2. Consumers, who want to maintain privacy and, at the same time, usability of services

3. The content industry, who wants to get fair payment for content creation and distribution. Their solution is again very restrictive, even violating the right of private copies, and characterized by a paranoia attitude: every customer is assumed to be a criminal pirate.

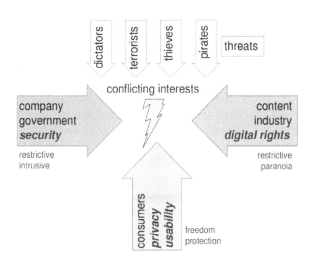

FIGURE 1.5
Conflicting interests in security [14].

All stakeholders are confronted with threats: pirates, thieves, terrorists, dictators, etc. The challenge is to find solutions that respect all the needs, not only the needs of one of the stakeholders. Another challenge is to make systems sufficiently secure, where "a little bit insecure" quickly means entirely insecure. Last but not least is the human factor, often the weakest link in the security chain.

1.2.5 Limitations of the Current Engineering Practices for System-on-Chip Design

One of the major limitations in the design of these systems is the dependence of the hardware and software models with the target implementation. Recent trends have focused on combining both language-based and synthesis-based approaches (hardware/software codesign) and on gaining, during the early design process, maximal independence from a specific implementation platform. We refer to these newer aproaches collectively as model-based, because they emphasize the separation of the design level from the implementation level, and they are centered around the semantics of abstract system descriptions (rather than on the implementation semantics). Consequently, much effort in model-based approaches goes into developing efficient code generators. Some examples can be found on the use of synchronous languages, such as Lustre and Esterel [5], and more recent modeling languages, such as United Modeling Language (UML) [16] and Architecture Analysis and Design Language (AADL) [2].

A common characteristic of all model-based designs is the existence of an effective theory of model transformations. Design often involves the use of multiple models that represent different views of a system at different levels of granularity. The traditional approaches for system design are the top-down, from the specifications to the implementation; and the bottom-up, that integrates library components until reaching the required system complexity. However, current practices are of a less directed fashion, and they iterate the model construction, model analysis, and model transformation. Some transformations between models can be automated; at other times, the designer must guide the model construction.

The ultimate success story in model transformation is the theory of compilation: today, it is difficult to manually improve on the code produced by a good optimizing compiler from programs (i.e., computational models) written in a high-level language. On the other hand, code generators often produce inefficient code from equation-based models: fixpoints of equation sets can be computed (or approximated) iteratively, but more efficient algorithmic insights and data structures must be supplied by the designer [7].

For extrafunctional requirements, such as timing, the separation of human-guided design decisions from automatic model transformations is even less well understood. Indeed, engineering practice often relies on a "trial-and-error" loop of code generation, followed by test, followed by redesign (e.g., priority tweaking when deadlines are missed). An alternative is to develop high-level programming languages that can express reaction constraints, together with compilers that guarantee the preservation of the reaction constraints on a given execution platform [6]. Such a compiler must mediate between the reaction constraints specified by the program, such as timeouts, and the execution constraints of the platform, typically provided in the form of worst-case execution times.

1.3 Open Issues

In spite of the rapid advances that have been incorporated in the design flow of systems-on-chip, there are still several open issues to be solved.

First of all, system-on-chip is a complex product and this is reflected in the design process. Moreover, the future SoC tend to be a Network-on-Chip (NoC), with much higher complexity [9]. NoC envisions systems with hundreds of processors with local and shared memory, communicating with each other via asynchronous communication links, similar to the switches and routers used in the Internet. In these systems, mechanisms are required to subdivide the application space among these processors and perform what-if scenarios to satisfy both performance and quality-of-service measures [8]. Also, it is necessary to create abstractions that model the process interaction at high levels to study their concurrency, communication, and synchronization issues [11]. And, finally, in the same environment, both software and hardware have to be modeled, using both top-down and bottom-up methods, and with verification techniques well integrated into the flow [15].

The second open issue in the design of systems-on-chip is the product development cost, which must be reduced. Increased cost prevents the consumer demand, that could be not sufficient to provide a healthy return on investment to the Electronic Design Automation (EDA) and the Systems (or high-tech) companies that justifies the investment in further innovation. Historically, system development tools of the past have been focused on software because the hardware platforms were standardized and stable. EDA tool companies may have shied away from this domain because of the low- cost, high- volume model, quite opposed to their modus operandi of high cost and low volume. On the other hand, today's embedded system development (ESD) companies may not have the expertise to deal with traditional hardware concepts of concurrency, communication, synchronization, hierarchy, and mixed mode design. System design companies may not appreciate the need for such wizardry by their tool vendors, and thus, may be unwilling to pay the price [18].

Moreover, certain visionary electronic system-level (ESL) tools developed by EDA companies failed because of some drawbacks: steep learning curve; lack of libraries of components; no back-end integration; and deficient reference designs and application notes. The past few years have also seen raging debates on the relative merits of SystemC and SystemVerilog, the two languages supported by the EDA industry for standardization. Development of efficient ESL tools will require the joint effort of EDA and ESD companies, a progress that currently is fairly slow.

Another factor to consider is that the integration issues are cross-disciplinary; that is, many changes made in one domain or discipline may effect the performance in another field. Analog and Radio Frequency (RF) designers are proficient in understanding and accounting for loading and interference effects, but this is a new concept for software and digital hardware developers. A delay in the time-to-market can happen frequently because a mismatch between both design groups generally leads to design iterations or a compromise on the product features.

The Systems Companies play an important role in the design flow of systems-on-chip. The hardware-centric model that led the industry scheme has changed to a software-centric model in such companies. Software development today occurs with ESD tools, which attain their fast simulation speed by abstracting hardware and ignoring technological and architectural variations. Concurrency issues may get ignored by software developers. Consequently, design iterations, integration, and verification may continue to be major productivity issues. Because of the pressure of quarterly deadlines, the engineers may be unwilling to invest time and effort to explore an immature technology. Further, systems companies may continue to view the technology issues as external variables to their domain; something to be taken care of by their intellectual property (IP) and application specific integrated circuit (ASIC) providers. However, the deep submicron (DSM) design, one essential actor in system-on-chip design, has plenty of analog characteristics in its behavior and cannot be ignored at the system level.

Also the universities present some specific issues to cope with in the near future. The universities had a great period during the past decade, with increasing enrollment in the computer science and engineering programs. Graduates got jobs soon after graduation. Thus, there was not any need to evaluate the future challenges and plan changes in the curriculum. Further, the U.S. university model of individualism makes cross-disciplinary team effort a challenge among research peers of equal stature. Federal funding, with its emphasis on research excellence in a single domain, creates silos and discourages faculty from venturing out to address general systems issues. The teaching-oriented faculty may not have the wherewithal to update the courses to include the latest topics. The practice-oriented faculty, typically adjunct faculty and affiliated with a local company, may be out of the loop with regard to curriculum development. Further, interdisciplinary communication is hampered by incompatible vocabulary, tools, and methodologies. These may lead to communication breakdowns and misunderstandings.

Finally, the Funding Agencies like SRC (Semiconductor Research Corporation) and NSF (National Science Foundation) have focused on mutually exclusive goals of technology and theory, and seem to have missed the all encompassing "practice" world of system-on-chip design [4]. However, there are many major cross-disciplinary intellectual challenges here and they should not be considered mere "practice" issues. The EDA and Systems companies may have put their relevant plans on hold because of the economic downturn. Further, many systems companies may have sought lateral opportunities to extend the reach of their current product portfolio, postponing the reckoning to a later quarter.

1.4 Conclusions

Systems-on-chip are a main component of today's market of electronic systems. The advances in integration technologies and software techniques have allowed their widespread use and application to the most diverse domains. Moreover, the successful integration of the hardware and software parts is key for the development of these systems.

The design flow of systems-on-chip is characterized by the need of models and abstractions at different levels, and a well-established theory that allows the efficient implementation, validation, and test of them. In particular, we need a mathematical basis for systems modeling, and analysis, which integrates both abstract-machine models and transfer-function models in order to deal with computation and physical constraints in a consistent, operative manner. However, the current trend of the systems technology has posed several contraints in the design that still have to be incorporated in the design flow.

In the following chapters of this book, we will review the modern and exciting design constraints and technologies that are being incorporated in this area. Moreover, the ways and mechanisms to alleviate the communication bottleneck in systems-on-chip will be reviewed, from the traditional communication buses, to the novel techniques based on RF and optical links.

1.5 Glossary

AADL: Architecture Analysis and Design Language

ABS: Anti-lock Braking System

EDA: Electronic Design Automation

ESB: Electricity Supply Board

ESD: Embedded System Development

ESL: Electronic System-Level

HDTV: High-Definition TV

LOC: Line of Code

MRI: Magnetic Resonance Imaging

SoC: System-on-Chip

1.6　Bibliography

[1] Y-K Chen and S. Y. Kung. Trend and challenge on system-on-a-chip designs. *J. Signal Process. Syst.*, 53(1-2):217–229, 2008.

[2] P.H. Feiler, B. Lewis, and S. Vestal. The SAE architecture analysis and design language (AADL) standard: A basis for model-based architecture-driven embedded systems engineering. In *RTAS Workshop on Model-driven Embedded Systems*, 2003.

[3] M. J. Flynn and P. Hung. Microprocessor design issues: Thoughts on the road ahead. *IEEE Micro*, 25(3):16–31, 2005.

[4] R. Gupta. EIC message: The neglected community. *IEEE Design and Test of Computers*, 19:3, 2002.

[5] N. Halbwachs. *Synchronous Programming of Reactive Systems*. Kluwer Academic Publishers, 1993.

[6] T. A. Henzinger, C. M. Kirsch, M. A. A. Sanvido, and W. Pree. From control models to real-time code using giotto. *IEEE Control Systems Magzine*, 23:50–64, 2003.

[7] T. A. Henzinger and J. Sifakis. The embedded systems design challenge. In *Proceedings of the 14th International Symposium on Formal Methods (FM), Lecture Notes in Computer Science*, 1–15. Springer, 2006.

[8] A. Jantsch. *Modeling Embedded Systems and SoCs: Concurrency and Time in Models of Computation*. Morgan Kaufmann, 2004.

[9] A. Jantsch and H. Tenhunen. *Networks on Chip*. Kluwer Academic Publishers, 2003.

[10] B. Lewis. SoC market is set for years of growth in the mainstream. Technical report, Gartner Market Report, October 2005.

[11] J. Magee and J. Kramer. *Concurrency: State Models and Java Programs*. John Wiley & Sons, 2000.

[12] G. E. Moore. Cramming more components onto integrated circuits. *Electronics*, 38(8):114–117, April 1965.

[13] G. E. Moore. No exponential is forever ... but we can delay. In *International Solid State Circuits Conference*, 2003.

[14] G. Muller. Opportunities and challenges in embedded systems. Technical report, Embedded Systems Institute, Eindhoven, The Netherlands, May 2010.

[15] W. Muller, W. Rosenstiel and J. Ruf, editors. *SystemC: Methodologies and Applications*. Kluwer Academic Publishers, 2003.

[16] J. Rumbaugh, I. Jacobson, and G. Booch. *The Unified Modeling Language Reference Manual*. Addison-Wesley, 2004.

[17] R. R. Schaller. Moore's law: Past, present, and future. *IEEE Spectrum*, 34(6):52–59, 1997.

[18] R. Shankar. Next generation embedded system design: Issues, challenges, and solutions. Technical report, Florida Atlantic University, 2008.

[19] Microwave Journal Staff. Smart appliances: Bringing the digital home closer to reality. *Microwave Journal*, 45(11):45, 2002.

2

Communication Buses for SoC Architectures

Pablo García del Valle

Complutense University of Madrid, Spain

José L. Ayala

Complutense University of Madrid, Spain

CONTENTS

2.1   Introduction

With powerful embedded devices populating the consumer electronics market, Multiprocessor Systems-on-Chip (MPSoCs) are the most promising way to keep on exploiting the high level of integration provided by the semiconductor technology while, at the same time, satisfying the constraints imposed by the embedded system market in terms of performance and power consumption.

A modern MPSoC usually integrates hundreds of processing units and storage elements in a single chip to provide both processing power and flexibility. In order to efficiently communicate the different elements, designers must define an interconnect technology and architecture. This is a key element, since it will affect both system performance and power consumption.

This chapter focuses on state-of-the-art SoC communication architectures, providing an overview of the most relevant system interconnects. Open bus specifications such as Open Core Protocol International Partnership Association (OCP-IP), Advanced Microcontroller Bus Architecture (AMBA), and CoreConnect will be described in detail, as well as the Software (SW) tools available to help designers in the task of defining a communication architecture.

The simple solution of one system bus interconnecting all the on-chip elements unfortunately suffers from power and performance scalability limitations. Therefore, a lot of effort is being devoted to the development of advanced bus topologies, like partial or full crossbars, bridged buses, and even packet-switched interconnect networks, some of them already implemented in commercially available products.

In this scenario, almost as important as the interconnect properties, is the existence of tools to help designers create a custom topology to meet their specific demands. Selecting and reconfiguring standard communication architectures to meet application specific performance requirements is a very time-consuming process, mainly due to the large exploration space created by customizable communication topologies, where many parameters can be configured at will (the arbitration protocols, clock speed, packet sizes, and so on).

Manufacturers have developed advanced tools to speed up the design process. They perform the automatic synthesis of communication architectures, helping designers to perform design space exploration and verification. Note that a system level model of the SoC and the communication architecture is highly desirable, since, when available, it can be used to estimate power consumption and get performance figures from the system. This is a difficult task, due to the high variability in data traffic exposed by these kind of systems.

We will also dedicate one section of the chapter to review other specific buses that present very different requirements, like those used in the automotive, avionics, and house automation industries. And, to conclude, we will take a look at the security options available to SoC designers, a recent concern.

2.1.1 Current Research Trends

Improvements in process technology have led to more functionality being integrated onto a single chip. This fact has also increased the volume of communication between the integrated components, requiring the design and implementation of highly efficient communication architectures [45].

However, selecting and reconfiguring standard communication architectures to meet application-specific performance requirements is a very time-consuming process. This is due to the large exploration space created by customizable communication topologies, arbitration protocols, clock speed, packet sizes, and all those parameters that significantly impact system performance. Moreover, the variability in data traffic exposed by these systems demands the synthesis of numerous buses of different types. This fact is also found in MPSoCs, where the complexity of the architecture and the broad diversity of constraints complicate their design [46]. Therefore, SoC design processes, which integrate early planning of the interconnect architecture at the system level [42] and the automatic synthesis of communication architectures [43], are a very active research area.

System designers and computer architects find it very useful to have a system-level model of the SoC and the communication architecture. This model can be used to estimate power consumption, get performance figures, and speed up the design process. Some communication models on multiple levels of abstraction have been proposed to overcome the simulation performance and modeling efficiency bottleneck [24, 44]. These models were not conceived under any standarization, until the Transaction-Level Modeling (TLM) [40] paradigm appeared, which allows to model system level bus-interfaces at different abstraction levels. TLM models employ SystemC as modelling language and, therefore, can rely on the support provided by a new generation of Electronic System Level (ESL) SoC design tools [48, 15].

2.1.2   Modeling and Exploring the Design Space of On-Chip Communication

The SoC functionality can be modeled at different abstraction levels, determined by the accuracy of communication, data, time and software modeling. From a detailed Hardware Description Language (HDL) model of the processor, up to an untimed functional specification of the system.

The basic model of a SoC using TLM includes a model of the processing unit (PU) with transactors as interfaces, as well as a model of the interconnection (also specified with transactors). Since the interconnection links have a strong impact on power consumption and power density, some of these interconnection models can include geometric parameters to define them, so that they are able to infer the power consumption of the simulated system. For the estimation of the power consumption, diverse approaches can be followed. At the system level, power consumption can be defined as a function of the number of messages that are sent through the interconnection between the PUs and the power wasted in the transfer of a single message (or instruction). At a lower abstraction level, the power consummed by the communication architecture will take into account the length and density of wires, the electrical response of the bus, and the distance between repeaters [50].

Once both the modeling of the processing elements and the interconnection architecture are completed, the resulting models are integrated in a single environment, which provides the simulation and design exploration capabilities.

With the detailed information provided by low-level models it is possible to obtain detailed reports of the communication infrastructure, in terms of power consumption or wire length [46]. In order to reduce the power consumption of the communication structures, several techniques have been devised, like applying voltage scaling to the buses that expose a higher-energy consumption, or restructuring the floorplan, so that the communications links are uniformly balanced.

2.1.3   Automatic Synthesis of Communication Architectures

Modern SoC and MPSoC systems have high bandwidth requirements that have to be addressed with complex communication architectures. Bus matrices and NoCs are some of the alternatives being considered to meet these requirements. Although using different approaches (a bus matrix consists of several buses in parallel, while the NoCs are made of multiple routing elements), both can support concurrent high bandwidth data streams. Regardless of their efficiency, the design complexity grows as the the number of processing units in the system increases.

In order to cope with this problem, automated approches for synthesizing advanced communication architectures are being developed. These approaches consider not only the topology of the system, but also the requirements of clock speed, interconnect area, and arbitration strategies [41].

2.2   The AMBA Interface

The Advanced Micro-Controller Bus Architecture (AMBA) is a bus standard that was originally conceived by ARM to support communication among ARM processor cores. However, nowadays AMBA is one of the leading on-chip busing systems because of its open access and its advanced features. it is licensed and deployed for use with third-party Intellectual Property (IP) cores [3].

The AMBA specification, introduced by ARM 15 years ago, provides standard bus interfaces for connecting on-chip components, custom logic, and specialized functions. These interfaces are independent of the ARM processor and generalized for different SoC structures.

The first AMBA buses (1996) were Advanced System Bus (ASB) and Advanced Peripheral Bus (APB) but, since its origins, the original AMBA specification has been significantly refined and extended to meet the severe demands of the quickly evolving SoC designs. In this way, the AMBA 2 Specification [3] appeared in 1999, adding the AMBA High-performance Bus (AHB), and, later on, AMBA 3 [4, 7, 6] was defined in 2003 to provide a new set of on-chip interface protocols able to interoperate with the existing bus technology defined in AMBA 2. As we will see in the next sections, the AMBA 3 specification defines four different interface protocols that target SoC implementations with very different requirements in terms of data throughput, bandwidth, or power.

The last version of the AMBA specification is AMBA 4 Phase One. Released in March, 2010, it provides more functionality and efficiency for complex, media-rich on-chip communication. Being an extension of the AMBA3 specification, the Advanced eXtensible Interface 4 (AXI4) protocol offers support for longer bursts and Quality of Service (QoS) signaling. It defines an expanded family of AXI interconnect protocols like AXI4, AXI4-Lite, and AXI4-Stream [9, 8].

Thanks to the collaboration with Xilinx, the new AMBA 4 specification and the AXI4 protocols meet the Field Programmable Gate Array (FPGA) implementation parameters. AXI4-Lite is a subset of the full AXI4 specification. This allows simple control register interfaces, thereby reducing SoC wiring congestion that facilitates implementation. The AXI4-Stream protocol provides a streaming interface for non-address-based, point-to-point communication like video and audio data.

The AMBA-4 specification has been written with contributions from 35 companies that include Original Equipment Manufactures (OEMs) and semiconductor and Electronic Design Automation (EDA) vendors. Some of the early adopters of these new specifications include Arteris, Cadence, Mentor, Sonics, Synopsys, and Xilinx.

AMBA 4 Phase Two, the next version of the AMBA specification, has already been announced, and will focus on simplifying Multicore SoC Design. It will aim at further maximizing performance and power efficiency by reducing traffic to the external memory adding hardware support for cache coherency and message-ordering barriers. By putting more capabilities in hardware, AMBA 4 will greatly simplify the software programmer's view of multicore systems.

Through the following sections we will briefly review the interface protocols proposed by AMBA. They can be summarized as:

- *AMBA AXI4 Interface*, that focuses on high performance high-frequency implementations. This burst-based interface provides the maximum of interconnect flexibility and yields the highest performance.

  AMBA AXI4-Lite Interface, is a subset of the full AXI4 specification for simple control register interfaces, reducing SoC wiring congestion and simplifying implementation.

  AMBA AXI4-Stream Interface, provides a streaming interface for non-address-based, point-to-point communication, such as video and audio data.

- *AMBA 3 AHB Interface*, that enables highly efficient interconnect between simpler peripherals in a single frequency subsystem where the performance of AMBA 3 AXI is not required. There are two flavors:

  AMBA 3 AHB-Lite, a subset of the whole AMBA 3 AHB specification, for designs that do not need all the AHB features.

  Multilayer AHB, a more complex interconnect that solves the main limitation of the AHB: enables parallel access paths between multiple masters and slaves in a system.

- *AMBA 3 APB Interface*, intended for general-purpose low-speed low-power peripheral devices, allowing isolation of slow data traffic from the high-performance AMBA interfaces.

- *AMBA 3 ATB Interface*, provides visibility for debug purposes by adding tracing data capabilities.

The following sections will describe these protocols in detail.

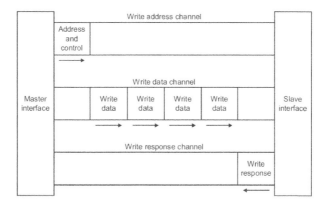

FIGURE 2.1
AXI write transaction.

2.2.1 AMBA 4 AXI Interface: AXI4

The Advanced eXtensible Interface (AXI^{TM}) [9] is designed to be used as a high-speed submicron interconnect. This high-performance protocol provides flexibility in the implementation of interconnect architectures while still keeping backward-compatibility with existing AHB and APB interfaces.

The AXI protocol is burst-based. Every transaction has address and control information on the address channel that describes the nature of the data to be transferred. The data is transferred between master and slave using a write data channel to the slave or a read data channel to the master. In write transactions (see Figure 2.1), in which all the data flow from the master to the slave, the AXI protocol has an additional write response channel that allows the slave to signal the master the completion of the write transaction.

AXI was first defined in AMBA 3. AXI4 is an incremental and backwards-compatible update to AXI3 that improves performance and interconnect utilization for multiple masters. It supports burst lengths of up to 256 beats (data packets), while previous support only went up to 16 beats. Long burst support enables integration of devices with large block transfers, and quality-of-service (QoS) signaling helps manage latency and bandwidth in complex multi-master systems. The new AXI4 protocol enables

- Simultaneous read and write transactions.

- Multiple transactions with out-of-order data completion.

- Burst transactions of up to 256 beats.

- Quality-of-service (QoS).

- Low-power operation (by including optional extensions).

- Pipelined interconnection for high-speed operation.

- Efficient bridging between frequencies for power management.

- Efficient support of high initial latency peripherals.

Two variations of the AXI4 exist, both of them with a clear focus on FPGA implementation: AXI4-Lite, a subset of the full AXI4 specification, and the AXI4-Stream protocol, for streaming communications.

2.2.1.1   AXI4-Lite

For simpler components, AMBA offers the AXI4-Lite version [9], a subset of the AXI4 protocol designed for communications between components with control registers. It greatly reduces the number of wires required to implement the bus. Designed with FPGAs in mind, it can also be used for custom chips.

In this protocol, all transactions have a burst length of one, all data accesses are the same size as the width of the data bus (32- or 64-bit), and exclusive accesses are not supported. Thanks to this specification, communication between AXI4 and AXI4-Lite interfaces is done through a single common conversion component.

2.2.1.2   AXI4-Stream

This protocol [8], unlike AXI4-Lite, is not a strict subset of AXI4. Designed also for FPGAs as the main target, it aims at greatly reducing signal routing for unidirectional data transfers from master to slave. The protocol lets designers stream data from one interface to another without needing an address. It supports single and multiple data streams using the same set of shared wires.

2.2.2   AMBA 3 AHB Interface

The AMBA 3 AHB interface specification enables highly efficient interconnect between simpler peripherals in a single frequency subsystem where the performance of AMBA 3 AXI is not required. Its fixed pipelined structure and unidirectional channels enable compatibility with peripherals developed for the AMBA 2 AHB-Lite specification. The main features of AMBA AHB can be summarized as follows:

- *Multiple bus masters.* Optimized system performance is obtained by sharing resources among different bus masters. A simple request-grant mechanism is implemented between the arbiter and each bus master. In this way, the arbiter ensures that only one bus master is active on the bus and, also, that when no masters are requesting the bus a default master is granted.

- *Pipelined and burst transfers.* Address and data phases of a transfer occur during different clock periods. In fact, the address phase of any transfer occurs during the data phase of the previous transfer. This over-lapping of address and data is fundamental to the pipelined nature of the bus and allows for high-performance operation, while still providing adequate time for a slave to provide the response to a transfer. This also implies that ownership of the data bus is delayed with respect to own-ership of the address bus. Moreover, support for burst transfers allows for efficient use of memory interfaces by providing transfer information in advance.

- *Split transactions.* They maximize the use of bus bandwidth by enabling high-latency slaves to release the system bus during dead time while they complete processing of their access requests.

- *Wide data bus configurations.* Support for high-bandwidth data-intensive applications is provided using wide on-chip memories. System buses support 32-, 64-, and 128-bit data-bus implementations with a 32-bit address bus, as well as smaller byte and half-word designs.

- *Non-tristate implementation.* AMBA AHB implements a separate read and write data bus in order to avoid the use of tristate drivers. In par-ticular, master and slave signals are multiplexed onto the shared com-munication resources (read and write data buses, address bus, control signals).

Like the original AMBA AHB system [3], it contains the following compo-nents:

AHB master: Only one bus master at a time is allowed to initiate and complete read and write transactions. Bus masters drive out the address and control signals and the arbiter determines which master has its signals routed to all the slaves. A central decoder controls the read data and response signal multiplexor. It also selects the appropriate signals from the slave that has been addressed.

AHB slave: It signals back to the active master the status of the pending transaction. It can indicate that the transfer completed successfully, that there was an error, that the master should retry the transfer, or indicate the beginning of a split transaction.

AHB arbiter: The bus arbiter serializes bus access requests. The arbitra-tion algorithm is not specified by the standard and its selection is left as a design parameter (fixed priority, round-robin, latency-driven, etc.), although the request-grant based arbitration protocol has to be kept fixed.

AHB decoder: This is used for address decoding and provides the select signal to the intended slave.

There are two variations of the AHB protocol: The AHB-lite, for simpler devices, and the Multilayer AHB, for more complex systems.

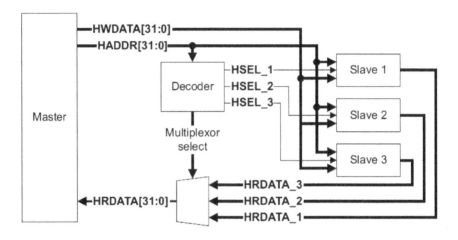

FIGURE 2.2
AHB-Lite block diagram.

2.2.2.1   AMBA 3 AHB-Lite

The AHB-Lite interface [6] is a subset of the whole AMBA 3 AHB specification. This simplified version also supports high bandwidth operation, and has the following features:

- Single bus master.

- Burst transfers.

- Single-clock edge operation.

- Non-tristate implementation.

- Wide data bus configurations, 64, 128, 256, 512, and 1024 bits.

The main components of the AHB-lite interface are the bus master, the bus slaves, a decoder and a slave-to master multiplexor, as can be seen in Figure 2.2.

The master starts a transfer by driving the address and control signals. These signals provide information about the address, direction, width of the transfer, and indicate if the transfer forms part of a burst. Transfers can be: (i) single, (ii) incrementing bursts that do not wrap at address boundaries, or (iii) wrapping bursts that wrap at particular address boundaries. The write data bus moves data from the master to a slave, and the read data bus moves data from a slave (selected by the decoder and the mux) to the master.

Given that AHB-Lite is a single master bus interface, if a multimaster system is required, an AHB-Lite multilayer structure is selected to isolate every master from each other.

2.2.2.2   Multilayer AHB Interface

The Multilayer AHB specification [5] emerges with the aim of increasing the overall bus bandwidth and provides a more flexible interconnect architecture with respect to AMBA AHB. This is achieved by using a more complex interconnection matrix that enables parallel access paths between multiple masters and slaves in a system.

Figure 2.3 shows a schematic view of the multilayer concept.

The multilayer bus architecture allows the interconnection of unmodified standard AHB or AHB-Lite master and slave modules with an increased available bus bandwidth. The resulting architecture becomes very simple and flexible: each AHB layer has only one master; hence, no arbitration and master-to-slave muxing is needed. Moreover, the interconnect protocol implemented in these layers can be very simple (AHB-Lite protocol, for instance): it does not have to support request and grant, nor retry or split transactions.

The additional hardware needed for this architecture with respect to the AHB is an interconnection matrix to connect the multiple masters to the peripherals. Point arbitration is also required when more than one master want to access the same slave simultaneously.

The interconnect matrix contains a decode stage for every layer in order to determine which slave is required during the transfer. A multiplexer is used to route the request from the specific layer to the desired slave.

The arbitration protocol decides the sequence of accesses of layers to slaves based on a priority assignment. The layer with lowest priority has to wait for the slave to be freed. Different arbitration schemes can be used, and every slave port has its own arbitration. Input layers can be served in a round-robin fashion, changing every transfer or every burst transaction, or based on a fixed priority scheme.

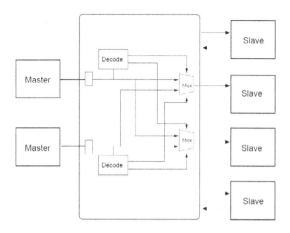

FIGURE 2.3
Schematic view of the multilayer AHB interconnect.

The number of input/output ports of the interconnect matrix is completely flexible and can be adapted to suit system requirements. However, as the number of masters and slaves implemented in the system increases, the complexity of the interconnect matrix can become significant. In this situation, different optimization techniques are suggested, in the AMBA specification to extend the capabilities of AHB-based designs: (i) defining multiple masters to share a single layer, (ii) making multiple slaves appear as a single slave to the interconnect matrix, or (iii) defining local slaves to a particular layer.

Because the multilayer architecture is based on the existing AHB protocol, previously designed masters and slaves can be totally reused.

2.2.3   AMBA 3 APB Interface

The Advanced Peripheral Bus interface (AMBA APB^{TM}) is intended for general-purpose low-speed low-power peripheral devices. It supports the low-bandwidth transactions necessary to access configuration registers and route data traffic among low bandwidth peripherals. Inside complex systems, the APB can appear as a subsystem connected to the main system bus via a bridge. All the bus devices are slaves, being the bridge the only peripheral bus master. Thus, it is able to isolate the data traffic from the high performance AMBA 3 AHB and AMBA 3 AXI interconnects.

The AMBA 3 APB interface is fully backwards compatible with the AMBA 2 APB. It is a static bus with unpipelined architecture, that provides a simple addressing, with latched address and control signals for easy interfacing. To ease compatibility with other design flows, all APB signal transitions only take place at the rising edge of the clock, requiring at least two cycles for every Read or Write transfer.

The main features of this bus are the following:

- Unpipelined architecture.

- Low gate count.

- Low power operation.

  - Reduced loading of the main system bus is obtained by isolating the peripherals behind a bridge.
  - Peripheral bus signals are only active during transfers.

The AMBA APB operation can be abstracted as a state machine like the one depicted in Figure 2.4, with three states. The default state for the peripheral bus is *IDLE*, that switches to the *SETUP* state when a transfer is required. The bus only remains in the SETUP state for one clock cycle, during which the peripheral select signal (PSELx) is asserted, and always moves to the *ACCESS* state on the next rising edge of the clock. The address, control, and data signals must remain stable during the transition from the *SETUP* to

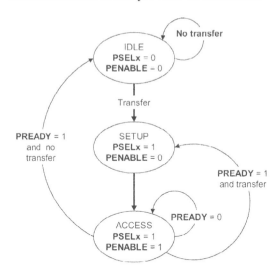

FIGURE 2.4
State diagram describing the operation of the AMBA APB bus.

the *ACCESS* state. The enable signal, PENABLE, is asserted in the *ACCESS* state. The bus remains in this state until the access is completed. At this moment, the PREADY signal is driven HIGH by the slave. Then, if other transfers are to take place, the bus goes back to the SETUP state, otherwise to IDLE.

As can be observed, AMBA APB should be used to interface to any peripherals that are low bandwidth and do not require the high performance of a pipelined bus interface. The simplicity of this bus results in a low gate count implementation.

2.2.4 AMBA 3 ATB Interface

The AMBA 3 ATB^{TM} interface specification adds a data diagnostic interface for tracing data in an AMBA system. The Advanced Trace Bus (ATB) is a common bus used by the trace components to pass format-independent trace data through the system. As depicted in Figure 2.5, both trace components and bus are placed in parallel with the peripherals and interconnects, providing visibility for debugging purposes.

The ATB interfaces can play two different roles depending on the sense of the trace data transfer: The interface is *Master* if it generates trace data, and *Slave* when it receives trace data. Some interesting features for debugging that are included in the ATB protocol are:

- Stalling of data, using valid and ready responses.

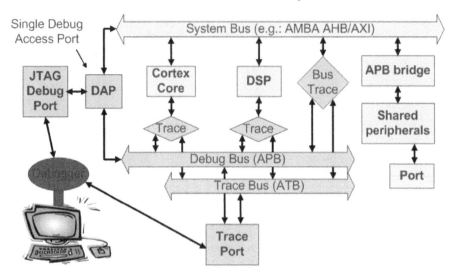

FIGURE 2.5
Schematic view of an AMBA system with trace and debug instrumentation.

- Control signals to indicate the number of bytes valid in a cycle.

- Labeling of the originating component since each data packet has an associated ID.

- Variable information formats.

- Identification of data from all originating components.

- Flushing.

2.2.5 AMBA Tools

To aid designers to build AMBA-based solutions, ARM and third party licensed companies make available advanced SW tools, as well as Hardware (HW) trace and debug elements aimed at rapid SoC prototyping.

2.2.5.1 ARM's CoreLink

ARM's CoreLink product includes interconnect, static and dynamic memory controllers, cache and DMA controllers, and AMBA design tools for easy and optimal SoC design. The CoreLink Interconnect products have been licensed by over 100 companies and shipped in millions of units to final consumers. Some of these CoreLink tools are:

- The AMBA Design Kit (ADK) is a mixed synthesizable and behavioral

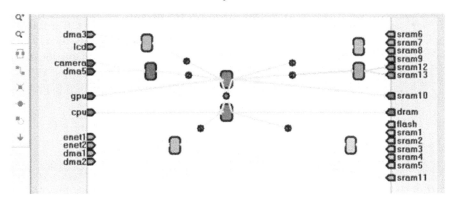

FIGURE 2.6
Snapshot of the tool configuring a CoreLink AMBA Network Interconnect.

package of AHB and APB components, test benches, example software, documentation, IP-XACT description files and microcontroller reference designs for AHB-based designs. For example, designers can use:

- The configurable AHB Bus Matrix element, to create an optimized multilayer AHB interconnect. The configuration and generation of the AHB Bus Matrix can be driven by a simple Perl script, or from the AMBA Designer tool.
- The File Reader Bus Master, can drive any bus transfer under the command of a text file. This can be used to test system integration, for example, to verify the memory map of a system by reading and writing to peripheral registers.

Besides these two examples, many other components are available in the AMBA Design Kit, such as Static Memory Controllers, Interrupt Controllers, Timers, AHB Downsizers, and several bridges (e.g., AHB Asynchronous Bridge, or AHB-to-APB Bridge).

- *CoreLink AMBA Designer,* makes it faster to configure and connect ARM Cortex^{TM} A, R, and M class processors, Mali^{TM} graphics hardware, and AMBA IP.

- *CoreLink Verification and Performance Exploration tool* is a tool that allows designers to optimize the performance of their AXI based System on chip, using generated traffic profiles running in RTL simulation.

- *CoreLink AMBA Network Interconnect.* For designs using AMBA 3 AXI, although AHB interfaces are also supported. Figure 2.6 shows a snapshot of the tool.

- *CoreLink Advanced Quality of Service*, is an option for the CoreLink^{TM} Network Interconnect product, and gives the SoC architect the essential tools to efficiently share the finite bandwidth of critical interfaces like Double Data Rate (DDR) memory. It performs traffic shaping to minimize average latency (best effort masters) or guarantee bandwidth and latency for real-time traffic.

2.2.5.2   ARM's CoreSight

For on-chip debug and trace, ARM has defined the open *CoreSight Architecture* to allow SoC designers to add debug and trace capabilities for other IP cores into the CoreSight infrastructure. ARM® CoreSight products include a wide range of trace macrocells for ARM processors, system and software instrumentation and a comprehensive set of IP blocks to enable the debug and trace of the most complex, multi-core SoCs (see Figure 2.5).

2.2.5.3   Third-Party Tools

Third-party tools are also available to be used independently, or itegrated with the ARM design tools, into one single design flow. To mention a couple of examples, the *Synopsys DesignWare IP solutions for AMBA® Interconnect protocol-based designs* include a comprehensive set of synthesizable and verification IP and an automated method for subsystem assembly with the Synopsys core Assembler tool. Mentor Graphics recently added AMBA 4 Verification IP to its *Questa Multi-View Verification Components Library*, and the *Platform Express* is supplied with an extensive range of example IP libraries that include a range of ARM processors, AMBA Peripherals, generators that create HDL for many of the AMBA System buses (and many others), and support for AXI and AMBA Verification IP.

2.3   Sonics SMART Interconnects

Sonics Inc. [27], is a premier supplier of system-on-chip interconnect solutions. Major semiconductor and systems companies including Broadcom, Samsung, Texas Instruments, and Toshiba have applied Sonics' Interconnect solutions in leading products in the wireless, digital multimedia, and communications markets, which include mobile phones, gaming platforms, HDTVs, communications routers, as well as automotive and office automation products. Sonics is a privately held company funded by Cadence Design Systems, Toshiba Corporation, Samsung Ventures and various venture capital firms among others.

The approach offered by this company is called SMART: *Sonics Methodology and Architecture for Rapid Time to Market*, and allows system designers to utilize predesigned, highly optimized and flexible interconnects to config-ure, analyze, and verify data flows at the architecture definition phase early in the design cycle. The SMART family offers different products to match designers' needs:

- *SNAP*: Sonics Network for AMBA Protocol designed for AMBA-based designs.

- *SonicsSX*: NoC communications for high-operating frequencies. It is a superset of the SonicsLX and SonicsMX solutions.

- *SonicsLX*: A crossbar-based structure for midrange SoC designs.

- *SonicsMX*: High-performance structure for multicore SoC.

- *S3220*: Interconnect devised for isolating low-speed peripherals.

- *SonicsExpress*: A high bandwidth bridge between two clock domains that allows connecting structures from other Sonics SMART intercon-nects.

All transactions in the SMART protocol can be seen as a combination of agents that communicate with each other using the interconnect. Agents isolate cores from each one and from the Sonics internal interconnect fabric. In this way, the cores and the interconnect are fully decoupled, allowing the cores to be re-used from system to system without rework. In the agent architecture we can distinguish *Initiators* and *Targets*:

Initiator: Who implements the interface between the interconnect and the master core (Central Processing Unit [CPU], Digital Signal Processing [DSP], Direct Memory Access [DMA]...). The initiator receives requests from the core, then transmits the requests according to the Sonics stan-dard, and finally processes the responses from the target.

Target: Who implements the interface between the physical interconnect and the target device (memories, Universal Asynchronous Receiver Transmitter [UARTs]...).

Intiator and target agents automatically handle any mismatch in data width, clock frequency, or protocol among the various SoC core interfaces and the Sonic interconnect with a minimum cost in terms of delay, latency, or gates.

As will be described in Section 2.3.6, Sonics' advanced SW tools allow designers to perform architectural exploration and configuration of the inter-connect to exactly match a particular SoC design.

Technologically, the SMART Interconnect solutions take advantage of the Practical Globally Asynchronous Locally Synchronous (GALS) approach. Practical GALS enables SoC developers to extend the capabilities to include

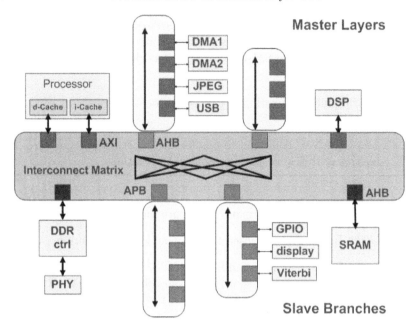

FIGURE 2.7
Typical SNAP Architecture.

voltage and domain isolation in an IP core or subsystem basis, which further facilitates low power and higher performance. Practical GALS do not disturb SoC design flows, which is typically the case when implementing asynchronous functionality.

2.3.1 SNAP: Sonics Network for AMBA Protocol

Sonics Network for AMBA Protocol (SNAP) [28], offers SoC developers a low-cost easy-to-configure bus solution specifically designed to simplify AMBA-based designs. The SNAP architecture consists of AHB master layers, AHB/APB slave branches and an interconnect matrix that provides better performance with efficient power consumption than traditional AHB architectures. An example of a SNAP architecture can be seen in Figure 2.7.

SNAP's interconnect matrix permits direct blending of AHB, AXI, and OCP masters and APB, AHB, AXI, and OCP slaves without the large latency penalties often associated with bridging. Thus, it is ideal for users looking to improve AHB bus performance, facing memory bottlenecks or needing to integrate AXI with AHB/APB.

SNAP suits embedded SoCs targeted for a wide range of applications, including both wireless and wired communications, consumer electronics (personal media players), and automotive control.

The main features of the SNAP architecture are:

- Naturally supports simultaneous accesses to different targets and multiple requests.

- Master layers can support up to 8 AHB masters per layer.

- Slave branches support up to 16 AHB/APB slaves per branch.

- Custom arbitration in the AHB master layers allows concurrent bus accesses for each core, and performance improvements.

- Low-latency crossbar simplifies multilayer AHB designs: High performance cores and memory can be connected directly to the interconnect matrix allowing the lowest latency and best performance.

- Supports a wide range of popular interfaces. No need for extra bridges or additional validation: All protocol, data widths, and clock conversions are handled automatically.

- Optional pipeline registers can be included to support high frequencies.

2.3.2 SonicsSX Interconnect

SonicsSX Smart Interconnect [33] is a superset of the current SonicsMX and SonicsLX solutions. It consists of silicon intellectual property (IP) and software tools that help users to build network-on-chip fabrics for systems-on-chip. Designed for SoCs requiring high quality, high definition, or HQHD, video support, SonicsSX accelerates video performance and eases global integration of intellectual property cores and subsystems onto a single chip.

First introduced in July 2008, SonicsSX is Sonics' highest performance product with up to 16GB/s bandwidth per port and it is ideal for solving the data throughput requirements of today's powerful SoCs (like real-time HQHD decoding). SonicsSX network-on-chip operates at high clock frequencies and provides native support for 2D data transactions and an expanded data bus width of 256 bits. With the addition of Sonics' Interleaved Multichannel Technology (IMT), enables SoC architectures to transition from single to multiple Dynamic Random Access (DRAM) channels, or multichannel, while automatically balancing the traffic among the channels.

SonicsSX has built-in security features that provide protection for digital media. Also, its low-power circuitry ensures extremely low active power for mobile battery-powered applications.

2.3.3 SonicsLX Interconnect

The SonicsLX SMART Interconnect [30] was conceived to target medium complexity SoCs. Based on a crossbar structure, SonicsLX provides a fully configurable interconnect fabric that supports transport, routing, arbitration, and translation functions. It is fully compatible with other interfaces, like the AMBA AXI/AHB or the Open Core Protocol, what ensures maximum reuse of cores regardless of their native configuration.

SonicsLX supports multithreaded, fully pipelined and nonblocking communications with a distributed implementation. This interconnect utilizes state-of-the-art physical structure design and advanced protocol management to deliver guaranteed high bandwidth together with fine grained power management.

Based on the agent structure, SonicsLX also presents decoupling of the functionality of each core from the communication interconnects required among the cores. It supports SoC cores running at different clock rates and establishes independent request and response networks to adapt to targets with long or unpredictable latency such as DRAM. The interconnect topology can be tuned to meet the specific constrains required by a particular application. As an example, SonicsLX supports high-definition video.

2.3.4 SonicsMX Interconnect

The SonicsMX SMART Interconnect [31] contains a full set of state-of-the-art fabric features and data flow services as dictated by the requirements of high-performance SoC designs. In particular, SonicsMX can perfectly face the design of low-power, cost-effective SoC devices powering multimedia-rich wireless and handheld products. It provides the physical structures, advanced protocols, and extensive power management capabilities necessary to overcome data flow and other design challenges of portable multicore SoCs. Its flexibility lets SoC developers fine-tune the interconnect specifically for their use and manage all communications and multimedia convergence data flow challenges. Compliance with OCP, AHB, and AXI interfaces is guaranteed, resulting in a maximum reuse.

SonicsMX supports crossbar, shared link, or hybrid topologies within a multithreaded and nonblocking architecture. It combines advanced features such as multiple clock rates, mixed latency requirements, Quality of Service (QoS) management, access security and error management, with low-power operation, high bandwidth and flexibility.

A typical example of application of the SonicsMX interconnect in a SoC is depicted in Figure 2.8. In this example, both crossbar and shared link structures are used in a system with seven cores accessing to the SonicsMX interconnect. Initiator and target agents are used to isolate the cores from the interconnect structures.

2.3.5 S3220 Interconnect

Sonics3220 SMART Interconnect [29] is perfect for low complexity SoCs because of its mature, low-cost structure. It is often used as an I/O interconnect

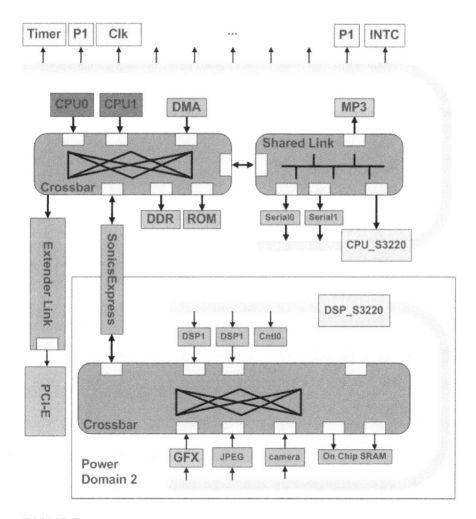

FIGURE 2.8
Example of system with a SonicsMX Interconnect including a crossbar and a shared link.

that off-loads slow transfers from the main system interconnect inside more complex SoCs.

Providing low latency access to a large number of low bandwidth, physically dispersed target cores, Sonics3220 is fully compatible with IP cores that support AMBA and OCP standards. Thus, providing the ability to decouple cores to achieve high IP core reuse. Using a very low die area interconnect structure facilitates a rapid path to simulation.

As other Sonics SMART Interconnects, the S3220 is a nonblocking peripheral interconnect that guarantees end-to-end performance by managing data, control, and test flows between all connected cores. By eliminating blocking, Sonics3220 allows multiple transfers to be in flight at the same time while there is no need for a multilayered bus architecture. For example, allowing latency-sensitive CPU traffic to bypass DMA-based I/O traffic.

2.3.6 Sonics Tools

2.3.6.1 SonicsStudio

All SMART solutions rely on the SonicsStudio^{TM} [32] development environment for architectural exploration, and configuration of the interconnect to exactly match a particular SoC design. The use of this tool significantly reduces the development time and schedule risks, that grow as the interconnect becomes more complex. The availability of precharacterization results enables reliable performance analysis and reduction of interconnect timing closure uncertainties. Furthermore, SonicsStudio can be useful to analyze data flows and execute performance verification testing.

SonicsStudio exploits the high scalability of the SMART Interconnects solutions to allow SoC developers to rapidly predict, implement, and validate the frequency, area, power, and application performance of a wide-range of potential product architectures. The ultimate goal is to avoid overdesigning interconnects.

This development environment accelerates every stage of System-On-Chip design; from architectural exploration to synthesis. SonicsStudio provides an array of graphical and command line tools that bring SoC architects and designers a single environment within which the entire SoC can be assembled, configured, simulated, and finally netlist generated. A typical integrated design flow using SonicsStudio can be observed in Figure 2.9. Nevertheless, they support as well industry standard simulation, verification, and synthesis tools.

Using SonicsStudio, designers can instantiate all of the SoC cores, configure the interconnect, create functional models for those IP cores not yet available, place monitors, stimulate the SoC components, and analyze the resultant performance of the interconnect using either SystemC or RTL. Within minutes, changes can be made to the interconnect, or the whole SoC can be re-architected for subsequent analysis. In this way, SoC architects can rapidly decide on the optimal SoC architecture that meets design and customer re-

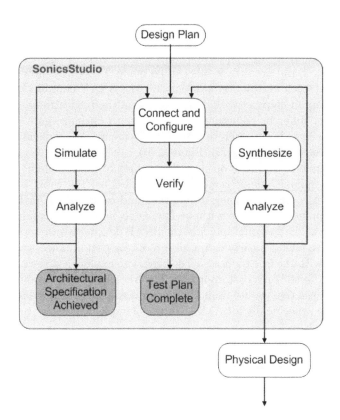

FIGURE 2.9
SonicsStudio^{TM} Development Flow.

quirements. The available SystemC model facilitates system modeling and verification using electronic system level (ESL) products from electronic design automation vendors. It can deliver outputs for ESL or Register Transfer Language (RTL) environments, so that the same source is used for development. In this manner, software engineers and system architects can work in parallel.

2.3.6.2   SNAP Capture Tool

For the SNAP interconnect, Sonics has created *The SNAP Capture Tool* [27], a design capture tool that provides a simplified and automated design flow to support architecture exploration of custom SNAP designs, so that performance, power and area trade-offs can be made. Designers enter a few parameters about each core in the system and the SNAP compiler automatically generates an on-chip network design based on these constrains. Next, designers can test different configurations in order to optimize gate counts, tweak performance needs and achieve power requirements. An snapshot of the SNAP graphical interface is shown in Figure 2.10.

Sonics makes available on its website a set of downloadable free tools for AMBA-compliant bus interconnect in SoC designs. The tools, a subset of the Sonics SNAP, allow designers to capture and evaluate an AMBA-compliant bus interconnect structure without having to purchase an evaluation license from Sonics. The free offer includes the SNAP design capture tool. In addition, the freeware includes early estimation tools for performance, gate-count, and power. So with the free tools a SoC designer can explore a range of interconnect options and settle on a best approach. The free capture and evaluation tools do not have locks that would limit the size or complexity of the design, but they are intended specifically for AMBA-protocol implementations, not proprietary bus protocols.

The code generation tool that actually creates an interconnect implementation is not included in the free package. Therefore, once a team settles on an implementation, they would need to license the full SNAP package from Sonics in order to complete the design with Sonics interconnect technology. A team could use the free tools to evaluate alternative AMBA architectures, and then implement the bus with ARM or in-house IP. But Sonics IP includes some concurrency and quality-of-service features that would not be trivial to implement in standard ARM IP.

2.4   CoreConnect Bus

CoreConnect is an IBM-developed on-chip bus architecture that eases the integration and reuse of processor, subsystem and peripheral cores within

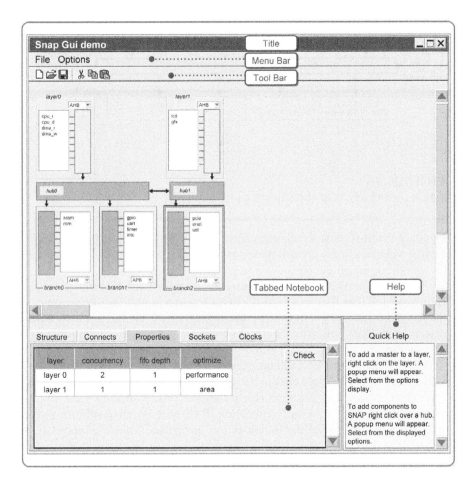

FIGURE 2.10
SNAP Capture Tool graphical interface.

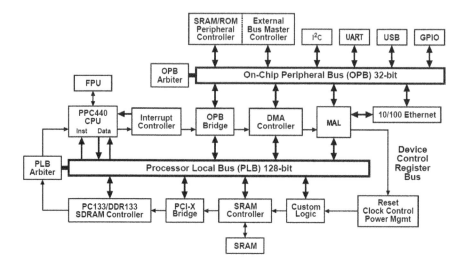

FIGURE 2.11
Schematic structure of the CoreConnect bus.

standard product platform designs. It is a complete and versatile interconnect clearly targeting high performance systems, and many of its features might be overkilling in simple embedded applications.

The CoreConnect bus architecture [25] serves as the foundation of IBM Blue Logic^{TM}. The Blue Logic ASIC/SOC design methodology is the approach proposed by IBM [18] to extend conventional ASIC design flows to current design needs: low-power and multiple-voltage products, reconfigurable logic, custom design capability, and analog/mixed-signal designs.

IBM's Blue Logic IP Collaboration Program promotes close working relationships with leading third-party Intellectual Property (IP) vendors and their ASIC customers. Through this program, IBM is working with suppliers of leading processor cores and application IP to integrate their solutions within IBM's Blue Logic design methodology and processes. This is intended to enable IBM's ASIC customers to more rapidly and reliably bring new products to market.

The CoreConnect bus is available as a no-fee, no-royalty architecture to tool vendors, core IP companies, and chip development companies. In the past decade, it has been freely licensed to over 1500 electronics companies such as Cadence, Ericsson, Lucent, Nokia, Siemens, and Synopsys.

Being an integral part of IBM's Power Architecture, it is used extensively in their PowerPC 4x0 based designs. Xilinx, for example, uses CoreConnect as the infrastructure for all of their embedded processor designs even though only a few are based on Power Architecture (more on Section 2.6, Xilinx FPGAs). To ease compatibility, with each Power Design Kit, IBM provides a bridge to the I/O core solutions of semiconductor IP supplier ARM. This enables ARM solutions to be used in Power Architecture technology-based designs. IBM and IBM Business Partners provide native CoreConnect I/Os for all latency-sensitive IP so that SoC designs can be optimized. Peripheral options include memory controllers, direct memory access (DMA) controllers, PCI interface bridges and interrupt controllers. When combined with SystemC models, available peripherals and design services, a complete PowerPC solution from system-level design to implementation can be realized.

As can be seen in Figure 2.11, the IBM CoreConnect architecture provides three buses for interconnecting cores, library macros, and custom logic:

- Processor Local Bus (PLB). The PLB bus is a high-performance bus that connects the processor to high-performance peripherals, such as memories, DMA controllers, and fast devices. The PLB on-chip bus is used in highly integrated systems. It supports read and write data transfers between master and slave devices equipped with a PLB interface and connected through PLB signals.

- On-Chip Peripheral Bus (OPB). Lower-performance peripherals are attached to the on-chip peripheral bus. A bridge is provided between the PLB and OPB to enable data transfer by PLB masters to and from OPB slaves.

- Device Control Register (DCR) Bus. It is a separate control bus that connects all devices, controllers, and bridges and provides a separate path to set and monitor the individual control registers. It is designed to transfer data between the CPU's general purpose registers and the slave logic's device control registers. It removes configuration registers from the memory address map, which reduces loading and improves bandwidth of the PLB.

Figure 2.11 illustrates how the CoreConnect architecture can be used to interconnect macros in a PowerPC 440 based SOC. High-performance, high-bandwidth blocks such as the PowerPC 440 CPU core, PCI-X Bridge and PC133/DDR133 SDRAM Controller reside in the PLB, while the OPB hosts lower data rate peripherals. The daisy-chained DCR bus provides a relatively low-speed data path for passing configuration and status information between the PowerPC 440 CPU core and other on-chip macros.

This architecture shares many high-performance features with the AMBA Bus specification. Both architectures allow split, pipelined and burst transfers, multiple bus masters and 32-, 64-, or 128-bits architectures. All Coreconnect buses are a fully synchronous interconnect with a common clock, but their devices can run with slower clocks, as long as all of the clocks are synchronized with the rising edge of the main clock.

2.4.1   Processor Local Bus (PLB)

The PLB is the main system bus targeting high performance and low latency on-chip communication. More specifically, PLB is a synchronous, multi-master, arbitrated bus. It supports *concurrent read and write transfers*, and implements *address pipelining*, that reduces bus latency by overlapping requests with an ongoing transfer [26]. Currently, two versions of the PLB co-exist: v4.x, called PLB4, and v6.0, also known as PLB6.

Access to PLB is granted through a central arbitration mechanism that allows masters to compete for bus ownership. This arbitration mechanism is flexible; with four levels of request priority for each master, allows for different *priority schemes* to be implemented. Additionally, an arbitration locking mechanism is provided to support master-driven atomic operations.

The PLB macro is the key component of PLB architecture, and consists of a bus arbitration control unit and the control logic required to manage the address and data flow through the PLB. Each *PLB master* is attached to the PLB through *separate address, read data and write data buses* and a plurality of transfer qualifier signals, while *PLB slaves* are attached through *shared, but decoupled, address, read data and write data buses* (each one with its own transfer control and status signals).

The separate address and data buses from the masters allow simultaneous transfer requests. The PLB arbitrates among them and sends the address, data and control signals from the granted master to the slave bus. The slave response is then routed back to the appropriate master. Up to 16 masters are supported by the arbitration unit, while there are no restrictions in the number of slave devices.

2.4.1.1   PLB6

With the purpose to support coherency in multiple-core designs, IBM released in October 2009 the CoreConnect PLB6 On-Chip System Bus specification, that includes the following key features:

- Support for concurrent traffic in read and write data buses.

- Support for concurrent transfers in all segments.

- Configurable bus controller that supports coherent and noncoherent masters and slaves.

- Symmetric multiprocessor (SMP) coherency with seven cache states (they include modified-exclusive-shared-invalid (MESI) states plus three cache intervention states).

- Out-of-order read data.

With each bus agent port including a 128-bit read data bus and a 128-bit write data bus, the point-to-point maximum data bandwidth is 25.6 GBps at 800 MHz. An optional coherency module can be instantiated, providing data transfers, cache, and translation lookaside buffer (TLB) coherency. The PLB6 bus controller core supports a maximum of 16 masters, with up to eight coherent masters. On the slave side, up to eight slave segments can be instantiated, with only one coherent slave segment. However, each slave segment supports up to four slaves.

The PLB6 can be interfaced to the PowerPC® 476FP embedded processor and the PowerPC 476FP L2 cache core, that supports SMP coherency. Of course, bridges to other CoreConnect cores are available. In particular, PLB6 is intended to cooperate with DMA controllers. They can be connected together through the DMA-to-PLB6 controller core. Compatibility with PLB4 is enabled with the PLB6-to-PLB4 and PLB4-to-PLB6 bridges.

2.4.2   On-Chip Peripheral Bus (OPB)

The On-Chip Peripheral Bus (OPB) is a secondary bus architected to alleviate system performance bottlenecks by reducing capacitive loading on the PLB. Frequently, the OPB architecture connects low-bandwidth devices such as serial and parallel ports, UARTs, timers, etc., and represents a separate, independent level of bus hierarchy. It is implemented as a multimaster, arbitrated bus. It is a fully synchronous interconnect with a common clock, but its devices can run with slower clocks, as long as all of the clocks are synchronized with the rising edge of the main clock.

The OPB supports multiple masters and slaves by implementing the address and data buses as a distributed multiplexer. This type of structure is suitable for the less data intensive OPB bus and allows adding peripherals to a custom core logic design without changing the I/O on either the OPB arbiter or existing peripherals. It also has support for burst transfers, and can transfer data between OPB bus master and OPB slaves in a single cycle. It can also overlap bus arbitration with last cycle of bus transfers.

2.4.3   Device Control Register Bus (DCR)

The DCR bus provides an alternative path to the system for setting the individual device control registers. Through the DCR bus, the host CPU can set up the device-control-registers without loading down the main PLB. This bus has a single master, the CPU interface, which can read or write to the individual device control registers. The DCR bus architecture allows data transfers among OPB peripherals to occur independently from, and concurrently with data transfers between processor and memory, or among other PLB devices. The DCR bus architecture is based on a ring topology to connect the CPU interface to all devices. The DCR bus is typicallly implemented as a distributed multiplexer across the chip such that each sub-unit not only has a path to place its own DCRs on the CPU read path, but also has a path that bypasses its DCRs and places another unit's DCRs on the CPU read path. DCR bus consists of a 10-bit address bus and a 32-bit data bus.

This is a synchronous bus, wherein slaves may be clocked either faster or slower than the master, although a synchronization of clock signals with the DCR bus clock is required. Finally, bursts are not supported by this bus.

2.4.4   Bridges

2.4.4.1   PLB-OPB Bridges

PLB masters gain access to the peripherals on the OPB bus through the PLB-to-OPB bridge macro. The OPB bridge acts as a slave device on the PLB and a master on the OPB. It supports bursts, word (32-bit), half-word (16-bit), and byte read and write transfers on the 32-bit OPB data bus, and has the capability to perform target word first line read accesses. The OPB bridge performs dynamic bus sizing, allowing devices with different data widths to efficiently communicate. When the OPB bridge master performs an operation wider than the selected OPB slave can support, the bridge splits the operation into two or more smaller transfers. Transactions from the OPB to the PLB under the direction of the OPB masters are also supported through the instantiation of an OPB-to-PLB bridge that works similar to the PLB-to-OPB, but inverting the roles: It is a slave on the OPB, and master on the PLB.

2.4.4.2   AHB Bridges

The AHB-to-PLB4 bridge core permits transfers of code and data between the Advanced Microcontroller Bus Architecture (AMBA) advanced high-performance bus (AHB) and the CoreConnect PLB. The AHB-to-PLB4 bridge is a slave on the AHB bus and a master on the PLB bus. The bridge also includes a DCR bus master interface. The counterpart bridge is the PLB4-to-AHB core. Moreover, according to IBM specification, they only provide compatibility with AHB rev 2.0. Some of the key features of these bridges are

- AMBA side: 32-bit AMBA AHB Specification, Version 2.0 compliant

- Coreconnect side: Support for versions 3 and 4 of the PLB specification: 64-bit and 128-bit PLB Architecture Specification, Version 4.4 compliant. DCR interface

- Clock and power-management capabilities

Figure 2.12 represents a typical IBM CoreConnect architecture with different buses and interconnect bridges.

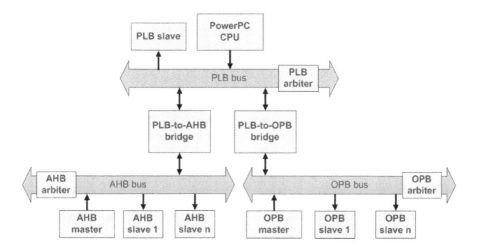

FIGURE 2.12
Typical IBM CoreConnect architecture with different buses and interconnect bridges.

2.4.5 Coreconnect Tools

2.4.5.1 Design Toolkits

Design toolkits are available for each of the on-chip buses. These toolkits contain master, slave and arbiter Bus Functional Models (BFM). They also provide a Bus Functional Compiler used to translate test cases written in a bus functional language into simulator commands executable by the master and slave models. The toolkits are available in Very-High-Speed Integrated Circuits (VHSIC), Hardware Description Language (VHDL), and Verilog such that they can be used in any simulation environment.

The Bus Functional Language (BFL) command definition is unique for each bus and allows the user to execute and respond to all allowable transactions on the particular bus. BFL is processed by the bus functional compiler into command sequences executable by a BFM running on an event-driven simulator such as VerilogXL, MTI, or VSS. Each bus toolkit provides a bus monitor that automatically performs architectural protocol checking on all masters and slaves attached to the bus. These checks verify that the masters and slaves under test adhere to the defined bus protocol and help to ensure compatibility when the macros are interconnected in a system environment. In addition to protocol checking, the master and slave models also perform read and write data checking. For example, when a master is programmed to perform a read transfer, the model checks the read data with the expected data as programmed in the BFL. The toolkits also support concurrently executing multiple master models. A model intercommunication scheme is provided to allow transaction synchronization between the masters. A very useful feature, for example, for creating bus contention testcases, often necessary to verify a macro.

2.4.5.2   PLB Verification: PureSpec

Under an agreement with IBM, Denali's PureSpec™ for Processor Local Bus (PLB) has been selected as the industry's primary protocol validation and planning solution for predictable verification of PLB designs. PureSpec-PLB enables coverage-driven verification closure and seamless integration via third-party verification planners. PureSpec features a hierarchical and configurable test plan, coverage, sequences and constraint libraries. PureSpec verification solution includes a configurable bus functional model, protocol monitor, and complete assertion library for all components in the topology.

PureSpec provides extensive coverage of the PLB 6.0 specification, including all high-speed PLB features:

- Full timing, bus functional models for PLB-4 or PLB-6 devices

- Complete assertion library with thousands of runtime checks, linked to configurable models supporting all valid topologies

- Controllable protocol checkers, monitors for interoperability testing

- Predefined traffic libraries with user-customizable packet generation

- Powerful error injection/detection capability

- Cumulative functional coverage reports

- Transaction logging

The framework includes a sophisticated data generation engine that helps on driving defined, pseudorandom bus traffic at all layers, to exercise the Device Under Test (DUT). Injected errors and error conditions are flagged and recovered according to PLB specifications. The Purespec framework organization is depicted in Figure 2.13.

2.5   STBus

STBus is an STMicroelectronics proprietary on-chip bus protocol. STBus is dedicated to SoC designed for high bandwidth applications such as audio/video processing [47]. The STBus interfaces and protocols are closely related to the industry standard VCI (Virtual Component Interface). The components interconnected by an STBus are either *initiators* (which initiate transactions on the bus by sending requests), or *targets* (which respond to requests). The interconnect allows for the instantiation of complex bus systems

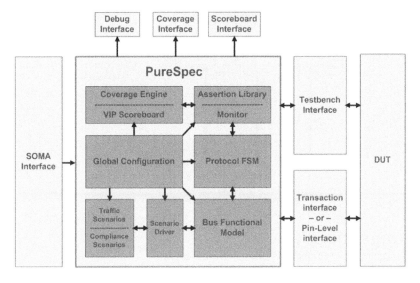

FIGURE 2.13
PureSpec validation framework.

such as heterogeneous multinode buses (thanks to size or type converters) and facilitates bridging with different bus architectures, provided proper protocol converters are made available (e.g., STBus and AMBA).

The STBus specification defines three different protocols that can be selected by the designer in order to meet the complexity, cost and performance constraints:

Type 1: Peripheral Protocol. This type is the low-cost implementation for low/medium performance. The peripheral STBus is targeted at modules that require a low complexity medium data rate communication path with the rest of the system. The supported operation subset is targeted at simple memory devices and supports read and write operations of up to 8 bytes. The interface is organized to minimize the need for the target device to interpret the operation, allowing the correct operation to be achieved by handling larger data quantities as if they were a series of smaller accesses.

Type 2: Basic Protocol. In this case, the limited operation set of the Peripheral Interface is extended to a full operation set, including compound operations, source labeling and some priority and transaction labeling. Moreover, this implementation supports split and pipelined accesses, and is aimed at devices that need high performance but do not require the additional system efficiency associated with the Advanced Protocol.

Type 3: Advanced Protocol. The most advanced implementation upgrades previous interfaces with support for out-of-order execution and

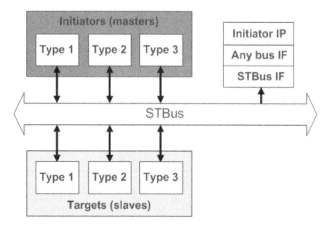

FIGURE 2.14
Schematic view of the STBus interconnect.

shaped packets, and is equivalent to the Advanced VCI protocol. Split
and pipelined accesses are supported. It allows performance improvements either by allowing more operations to occur concurrently, or by
rescheduling operations more efficiently.

An STBus system includes three generic architectural components. The
node arbitrates and routes the requests and, optionally, the responses. The
converter is in charge of converting the requests from one protocol to another
(for instance, from basic to advanced). Finally, the *size converter* is used between two buses of the same type but of different widths (8, 16, 32, 64, 128
or 256 bits). It includes buffering capability.

In Figure 2.14 we can see a schematic view of the STBus interconnect.
STBus can instantiate different bus topologies, trading-off communication
parallelism with architectural complexity. From simple Single Shared Bus
(minimum wiring area) suitable for simple low-performance implementations,
through Partial Crossbar, and up to a Full Crossbar, for complex high-
performance implementations.

A wide variety of arbitration policies is also available to help system inte-
grators meet initiators/system requirements. These include bandwidth limitation, latency arbitration, Least Recently Used (LRU), priority-based arbitration and others. A dedicated block, the node, is responsible for the arbitration
and routing and allows to change them dynamically.

The STBus signals present at any STBus module interface can be grouped
into three sets:

- Control Signals: The control signals determine when a transaction starts,
  when it finishes, and if two or more consecutive transactions are linked
  together to build more complex transactions.

- Transaction Signals: Contain the address, data, etc...

- Service Signals: Such as the clock, the reset, power down, or security mode signals.

The STBus interconnect is fully scannable for testability reasons. However, the test signals do not appear explicitly at STBus components or interconnect system interfaces, since application of test methodology is left to the system integration process and is neither performed at block level nor at interconnect level. The test signals to be added to STBus interface are standardized in the specification document.

2.5.1   STBus Verification Tools

The STBusGenkit toolkit, developed by STMicroelectronics, enables a graphical interface-based development flow. It can automatically generate top level backbone, cycle accurate high-level models, way to implementation, bus analysis (latencies, bandwidth) and bus verification (protocol and behavior). The "STBus GenKit" toolkit allows getting the maximum benefit from the STBus highly parametric features, giving an instrument for a quick convergence of critical design constraints and reducing the STBus design effort. STBus GenKit allows analyzing, at an early stage of the design flow, the power, the area and the performance of any STBus topology and configuration. Thus, it helps in getting quickly to a trade-off of those constraints for any SoC application.

Combining the STBus GenKit with other existing tools, designers can benefit from a complete development environment, as, for example, the Fully Integrated Vera based System Verification Environment (SVE) explained in [19]. The objective of this verification methodology is to yield a production gain, integrating different tools into one single design flow: "STBus GenKit" is an in-house tool where the backbone design entry and configuration are accomplished through a user-friendly Graphical User Interface (GUI). Still within a unique framework, all the steps from the system level (TLM and Bus-Cycle-Accurate (BCA)) analysis to the gate level implementation are accomplished. The RTL generation and verification is automated thanks to the Synopsys coretools suite (coreBuilder, coreConsultant, and coreAssembler) and Vera based verification environment STBus-SVE. The gate-level synthesis follows a similar approach being based on Synopsys Design compiler and is automated thanks to coreAssember. In the platform several add-on tools are offered for each step of the design flow. The final quality of the design is ensured by properties and protocol automatic check capability.

This gain is achieved utilizing preverified SVE based on complex protocol, higher-abstraction level, ability to interact with other SVEs, modular independent components, built-in functional coverage, built-in error injection, automated generation of highly configurable RTL and SVE, built-in path to SystemC modeling (TLM, BCA).

2.5.2 Towards NoC

In 2006, STMicroelectronics announced details of an innovative on-chip interconnect technology that the company was developing to meet the increasingly demanding needs of current and future SoC designs. The new technology, called ST Network on Chip (STNoC), was a flexible and scalable packet-based on-chip micronetwork designed according to a layered methodology, that extended the STbus. The network topology was defined depending on the application traffic, from the tree, to the simple ring and up to the patented Spidergon topology [21]. STNoC technology had Quality of Service (QoS) support, and STNoC's Network Interface allowed any kind of IP protocol such as AXI, OCP, or STBus to be converted into communication packets. Some time later, it was renamed to VSTNoC (Versatile STNoC).

However, in March, 2006, Arteris, a provider of network on chip (NoC) solutions, announced that STMicroelectronics had selected its patented Arteris NoC architecture for advanced on-chip communication in one of their SoC designs. Since then, STMicroelectronics dropped plans regarding STNoC, and now relies on Arteris solutions as network-on-chip interconnect.

2.6 FPGA On-Chip Buses

Modern FPGAs provide the silicon capacity to implement full multiprocessor systems containing several processors (hard and soft cores), complex memory systems and custom IP peripherals. These platforms require high-performance on-chip communication architectures for efficient and reliable interprocessor communication. With this goal in mind, some on-chip buses have been created, or adapted to the FPGA market.

A good FPGA interconnect must efficiently map into the taget FPGA architecture. Some buses, like Wishbone, are quite FPGA-friendly, since they offer a hardware implementation of bus interfaces simple and compact. While Altera designed the Avalon interconnect from scratch targeting FPGAs, Xilinx worked together with IBM to incorporate the Coreconnect technology into Xilinx's FPGAs. Many academic institutions have succesfully mapped networks on chip into FPGAs [49] and reported good performance and acceptable resource utilization so, chances are high that, in the near future, they will be incorporated into commercial solutions. As FPGAs grow in size, more complex elements will be embedded, like the ARM Cortex^{TM}-M1 processor. It is the first ARM processor designed specifically for implementation in FPGAs. The Cortex-M1 processor targets all major FPGA devices and includes support for leading FPGA synthesis tools.

Next, we will review the interconnect solutions from Xilinx and Altera, the two main FPGA manufacturers. Both aim at providing designers an adequate integrated design flow for rapid system prototyping.

2.6.1   Altera: Avalon

Altera's interconnect solution is called Avalon [2]. Avalon interfaces simplify system design by allowing easily connect components in the FPGA. Altera's tools (the System on a Programmable Chip Builder [SOPC Builder]) and the MegaWizard Plug-In Manager) allow users to create complex SoCs by symply instantiating and interconnecting elements from the library of components. Nios II is a 32-bit embedded-processor architecture designed specifically for the Altera family of FPGAs, licensable through a third-party IP provider, Synopsys Designware, for ports to a mass production ASIC-device. In addition to this, and many other cores that offer standard Avalon interfaces, designers can create custom components compliant with the Avalon standard, and make them available for future reuse. The Avalon interface family defines interfaces for use in both high-speed streaming and memory-mapped applications. The three main interface types are:

- Avalon Memory Mapped Interface (Avalon-MM) an address-based read/write interface typical of master/slave connections.

- Avalon Streaming Interface (Avalon-ST) an interface that supports the unidirectional flow of data, including multiplexed streams, packets, and DSP data.

- Avalon Memory Mapped Tristate Interface an address-based read/write interface to support off-chip peripherals.

Three more interfaces are defined aimed at delivering signals to on-chip components, such as the clock and reset (Avalon Clock) and interrupts (Avalon Interrupt), or exporting them to external FPGA pins (Avalon Conduit).

2.6.2   Xilinx

2.6.2.1   IBM's CoreConnect

Xilinx offers the IBM CoreConnect license to all its embedded processor customers since CoreConnect technology serves as the infrastructure for all Xilinx embedded processor designs.

IBM optimized a PowerPC hard core for the Xilinx architecture. It is included in some FPGA models. The Microblaze is a synthesizable 32-bit Reduced Instruction Set Computing (RISC) processor designed by Xilinx that can be instantiated as desired. Initially, the PowerPC allowed connection to the PLB (with optional DCR interface), while the Microblaze offered the OPB interface. Intercommunication between PLB and OPB subsytems was enabled through Opb2Plb and Plb2Opb bridges. However, in the last version of the Xilinx design tools (Integrated Software Environment [ISE] suite), the OPB bus is not available, although still supported for legacy designs through the use of adaptation bridges. Now, both the PowerPC and the Microblaze connect to the PLB.

In addition to the Processor Local Bus (version PLBv46) [53], the Embedded Development Kit tool (EDK), from Xilinx, offers the Fast-Symplex Link (FSL) [54] (point-to-point) and Local Memory buses [52], for tightly coupled memories and modules. A library of components is available with modules that offer PLB/FSL/Local Memory Bus (LMB) interfaces. Through the "Create peripheral Wizard" tool, EDK generates a template that will guide designers in the task of creating PLB/FSL peripherals.

Now, with Xilinx possibly looking to embed ARM processors in its products, they tightly collaborated with ARM to elaborate the new AMBA4 standard, that offers the AXI4-Lite version, a subset of the AXI4 protocol that greatly reduces the number of wires required to implement the bus. Designed with FPGAs in mind, can also be used for custom chips. The AXI4-Stream protocol, for streaming communications, is also very FPGA-oriented.

2.6.2.2   Extensible Processing Platform

Xilinx's announced at the beginning of 2010 the "Extensible Processing Platform." A new architecture that combines a high-performance, low-power, low-cost processor system, based on ARM's dual-core Cortex^{TM}-A9 MPCore, with Xilinx's high-performance, low-power, 28 nm programmable logic. The adoption of ARM technology is part of Xilinx's move to embrace open industry standards and to foster development of system solutions.

With a processor-centric approach, the platform behaves like a typical processor solution that boots on reset and provides Software Developers a consistent programming model. To achieve this, the Processing System (see Figure 2.15) is fully integrated and hardwired, using standard design methods, and includes caches, timers, interrupts, switches, memory controllers, and commonly used connectivity and I/O peripherals.

Extensibility is enabled by the use of high-bandwidth AMBA-AXI interconnects (Figure 2.15) between the Processing System and programmable logic for control, data, I/O, and memory. In the March 2010 release of version 4 of its AMBA Advanced Extensible Interface (AXI), ARM included an extension of the AXI specification optimized for programmable logic, co-developed with Xilinx (see Section 2.2).

Programmable logic devices such as Xilinx's FPGAs are well supported by sophisticated tools suites. These tools provide the Logic Designers an environment rich in features to optimize their IP solutions and render them onto the silicon devices. In this environment, the programming languages are hardware-oriented such as Verilog and VHDL. Software Developers work almost exclusively in high-level languages, such as C/C++, which are also well supported by today's processor-based solutions. In the Extensible Processing Platform, Xilinx will enable support for familiar software development and debug environments, using tools such as ARM Real View and related third-party tools, Eclipse-based Integrated Development Environment (IDEs), GNU toolchains, Xilinx® Software Development Kit, and others. The programmable logic por-

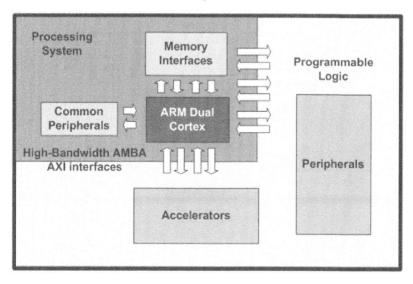

FIGURE 2.15
System showing Wrapped Bus and OCP Instances.

tion can be developed and debugged using the standard ISE® Design Suite, and other third-party HDL and algorithmic design tools.

Because the Extensible Processing Platform takes a processor-centric approach (it boots the Processing System at reset and then manages the programmable logic configuration), a more software-centric development flow is enabled (Figure 2.16).

This flow enables the System Architect, Logic Designer, and Software Developer to work in parallel, using their familiar programming environments, then merge the final releases into the software baseline. As a result, key partitioning decisions on system functions/performance can be made early and throughout the development process. This is critical for embedded systems where application complexity is driving tremendous levels of system performance against tightly managed cost, schedule, and power budgets. System Architects and Software Developers typically define the system initially from the software perspective and then determine what functions they need to offload or accelerate in hardware. This allows them to trial fit their design against the performance, cost, and power targets of the application. At this proof-of-concept stage, System Architects and Software Developers are most concerned with having flexibility over what can be performed in hardware or run in software to meet the specific application requirements. Iteratively, they converge on the optimal partitioning of hardware and software, and then refine both to fit the system requirements. The Extensible Processing Platform is ideal for this process as it will accelerate convergence on a more idealized programming platform. It is important to note that the AMBA-AXI interfaces

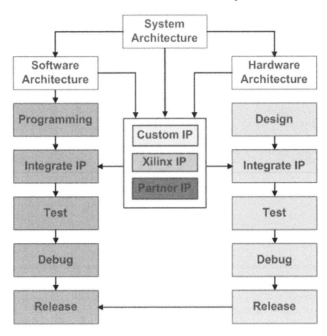

FIGURE 2.16
System showing Wrapped Bus and OCP Instances.

are key in enabling the software-centric flow because they present a seamless, common, and well-defined environment for the hardware extensions. While the Logic Designer will need to deeply understand this technology, for the Software Developer, the AMBA interfaces abstract the extended logic as memory mapped calls. This allows for a straightforward interplay of hardware and software programming in a parallel state of development.

2.7   Open Standards

The VSIA (Virtual Socket Interface Alliance), an open, international organization that included representatives from all segments of the SoC industry was founded in 1996 and dissolved in 2008. It was the first attempt to enhance the productivity of the SoC design community by providing reuse of IP development and promotion of business solutions and open standards used in the design of Systems-on-Chip.

With the goal to standarize IP protection, IP transfer, IP integration and IP reuse processes, the VSIA's (VSIA) Design Working Group on On-Chip Buses (DWGOCB) specified a bus wrapper to provide a bus-independent

Transaction Protocol-level interface to IP cores. They defined different Virtual Component interfaces (VCI): the BVCI (basic VCI), the PVCI (peripheral VCI), and AVCI (advanced VCI). These standards were not complete. They addressed only data flow, not issues of test and control.

The On-Chip Bus (OCB) documents (OCB 1 2.0 and OCB 2 2.0) are early bus specifications and standards developed at VSIA, describing the On-Chip Bus Attributes Specification and the Virtual Component Interface Standard.

VSIA also developed an international standard, the Quality Intellectual Property (QIP) Metric for measuring Semiconductor Intellectual Property (SIP) quality and examining the practices used to design, integrate, and support the IP. As described in the latest revision (4.0), the QIP is an objective way to compare IP for reuse on an "apples" to "apples" basis. The QIP consists of interactive Microsoft Excel^{TM} spreadsheets with sets of questions to be answered by the IP vendor. The QIP examines the practices used to design, integrate, and support the IP. Among similar IPs, the higher the QIP score, the higher will be the probability of successful reuse. Thus, the tool is like a "virtual advisor" that will guide designers in the process of making the right decisions. The VSI Alliance's Quality Pillar developed the QIP to help both IP users evaluate similar IPs and IP vendors clearly display their quality standards.

According to an agreement between VSIA and OCP International Partnership, OCP-IP is currently providing archival access to all these specifications. OCP-IP is the successor of the VSIA.

2.7.1   Introducing the OCP-IP

The Open Core Protocol International Partnership Association [10] (OCP-IP) was formed in December 2001 to promote and support the open core protocol (OCP) as the complete socket standard that ensures rapid creation and integration of interoperable virtual components.

OCP-IP is a nonprofit corporation focused on delivering the first fully supported, openly licensed core-centric protocol that comprehensively fulfills system-level integration requirements. The OCP facilitates IP core reusability and reduces design time and risk, along with manufacturing costs for SoC designs.

Sonics Inc., the inventor of the OCP technology donated and legally provided it to OCP-IP. This ensured the widest availability of the only complete, fully-supported and vendor neutral socket solution in the industry.

The industry strongly supports the Open Core Protocol as the universal complete socket standard. Among OCP-IP's founding members and initial Governing Steering Committee participants, Texas Instruments (TI), demonstrating its support for open standards and specifications developed its open media access platform Open Multimedia Application Platform (OMAP) 2 chips in compliance of the OCP specification. Nokia, Intel, Broadcom and PMC-Sierra have used the spec in production designs. Sonics' foundry part-

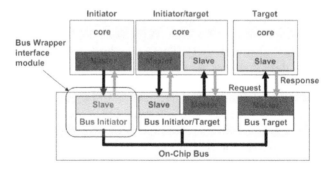

FIGURE 2.17
System showing Wrapped Bus and OCP Instances

ner United Microelectronics Corp., ARM Ltd. and MIPS Technologies Inc. are part of the effort, too.

2.7.2 Specification

OCP-IP is dedicated to proliferating a common standard for intellectual property (IP) core interfaces, or sockets, that facilitate "plug and play" System-on-Chip design.

The Open Core Protocol (OCP) version 1.0 defines the basis of this high-performance, bus-independent interface between IP cores. An IP core can be a simple peripheral core, a high-performance microprocessor, or an on-chip communication subsystem such as a wrapped on-chip bus.

OCP separates the computational IP core from its communication activity defining a point-to-point interface between two communicating entities, such as IP cores and bus interface modules (bus wrappers). One entity acts as the master of the OCP instance and the other as the slave. Only the master can present commands and be the controlling entity. The slave responds to commands presented to it. For two entities to communicate in a peer-to-peer fashion, two instances of the OCP connecting them are needed.

Figure 2.17 shows a simple system containing a wrapped bus and three IP core entities: a system target, a system initiator, and an entity that behaves as both.

A transfer across this system occurs as follows. A system initiator (as the OCP master) presents command, control, and possibly data to its connected slave (a bus wrapper interface module). The interface module plays the request across the on-chip bus system. The OCP does not specify the embedded bus functionality. Instead, the interface designer converts the OCP request into an embedded bus transfer. The receiving bus wrapper interface module (as the OCP master) converts the embedded bus operation into a legal OCP command. The system target (OCP slave) receives the command and takes the requested action.

The Open Core Protocol version 2.0, released in September 2003 adds many enhancements to the 1.0 specification, including a new burst model, the addition of in-band signaling, endianness specification, enhanced threading features, dual reset facilities, lazy synchronization, and additional write semantics. In November 2009 the Specification Working Group released the OCP 3.0 Specification [38]. This latest version contains extensions to support cache coherence and more aggressive power management, as well as an additional high-speed consensus profile and other new elements.

Since version 1.0, all signaling is synchronous with reference to a single interface clock, and all signals except for the clock are unidirectional, point-to-point, resulting in a very simple interface design, and very simple timing analysis. However, given the wide range of IP core functionality, performance and interface requirements, a fixed-definition interface protocol cannot address the full spectrum of requirements. The need to support verification and test requirements adds an even higher level of complexity to the interface. To address this spectrum of interface definitions, the OCP defines a highly configurable interface. The OCP's structured methodology includes all of the signals required to describe an IP cores' communication including data flow, control, verification, and test signals.

Next, we describe some of the main characteristics of the OCP protocol:

- Flexibility: Each instance of the OCP is configured independently of the others. For instance, system initiators may require more address bits in their OCP instances than do the system targets; the extra address bits might be used by the embedded bus to select which bus target is addressed by the system initiator. Only those addressing bits that are significant to the IP core should cross the OCP to the slave.

- IP design reuse: The OCP transforms IP cores, making them independent of the architecture and design of the systems in which they are used.

- Bus Independence: A core utilizing the OCP can be interfaced to any bus.

- Optimizes die area by configuring into the OCP interfaces only those features needed by the communicating cores.

- Simplifies system verification and testing, by providing a firm boundary around each IP core that can be observed, controlled, and validated.

The OCP provides the option of having responses for Write commands, or completing them immediately without an explicit response (posted write commands). The OCP protocol provides some advanced features, like burst transfers or multiple-cycle access models, where signals are held static for several clock cycles to simplify timing analysis and reduce implementation area. According to the OCP specifications, there are two basic commands (types of transactions): Read and Write, and five command extensions (WriteNonPost, Broadcast, ReadExclusive, ReadLinked, and WriteConditional).

Separating requests from responses, permits pipelining, that improves bandwidth and latency characteristics. Also, out-of-order request and response delivery can be enabled using tags and multiple threads. Threads can have independent flow control (requires independent buffering for each thread). Tags, on the other hand, exist within a single thread and are restricted to shared flow control.

OCP supports word sizes of power-of-two and nonpower-of-two (as would be needed for a 12-bit DSP core). The OCP address is a byte address that is word aligned. Transfers of less than a full word of data are supported by providing byte enable information. Byte lanes are not associated with particular byte addresses. This makes the OCP endian neutral, able to support both big and little-endian cores.

The OCP is equivalent to VSIA's Virtual Component Interface (VCI). While the VCI addresses only data flow aspects of core communications, the OCP is a superset of VCI that additionally supports configurable control and test signaling. The OCP is the only standard that defines protocols to unify all of the intercore communication.

While moving data between devices is a central requirement of on-chip communication systems, other types of communications are also important like reporting events (Interrupts and Errors), performing high-level flow control (sending a global reset), and other signaling (e.g., cacheable information, or data parity). The OCP provides support for such signals through the "Sideband Signaling" extensions.

2.7.3   Custom Cores

In order to design an OCP-compliant core, each OCP interface present in the core (there must be, at least, one) must comply with all aspects of the OCP interface specification. In addition to this, both the core and OCP interfaces must be described using an RTL configuration file, and have their timing described using a synthesis configuration file. The specific syntax of these configuration files is described in detail in the OCP documents. They also provide detailed timing diagrams to help understand the protocol. As can be seen in Figure 2.18, it shows three read transfers using pipelined request and response semantics. In each case, the request is accepted immediately, while the response is returned in the same or a later cycle.

The sequence of operations observed is the following:

- A) The master starts the first read request, driving RD on MCmd and a valid address on MAddr. The slave asserts SCmdAccept, for a request accept latency of 0.

- B) Since SCmdAccept is asserted, the request phase ends. The slave

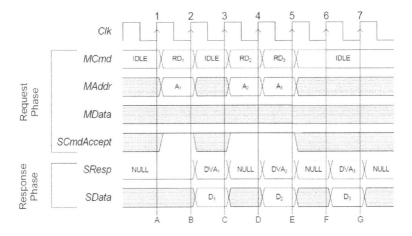

FIGURE 2.18
Pipelined OCP Request and Response.

responds to the first request with DVA on SResp and valid data on SData.

- C) The master launches a read request and the slave asserts SCmdAccept. The master sees that SResp is DVA and captures the read data from SData. The slave drives NULL on SResp, completing the first response phase.

- D) The master sees that SCmdAccept is asserted, so it can launch a third read even though the response to the previous read has not been received. The slave captures the address of the second read and begins driving DVA on SResp and the read data on SData.

- E) Since SCmdAccept is asserted, the third request ends. The master sees that the slave has produced a valid response to the second read and captures the data from SData. The request-to-response latency for this transfer is 1.

- F) The slave has the data for the third read, so it drives DVA on SResp and the data on SData.

- G) The master captures the data for the third read from SData. The request-to-response latency for this transfer is 2.

2.7.4   Tools

One of the advantages of the OCP protocol are the multiple tools available to perform design space exploration and verification of OCP-based systems. A big effort from the community has been made in this direction, including the OCP-IP University Program. Members of this program are entitled to receive free software tools, technical support, and training that is packaged and ready for incorporation into a course or immediate independent use by students. OCP-IP members also receive free training and support, and software tools, enabling them to focus on the challenges of SoC design. We next describe some of these tools: the CoreCreator, The OCP TLM Modeling Kit, and the OCP conductor:

2.7.4.1   CoreCreator

Corecreator is a complete OCP Verification Environment. The last version, CoreCreator II [37], provides capabilities to simulate OCP Cores and OCP-based Systems. It includes verification IP to generate and respond to OCP stimulus, an OCP checker to ensure protocol compliance, a performance analyzer to measure system performance and a disassembler, which helps to view the behavior of OCP traffic (see Figure 2.19). CoreCreator II can be used with

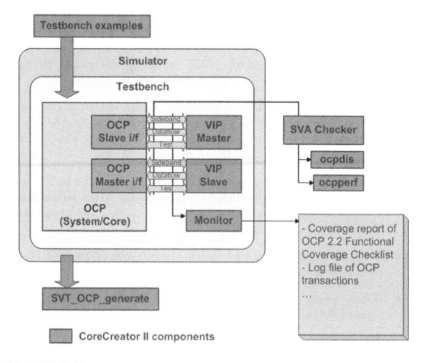

FIGURE 2.19
CoreCreator components.

traditional Verilog and VHDL testbench environments to create directed tests for OCP designs. It provides the Verification IP and debugging tools necessary for validating Open Core Protocol implementations, reducing design time and risk, ensuring rapid time to market.

CoreCreator integrates many features. Based on the OCP library for verification, it can be interfaced with several standard simulators: Questa (Mentor Graphics), NC-Sim (Cadence Design Systems), and VCS and VCS-MX (Synopsys Inc).

The OCP Library for Verification (SOLV), developed by Sonics, is designed to enable debugging, performance tuning and validation of complex intellectual property (IP) cores using OCP interfaces. This suite of tools enables core designers to validate their OCP socket interfaces and maximize performance. Sonics OCP Library for Verification (SOLV) is a package of three components:

- The first component in the SOLV package is the Sonics SVA OCP Checker. It validates OCP sockets for protocol compliance during simulation and generates OCP trace files for use by postprocessing tools. The checker sits in between the Master and the Slave, as depicted in Figure 2.20, and captures values of OCP signals at each OCP clock cycle and compares them to OCP protocol requirements as defined within the official OCP specification. This tool enables users to quickly and efficiently identify OCP protocol violations within individual sockets thereby reducing debug and validation workloads.

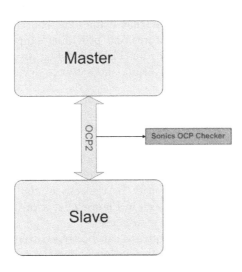

FIGURE 2.20
An instantiation of the Sonics OCP Checker (example of the Sonics OCP Library for Verification: SOLV) monitors the OCP2 protocol compliance during the simulation.

- The second tool within the SOLV package is the OCP Disassembler (OCPDis2). OCPDis2 is a command line tool that allows for the display of OCP connection activity in a convenient report format. During simulation, OCP connection activity can be logged into an OCP trace file. These trace files are essentially tables of hexadecimal values until OCPDis2 disassembles them into human readable data.

- The OCP Performance Analyzer (OCPPerf2), the third component of the SOLV package, is a command line tool that processes OCP trace files to measure the performance of OCP transfers and burst transactions. OCPPerf2 can process and interpret OCP trace files, making them useful for determining performance results within an OCP socket. Using this knowledge, a designer can individually tune OCP socket performance to optimal levels for any given application.

2.7.4.2 OCP SystemC TLM Kit

The OCP SystemC TLM Kit allows designers to create Interoperable Transaction Level Models of OCP-Based Components. The OCP TLM Modeling Kit is available to members of OCP-IP and can be downloaded at www.ocpip.org. A version of the Kit, fully functional but containing no monitor components, is available to nonmembers via click-thru research license agreement at www.ocpip.org.

The OCP Modeling Kit provides a full interoperability standard for SystemC models of SOC components with OCP interfaces. The Kit is built on top of OSCI's TLM 2.0 technology (that defines the concept of sockets), adding support for OCP protocol features and providing support for code development and testing. All use cases for TLM modeling are supported, including verification, architecture exploration, and software development.

2.7.4.3 OCP Conductor

OCP Conductor is an innovative, detailed OCP transaction viewer and debugger that allows very granular analysis of bus transactions. With this tool, system designers can get instant abstraction of OCP signalling into a high-level format to allow enhanced OCP debugging, so that they can easily check for protocol violations, or identify bottlenecks in the system. With a nice graphical interface, as can be seen in Figure 2.21, it provides powerful unique analysis of every type of OCP transaction.

A complete transaction sequence can be traced from request to response along with a host of related information about the transaction, permitting instant, high-level OCP transaction analysis. The tool permits intuitive visualisation of OCP transaction activity, provides functionality for searching for transactions by type, thread, latency, and bus occupancy.

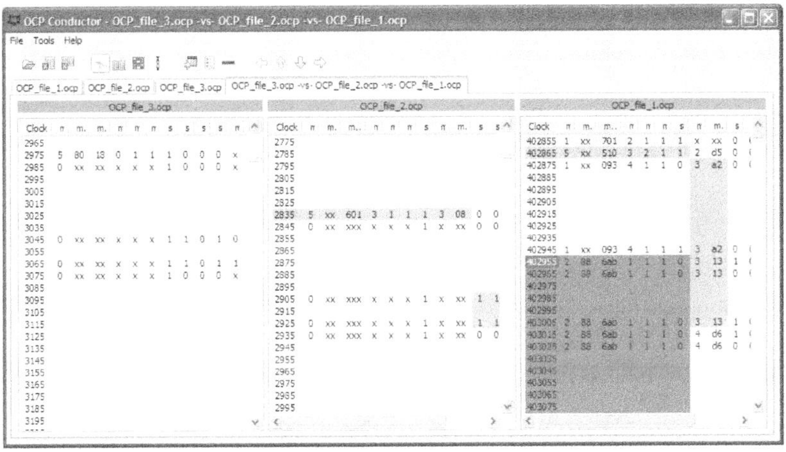

FIGURE 2.21
System showing Wrapped Bus and OCP Instances.

2.8 Wishbone

The WISHBONE System-on-Chip interconnect [51] is an open source hardware computer bus. It is an attempt to define a standard interconnection scheme for IP cores so that they can be integrated more quickly and easily by the end user.

A large number of open-source designs for CPUs, and auxiliary computer peripherals have been released with Wishbone interfaces. Many can be found at OpenCores [39], a foundation that attempts to make open-source hardware designs available.

It uses a *Master/Slave* architecture. Cores with Master interfaces initiate data transactions to participating Slave interfaces. All signals are synchronous to a single clock but some slave responses must be generated combinatorially for maximum performance. Some relevant Wishbone features that are worth mentioning are the multimaster capability which enables multiprocessing, the arbitration methodology defined by end users attending to their needs, and the scalable data bus widths (8-, 16-, 32-, and 64-bit) and operand sizes. Moreover, the hardware implementation of bus interfaces is simple and compact (suitable for FPGAs), and the hierarchical view of the Wishbone architecture supports structured design methodologies [23]. Wishbone permits addition of a "tag bus" to describe the data.

The hardware implementation supports various IP core interconnection schemes, including: point-to-point connection, shared bus, crossbar switch implementation, data-flow interconnection and even off-chip interconnection. The crossbar switch interconnection is typically used when connecting two or more masters together so that every one can access two or more slaves (in a point-to-point fashion).

The Wishbone specification does not require the use of specific development tools or target hardware. Furthermore, it is fully compliant with virtually all logic synthesis tools.

2.8.1　The Wishbone Bus Transactions

The Wishbone architecture defines different transaction cycles attending to the action performed (read or write) and the blocking/nonblocking access— namely, single read/write transfers, and block read/write transfers. In addition to those, the read-modify-write (RMW) transfer is also specified, which can be used in multiprocessor and multitasking systems in order to allow multiple software processes to share common resources by using semaphores. This is commonly done on interfaces for disk controllers, serial ports, and memory. The RMW transfer reads and writes data to a memory location in a single bus cycle.

2.9　Other Specific Buses

This section is dedicated to a special set of buses: those used in the automotive, avionics, and house automation industries. As we will see, they have different requirements compared to the ones we have reviewed through this chapter.

Although difficult to imagine, some modern automobiles can easily contain three miles of cabling. Because this number is liable to rise as components grow more intelligent, a bus architecture is the only way to keep the volume of wiring from becoming unmanageable. Cars are not the only application domain affected by wiring weight and complexity. In some luxury yachts, manufacturers add concrete blocks to one side of the boat to compensate for the heavy wiring loom on the far side.

As opposed to on-chip buses, a bus in an automotive, avionics, or house automation environment, will be, first of all, connecting together elements from different chips. Thus, the delays, size constrains, cost, etc... are in other range of magnitudes. However, the main purpose of the bus still remains: to interconnect together system elements. Once again, the topology of the bus will be affected depending on the services it needs to provide: optimum performance, minimum wiring, lowest cost, etc. As we will see, real-time performance and reliability are usually a must. If there is a nondeterministic component in the bus architecture, it will be impossible to compensate for it at higher levels.

We first take a look at the automotive industry. Next, we overlook the avionics standards, and will conclude with the home automation market.

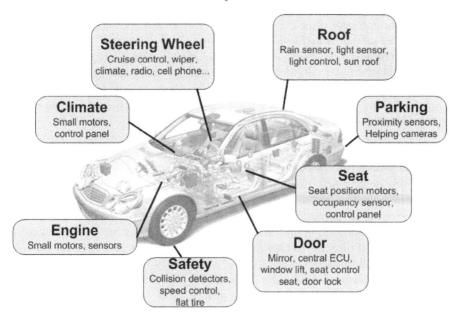

FIGURE 2.22
In-car intelligent elements.

2.9.1 Automotive

The most widely used automotive bus architecture is the CAN bus. The CAN (Controller Area Network) is a multimaster broadcast serial bus standard for connecting electronic control units (ECUs). Originally developed by Robert Bosch GmbH in 1986 for in-vehicle networks in cars. Bosch published the CAN 2.0 specification in 1991 [13]. It currently dominates the automotive industry, also having considerable impact in other industries where noise immunity and fault tolerance are more important than raw speed, such as factory automation, building automation, aerospace systems, and medical equipment. Hundreds of millions of CAN controllers are sold every year and most go into cars. Typically, the CAN controllers are sold as on-chip peripherals in microcontrollers. Bosch holds patents on the technology, and manufacturers of CAN-compatible microprocessors pay license fees to Bosch.

The applications of CAN bus in automobiles include window and seat operation (low speed), engine management (high speed), brake control (high speed), and many other systems. Figure 2.22 shows some of the typical in-car elements.

Choosing a CAN controller defines the physical and data-link portions of your protocol stack. In a closed system, designers can implement their own higher-level protocol. If they need to interoperate with other vehicle components, though, the vehicle manufacturer will most likely mandate the use of one of the standard higher-level protocols.

For the physical layer, a twisted pair cable (with a length ranging from 1,000m at 40Kbps to 40m at 1Mbps) carries the information on the bus as a voltage difference between the two lines. The bus is therefore immune to any ground noise and to electromagnetic interference, which in a vehicle can be considerable.

Each node is able to send and receive messages, but not simultaneously. A message consists primarily of an ID; usually chosen to identify the message-type or sender (interpreted differently depending on the application or higher-level protocols used), and up to eight data bytes. It is transmitted serially onto the bus. This signal pattern is encoded in Nonreturn to Zero (NRZ) and is sensed by all nodes. The devices that are connected by a CAN network are typically sensors, actuators, and other control devices. These devices are not connected directly to the bus, but through a host processor and a CAN controller. All messages carry a cyclic redundancy code (CRC).

CAN features an automatic "arbitration free" transmission. A CAN message that is transmitted with highest priority will "win" the arbitration, and the node transmitting the lower priority message will sense this and back off and wait. This is achieved by CAN transmitting data through a binary model of "dominant" bits and "recessive" bits where dominant is a logical 0 and recessive is a logical 1. This means open collector, or "wired or" physical implementation of the bus (but since dominant is 0 this is sometimes referred to as wired-AND). If one node transmits a dominant bit and another node transmits a recessive bit then the dominant bit "wins" (a logical AND between the two).

If the bus is free, any node may begin to transmit. If two or more nodes begin sending messages at the same time, the message with the more dominant ID (which has more dominant bits, i.e., zeroes) will overwrite other nodes' less dominant IDs, so that eventually (after this arbitration on the ID) only the dominant message remains and is received by all nodes.

There are several high-level communication standards that use CAN as the low level protocol implementing the data link and physical layers. For example, The time-triggered CAN (TTCAN) protocol [12] (standardized in ISO 11898-4). Time-triggered communication means that activities are triggered by the elapsing of time segments. In a time-triggered communication system all points of time of message transmission are defined during the development of a system. A time-triggered communication system is ideal for applications in which the data traffic is of a periodic nature. Other high-level protocols are CANopen, DeviceNet, and J1939 (*http://www.can-cia.org/*). For engine management, the J1939 protocol is common, while CANopen is preferred for body management, such as lights and locks. Both buses run on the same hardware; different application-specific needs are met by the higher-level protocols.

CAN is a relatively slow medium and cannot satisfy all automotive needs [34]. For example in-car entertainment requires high-speed audio and video streaming. These needs are being addressed by Media-Oriented Systems Transport (MOST) and IDB-1394b, which is based on Firewire. Recently, after MOST and Firewire, Ethernet is emerging as a third option for in-car information and entertainment networks. Due to its inherent openness, the Ethernet/IP combination facilitates the integration of mobile devices such as cell phone or portable music players into the automotive environment.

Another standard, the Local Interconnect Network (LIN) [35], is a broadcast serial network comprising one master and many (up to 16) slaves. No collision detection exists, therefore all messages are initiated by the master with at most one slave replying for a given message identifier. The LIN bus is an inexpensive serial communications protocol, which effectively supports remote application within a car's network. It is particularly intended for mechatronic nodes in distributed automotive applications, but is equally suited to industrial applications. Used together with the existing CAN network leads to hierarchical networks within cars.

2.9.2 Avionics

With the purpose of reducing integration costs and provide distributed data collection and processing, the most relevant features of avionics buses include deterministic behavior, fault tolerance, and redundancy. Most avionics buses are serial in nature. A serial bus using only a few sets of wires keeps the point to point wiring and weight down to a minimum. Newest bus standards use fiber optics to provide a significantly reduced weight and power solution to spacecraft subsystem interfacing. Another benefit from the technology advances is that small-scale embedded subsystems can now be implemented as a System-on-Chip using the latest generations of FPGAs. This creates opportunities for Commercial, Off-the-Shelf (COTS) vendors to offer IP cores for single or multiple remote terminals, bus controller, and bus monitor functions.

Avionics architectures typically separate the flight safety-critical elements such as primary flight control, cockpit, landing gear, and so on from less critical elements such as cabin environment, entertainment, and, in the case of military aircraft, the mission systems. This separation offers less onerous initial certification and allows incremental addition, as is often required for regulatory reasons, without the need for complete recertification. Significant savings in weight and power could be made with an integrated systems approach, using centralized computing supporting individual applications running in secure partitions with critical and noncritical data sharing the same bus.

While it appears that avionics buses are being left behind by the pace of technological change, there are sound economic and safety reasons why avionics architectures cannot change so rapidly. Avionics buses are traditionally slow to evolve, partly because requirements change so slowly and partly because of the costs of development, certification, and sustainment. It is with the development of new airplanes that the demand for new bus architectures evolves. This can be seen in the adoption of Fibre Channel for JSF and AR-

INC 664, also known as AFDX (Avionics Full-Duplex Switched Ethernet) for new Boeing and Airbus airplane types. Some buses, although ideally suited technically such as Time-Triggered Protocol (TTP), have been sluggish to be adopted and might only find use in niche applications. However, although the rate of change might be slow, the nature of the market still leaves room for much innovation in packaging, soft cores, and test equipment by embedded computing vendors.

Next, we will review some of the most widespread standards, divided into military and civil solutions and, then, we dedicate a small section to the task of debugging this type of systems.

2.9.2.1   Military

MIL-STD-1553 (rev B) [16] is the best known military example, developed in the 1970s by the United States Department of Defense. It was originally designed for use with military avionics, but has also become commonly used in spacecraft on-board data handling (OBDH) subsystems, both military and civil. It features a dual redundant balanced line physical layer, a (differential) network interface, time division multiplexing, half-duplex command/response protocol, and up to 31 remote terminals (devices). A version of MIL-STD-1553 using optical cabling in place of electrical is known as MIL-STD-1773. It is now widely used by all branches of the U.S. military and has been adopted by NATO as STANAG 3838 AVS. MIL-STD-1553 is being replaced on some newer U.S. designs by FireWire.

Another specification, the MIL-STD-1760 (rev D) [17] Aircraft/Store Electrical Interconnection System, is a very particular case. It defines an electrical interface between a military aircraft and its carriage stores. Carriage stores range from weapons, such as GBU-31 JDAM, to pods, such as AN/AAQ-14 LANTIRN, to external fuel tanks. Prior to adoption and widespread use of MIL-STD-1760, new store types were added to aircraft using dissimilar, proprietary interfaces. This greatly complicated the aircraft equipment used to control and monitor the store while it was attached to the aircraft: the stores management system, or SMS. The specification document defines the electrical characteristics of the signals at the interface, as well as the connector and pin assignments of all of the signals used in the interface. The connectors are designed for quick and reliable release of the store from the aircraft. Weapon stores are typically released only when the aircraft is attacking a target, under command of signals generated by the SMS. All types of stores may be released during jettison, which is a nonoffensive release that can be used, for example, to lighten the weight of the aircraft during an emergency.

2.9.2.2   Civil

SpaceWire [1] is a spacecraft communication network based in part on the IEEE 1355 standard of communications. SpaceWire is defined in the European Cooperation for Space Standardization ECSS-E50-12A standard, and is used worldwide by the European Space Agency (ESA), and other international space agencies including NASA, JAXA, and RKA. Within a SpaceWire network the nodes are connected through low-cost, low-latency, full-duplex, point-to-point serial links and packet switching wormhole routing routers. SpaceWire covers two (physical and data-link) of the seven layers of the OSI model for communications.

AFDX [11] is defined as the next-generation aircraft data network (ADN). It is based upon IEEE 802.3 Ethernet technology, and utilizes COTS components. AFDX is described specifically by Part 7 of the ARINC 664 Specification, as a special case of a profiled version of an IEEE 802.3 network. This standard was developed by Airbus Industries for the A380. It has been accepted by Boeing and is used on the Boeing 787 Dreamliner. AFDX bridges the gap on reliability of guaranteed bandwidth from the original ARINC 664 standard. It utilizes a cascaded star topology network, where each switch can be bridged together to other switches on the network. By utilizing this form of network structure, AFDX is able to significantly reduce wire runs, thus reducing overall aircraft weight. Additionally, AFDX provides dual link redundancy and Quality of Service (QoS). The six primary aspects of AFDX include full duplex, redundancy, deterministic, high-speed performance, switched and profiled network.

Prior to AFDX, ADN utilized primarily the ARINC 429 standard. The ARINC 429 Specification [22] establishes how avionics equipment and systems communicate on commercial aircrafts. This standard, developed over thirty years ago, has proven to be highly reliable in safety critical applications, and is still widely used today on a variety of aircrafts from both Boeing and Airbus, including the B737, B747, B757, B767, and Airbus A330, A340, A380, and the A350. The specification defines electrical characteristics, word structures, and protocols necessary to establish the bus communication. Hardware consists of a single transmitter, or source, connected to 1-20 receivers, or sinks, on one twisted wire pair. A data word consists of 32 bits communicated over the twisted pair cable using the Bipolar Return-to-Zero Modulation. There are two speeds of transmission: high speed operates at 100 kbit/s and low speed operates at 12.5 kbit/s. Data can be transmitted in one direction only (simplex communication), with bidirectional transmission requiring two channels or buses. ARINC 429 operates in such a way that its single transmitter communicates in a point-to-point connection, thus requiring a significant amount of wiring that amounts to added weight.

Another standard, ARINC 629 [36], introduced by Boeing for the 777 provides increased data speeds of up to 2 Mbit/s and allows a maximum of 120 data terminals. This ADN operates without the use of a bus controller, thereby increasing the reliability of the network architecture. The drawback of this system is that it requires custom hardware that can add significant cost to the aircraft. Because of this, other manufacturers did not openly accept the ARINC 629 standard.

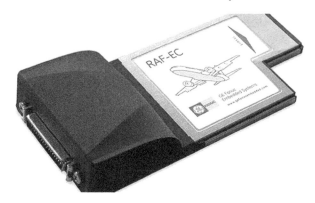

FIGURE 2.23
In-car intelligent elements.

2.9.2.3 Debugging in Avionics

With the continued development of new applications for avionics interfaces, bus analyzers play an important role in the testing and verification process. Similar to many market sectors, the laptop computer has become the ubiquitous test vehicle, easily supporting the four basic analyzer functions of display, logging and analysis, simulation, and playback when hooked up to a test rig or aircraft system. The laptop's PCMCIA slot has, until recently, been used to connect to the system under test. However, PCMCIA is rapidly being displaced by the new ExpressCard standard. This replaces the parallel interface of PCMCIA with PCI Express, offering significant improvements in performance and bandwidth plus compatibility with many other forms of embedded computing technology.

Despite its small size, an ExpressCard can support a dual-redundant AFDX port, logging all bus traffic, complete with 64-bit time tagging and IRIG-B for synchronization with external time sources. As an example, the RAF-EC and MIL-STD-1553B soft cores are produced by GE Fanuc Intelligent Platforms. They are designed for use in avionics bus analyzers, along with its sibling ExpressCard products for MIL-STD-1553B and ARINC 429 (see Figure 2.23).

2.9.3 Home Automation

Home automation (also called *domotics*) is becoming more and more popular around the world and is becoming a common practice. Home automation takes care of a lot of different activities in the house, as can be observed in Figure 2.24. Some of them are as simple as controlling your lights, appliances, home theater system, or turning on the sprinklers at a certain time every day. But some others are critical tasks, like fire sensors, or detecting burglars in

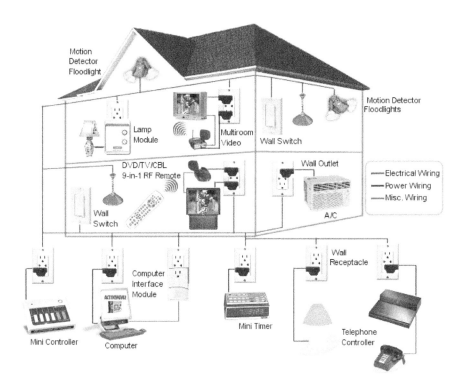

FIGURE 2.24
Automated House.

the middle of the night. Automating your house means interconnecting many sensors, actuators, and controls together to provide a centralized service. For this reason, multiple standards have been developped aiming at efficiently intercommunicating the elements of an automated environment. Currently, the two most succesful home automation technologies are X10 and C-Bus.

X10 [20] is an international and open industry standard for communication among electronic devices used for home automation. It primarily uses power line wiring for signaling and control (a signal imposed upon the standard AC power line). A wireless radio based protocol transport is also defined. X10 was developed in 1975 by Pico Electronics of Glenrothes, Scotland, in order to allow remote control of home devices and appliances. It was the first general purpose domotic network technology and remains the most widely available due to its simplicity (can be installed without re-cabling the house) and low price. An X-10 command usually includes two actions: activate a particular device (message code indicating device), and then send the function to be executed (message with the function code). Table 2.1 describes all commands supported by the standard X-10.

TABLE 2.1
X10 commands

Code	Function	Description
0 0 0 0	All units Off	Switch off all devices with the house code indicated in the message
0 0 0 1	All lights On	Switches on all lighting devices (with the ability to control brightness)
0 0 1 0	On	Switches on a device
0 0 1 1	Off	Switches off a device
0 1 0 0	Dim	Reduces the light intensity
0 1 0 1	Bright	Increases the light intensity
0 1 1 1	Extended Code	Extension code
1 0 0 0	Hail Request	Requests a response from the device(s) with the house code indicated in the message
1 0 0 1	Hail Acknowledge	Response to the previous command
1 0 1 x	Pre-Set Dim	Allows the selection of two predefined levels of light intensity
1 1 0 0	Extended Data	Additional data
1 1 0 1	Status is On	Response to the Status Request indicating that the device is switched on
1 1 1 0	Status is Off	Response indicating that the device is switched off
1 1 1 1	Status Request	Request requiring the status of a device

C-Bus [14] is a proprietary communications protocol created by Clipsal's Clipsal Integrated Systems division for use with its brand of home automation and building lighting control system. C-Bus uses a dedicated low-voltage cable to carry command and control signals. This improves the reliability of command transmission and makes C-Bus far more suitable for large, commercial applications than X10 but, on the other hand, it is the major obstacle to widespread use of wired C-Bus, since it will not work with a standard mains wire installation. A completely new wiring system must be installed for a wired C-Bus system which means that it is normally only used for new builds. A cable-less version of the protocol is also defined, where the different components communicate in a two-way wireless network.

There are some key features to be observed in this kind of systems that may result more or less relevant, depending on the particularities of the controlled system:

- Physical medium: Typically, it is easier to outfit a house during construction due to the accessibility of the walls, outlets, and storage rooms, and the ability to make design changes specifically to accommodate certain technologies. Wireless systems are commonly installed when outfitting a preexisting house, as they obviate the need to make major structural changes. These communicate via radio or infrared signals with a central controller.

- Scalability: The X10 protocol, for example, can only address 256 devices; each controlled device is configured to respond to one of the 256 possible addresses (16 house codes × 16 unit codes); each device reacts to commands specifically addressed to it, or possibly to several broadcast commands.

- Security and error control features: Key factors when automating security-critical elements. Both the standard X10 power line and Radio Frequency (RF) protocols lack support for encryption. Unless filtered, power line signals from close neighbors using X10 may interfere with each other if the same device addresses are used by each party. Interfering RF wireless signals may similarly be received, with it being easy for anyone nearby with an X10 RF remote to wittingly or unwittingly cause mayhem if an RF to a power line device is being used on a premises.

In addition to these two protocols, there are other solutions, like KNX, that is the successor to, and convergence of, three previous standards: the European Home Systems Protocol (EHS), BatiBUS, and the European Installation Bus (EIB or Instabus). Some other protocols and technologies used for home automation are LonWorks, Universal powerline bus (UPB), BACnet, INSTEON, SCS BUS with OpenWebNet, ZigBee, Z-Wave, and EnOcean, to mention some. However, despite their oldness, the most widespread standards used today are still X10 and CBus.

2.10 Security

It is clear that programmability and high performance will increasingly take center stage as desirable elements in future SoC designs. However, there is another element, security, that is likely to become crucial in most systems. In the last years, SoC designers have become increasingly aware of the need for security functionality in consumer devices. Security has emerged as an increasingly important consideration, protecting both the device and its content from tampering and copying. A system is only as secure as its weakest link, and security becomes ever more important as more equipment moves to a system-on-chip approach.

With more and more e-commerce applications running on phone handsets today, mobile systems are now looking to adopt security measures. While mobile processors have previously relied on Subscriber Identity Module (SIM) cards as the secure element, the processor architecture and integration architecture are now critical to the security of the whole system as more and more of the peripherals are being integrated into a single chip.

One way to provide multiple levels of security inside a SoC is to create a Hypervisor, a thin layer of software that has greater priority than the supervisor in a system, and checks all the transactions. Another possibility is to use cryptographic acceleration, or a trusted environment to run trusted software and store sensitive data in a secure area. Looking to the future, a new generation of standards is enabling a wide range of secure peripherals with individualized access levels, avoiding the problems that can occur with a single trusted environment. In that situation, if one peripheral is breached, it can be used to access all the others. With multiple levels of access, the peripherals and assets that need to be most secure, such as those handling certificates and credit card numbers, can still be kept secure from other peripherals.

Here, we will focus on the later alternative: the implementation of secure peripherals that prevent unauthorized access, ideally with multiple levels of access. To support complex system designs, in addition to these secure peripherals, it is necessary to provide protection against illegal transactions at the interconnect level. For this purpose, dedicated signals are added to the bus, in order to qualify the secure accesses.

The OCP-IP specification (Open Core Protocol International Partnership) is defining a standard way of building secure peripherals based on a signal on the bus that can be defined by an arbitrary number of bits to set different levels of access. This will allow processor cores to build in capabilities similar to that of Secure Machines and chip designers to use a wide range of peripherals and easily build more secure systems. The OCP interface can be used to create secure domains across a system. Such domains might include a CPU, shared memory, or I/O subsystems. The OCP master drives the security level for a request by using a special set of signals: the MSecure signal to define security access modes, along with using MConnID to identify a specific target in the system for the request. This MSecure signal is a user-defined subnet of the MReqInfo signal.

A similar approach is present in some of Sonics' interconnect solutions (Sonics3220^{TM}, SonicsMX, and SonicsSX). They have built-in access security hardware that protects designated cores from access by unauthorized initiators, protecting media content and intellectual property. The mechanism, similar to a firewall, allows dynamically configured access protection for cores or memory regions against access by specific initiating cores or processes, or defining protected regions within address space.

The AMBA bus specifications also add extra signals to provide a similar protection control mechanism primarily intended for use by any module that wants to implement some level of protection. As described in the AMBA 3 AHB lite specification, the protection control signals provide additional information about a bus access. The signals indicate if the transfer is an opcode fetch or data access, and distinguish a privileged mode access or user mode access. For masters with a memory management unit these signals also indicate whether the current access is cacheable or bufferable. Compatibility with masters that are not capable of generating accurate protection information, is achived by qualifying all transactions as a noncacheable, nonbuffer-able, privileged, data access. The AXI protocol uses two signals (The Address Read/Write Protection Signals: AWPROT and ARPROT, respectively) signals to provide three levels of protection unit support: normal or privileged, secure or nonsecure, and an indication if the transaction is an instruction or a data access.

For complex systems, it is often necessary for both the interconnect and other devices in the system to provide protection against illegal transactions. It should be noted that designing a secure system takes a chipwide approach. As we said at the beginning of the section, any system is only as secure as its weakest element, and retrofitting elements for security to a system that has not been designed with security in mind is only a temporary fix.

2.11 Conclusions

MPSoCs designs are increasingly being used in today's high-performance systems. The choice of the communication architecture in such systems is very important because it supports the entire interconnect data traffic and has a significant impact on the overall system performance. Through this chapter, we have provided an overview of the most widely used on-chip communication architectures.

Starting from the simple solution of the system bus that interconnects the different elements, we have seen how, as technology scales down transistor size and more elements fit inside a chip, scalability limitations of bus-based solutions are making on-chip communications manufacturers rapidly evolve to multilayer hierarchical buses or even packet-switched interconnect networks.

Selecting and configuring communication architectures to meet application specific performance requirements is a very time-consuming process that cannot be solved without advanced design tools. Such tools should be able to automatically generate a topology, guided by the designer-introduced constraints, and report estimated power consumption and system performance, as well as generate simulation and verification models.

New design methodologies allow simple system creation, based on the instantiation of different IP components included in a preexisting library. Thanks to the standarization of system interconnects, the designer can select the components, and interconnect them in a plug-and-play fashion. If needed, new components can be easily created, thanks to the existence of tools for validation of the different protocols.

In this heterogeneous context, Open Standards, such as the OCP-IP, are key towards unification. Bridges, adapters, and protocol converters are provided by manufacturers to ease interoperability among the different protocols, so that designers are not restricted to one proprietary solution. They can trade off performance and complexity to meet design requirements.

To conclude, we have visited some specific buses, present in the home automation, avionics and automotive areas, showing their different inherent characteristics.

The new big issue, for upcoming generations of chips, will be security, and interconnect support is vital to provide systemwide protection.

2.12   Glossary

ADK: AMBA Design Kit

ADN: Aircraft Data Network

AFDX: Avionics Full-Duplex Switched Ethernet

AHB: AMBA High-performance Bus

AMBA: Advanced Microcontroller Bus Architecture

APB: Advanced Peripheral Bus

ASB: Advanced System Bus

ASIC: Application-Specific Integrated Circuit

ATB: Advanced Trace Bus

AVCI: Advanced VCI

AXI: Advanced eXtensible Interface

BCA: Bus-Cycle Accurate

BFL: Bus Functional Language

BFM: Bus Functional Model

BVCI: Basic VCI

CAN: Controller Area Network

COTS: Commercial, Off-the-Shelf

CRC: Cyclic Redundancy Code

DCR: Device Control Register Bus

DMA: Direct Memory Access

DSP: Digital Signal Processor

DUT: Device under test

DWGOCB: Design Working Group on On-Chip Buses

ECU: Electronic Control Unit

EDK: Embeded Development Kit

EHS: European Home Systems

EIB: European Installation Bus

FPGA: Field Programmable Gate Array

ESL: Electronic System Level

FSL: Fast Simplex Link

GALS: Globally-Asynchronous Locally-Synchronous

GNU: General Public License

GUI: Graphical User Interface

HQHD: High Quality, High Definition

IDE: Integrated Development Environment

IMT: Interleaved Multichannel Technology

IP core: Intellectual Property Core

ISE: Integrated Software Environment

LIN: Local Interconnect Network

LMB: Local Memory Bus

MOST: Media-Oriented Systems Transport

MPSoC: Multiprocessor System-on-Chip

NoC: Network-on-Chip

NRZ: Nonreturn to Zero

OBDH: On-Board Data Handling

OCP-IP: Open Core Protocol International Partnership Association

OMAP: Open Multimedia Application Platform, from Texas Instrument

OPB: On-Chip Peripheral Bus

PLB: Processor Local Bus

PU: Processing Unit

PSELx: Peripheral Select Signal

PVCI: Peripheral VCI

QIP metric: Quality Intellectual Property Metric

QoS: Quality of Service

RMW: Read-Modify-Write

RTL: Register Transfer Language

SMART: Sonics Methodology and Architecture for Rapid Time to Market

SMP: Symmetric Multiprocessing

SMS: Stores Management System

SNAP: Sonics Network for AMBA Protocol

SoC: System-on-a-Chip

SOLV: Sonics' OCP Library for Verification

SOPC Builder: System on a Programmable Chip Builder

STNOC: ST Network on Chip

SVE: System Verification Environment

TLB: Translation Lookaside Buffer

TLM: Transaction-Level Modeling

TTCAN: Time-Triggered CAN

TTP: Time-Triggered Protocol

UPB: Universal Powerline Bus

VCI: Virtual Component Interface

VSIA: Virtual Socket Interface Alliance

2.13 Bibliography

[1] http://spacewire.esa.int.

[2] Altera. *Avalon Interface Specifications v1.2.*, 2009.

[3] ARM. AMBA Specification v2.0. 1999.

[4] ARM. AMBA 3 APB Protocol Specification v1.0. 2004.

[5] ARM. AMBA Multilayer AHB Overview. 2004.

[6] ARM. AMBA 3 AHB-Lite Protocol Specification v1.0. 2006.

[7] ARM. AMBA 3 ATB Protocol Specification v1.0. 2006.

[8] ARM. AMBA 4 AXI4-Stream Protocol Specification v1.0. 2010.

[9] ARM. AMBA AXI Protocol Specification v2.0. 2010.

[10] The Open Core International Partnership Association. http://www.ocpip.org/.

[11] SBS Technologies. Bob Pickles. Avionics Full Duplex Switched Ethernet (AFDX), White paper.

[12] Bosch. *Time-Triggered Communication on CAN (Time Triggered CAN-TTCAN).*

[13] Bosch. *CAN Specification, Version 2.0*, 1991.

[14] Clipsal. http://www.clipsal.com.au/.

[15] CoWare. ConvergenSC. http://www.coware.com.

[16] United States of America. Department of Defence. MIL-STD-1553B specification document.

[17] United States of America. Department of Defence. MIL-STD-1760D specification document.

[18] G. W. Doerre and D. E. Lackey. The IBM ASIC/SoC methodology. a recipe for first-time success. *IBM Journal Research & Development*, 46(6):649–660, November 2002.

[19] G. Falconeri et al. from ST Microelectronics and Synopsys. A powerful development environment and its fully integrated VERA based system verification environment, for the highly parametric STBus communication system. *IBM Journal Research & Development*, 2005.

[20] EuroX10. http://www.eurox10.com/Content/x10information.htm.

[21] G. Palermo et al. Mapping and topology customization approaches for application-specific STNoC designs.

[22] AIM GmbH. ARINC429 Specification Tutorial Ver 1.1.

[23] R. Herveille. *WISHBONE System-on-Chip (SoC) Interconnection Architecture for Portable IP Cores. Specification*, 2002.

[24] K. Hines and G. Borriello. Dynamic communication models in embedded system co-simulation. In *IEEE DAC*, 1997.

[25] IBM Microelectronics. *CoreConnect Bus Architecture Overview*, 1999.

[26] IBM Microelectronics. *The CoreConnect Bus Architecture White Paper*, 1999.

[27] Sonics Inc. SMART Interconnect Solutions. http://www.sonicsinc.com.

[28] Sonics Inc. SNAP—Sonics Network for AMBA Protocol.

[29] Sonics Inc. Sonics3220 SMART Interconnect Solution.

[30] Sonics Inc. SonicsLX SMART Interconnect Solution.

[31] Sonics Inc. SonicsMX SMART Interconnect Solution.

[32] Sonics Inc. SonicsStudio Development Environment Product Brief.

[33] Sonics Inc. SonicsSX SMART Interconnect Solution.

[34] Atmel. Markus Schmid. Automotive Bus Systems.

[35] Motorola. LIN protocol description.

[36] Aerospace Airbus GmbH Hamburg Germany. N. Rieckmann, Daimler-Benz. ARINC 629 data bus physical layer technology.

[37] OCP-IP. *OCP CoreCreator II.*, 2008.

[38] OCP-IP. *Open Core Protocol Specification, Release 3.*, 2009.

[39] OpenCores. http://opencores.org/.

[40] S. Pasricha, N. Dutt, and M. Ben-Romdhane. Using TLM for Exploring Bus-based SoC Communication Architectures. In *IEEE ASAP*, 2005.

[41] S. Pasricha, N. Dutt, and M. Ben-Romdhane. Constraint-driven bus matrix synthesis for MPSoC. In *IEEE ASP-DAC*, 2006.

[42] S. Pasricha, N. Dutt, and M. Ben-Romdhane. BMSYN: Bus Matrix Communication Architecture Synthesis for MPSoC. *IEEE Trans. on CAD*, 8(26):1454–1464, Aug 2007.

[43] S. Pasricha, N. Dutt, E. Bozorgzadeh, and M. Ben-Romdhane. Floorplan-aware automated synthesis of bus-based communication architectures. In *IEEE DAC*, 2005.

[44] J.A. Rowson and A. Sangiovanni-Vincentelli. Interface-based design. In *IEEE DAC*, 1997.

[45] J.A. Rowson and A. Sangiovanni-Vincentelli. Getting to the bottom of deep sub-micron. In *IEEE ICCAD*, 1998.

[46] Y. Sheynin, E. Suvorova, and F. Shutenko. Complexity and Low Power Issues for On-chip Interconnections in MPSoC System Level Design. In *IEEE ISVLSI*, 2006.

[47] STMicroelectronics. *STBus communication system concepts and definitions*, 2007.

[48] Synopsys. CoCentric System Studio. http://www.synopsys.com.

[49] P. Y. K. Cheung T. S. T. Mak, P. Sedcole, and W. Luk. On-FPGA communication architectures and design factors.

[50] A. Wieferink, T. Kogel, R. Leupers, G. Ascheid, H. Meyr, et al. A system level processor/communication co-exploration methodology for multi-processor system-on-chip platforms. In *IEEE DATE*, 2004.

[51] www.opencores.org. *Specification for the: WISHBONE System-on-Chip (SoC) Interconnection Architecture for Portable IP Cores, Revision B.3.*, 2002.

[52] Xilinx. *Local Memory Bus (LMB) V10 (v1.00a)*, 2009.

[53] Xilinx. *Processor Local Bus (PLB) v4.6 (v1.04a) Data Sheet.*, 2009.

[54] Xilinx. *Fast Simplex Link (FSL) Bus (v2.11c) Data Sheet.*, 2010.

3

NoC Architectures

Martino Ruggiero

Ecole Polytechnique Federale de Lausanne (EPFL), Switzerland

CONTENTS

3.1   Introduction

Constant technology scaling enables the integration and implementation of increasingly complex functionalities onto a single chip. Nowadays advancements in chip manufacturing technology allow the integration of several hardware components into a single integrated circuit reducing both manufacturing costs and system dimensions. System-on-a-Chip (SoC) can integrate several Intellectual Property (IPs) and industry chip vendors are releasing multicore products with increasing core counts [18]. This multicore trend may lead to hundreds and even thousands of cores integrated on a single chip. As core counts increase, there is a corresponding increase in bandwidth demand to facilitate high core utilization. The basic performance of the processing elements are of no use unless the data can be fed to them at the appropriate rates. System-on-chip (SoC) architectures are indeed getting communication-bound and a scalable and high-bandwidth communication fabric becomes critically important [29]. It is clear that there is a critical need for scalable interconnection fabrics.

Considering the level of integration enabled by recent silicon technology advances, a reliable communication of the system components is also becoming a major concern. New challenges have to be faced by on-chip communication in a billion-transistor SoC paradigm: not only scalability and performance, but also reliability and energy reduction are issues. Deep submicron effects, like crosstalk effects, capacitance coupling, and wire inductance caused by feature size shrinking of transistors, will be more and more an issue. Process, thermal, and voltage variations introduce unpredictable performance and errors at every architectural level, mainly due by uncertainties in fabrication and run-time operations [20]. Implementation constraints are also increasingly complicating the design process. While silicon area still requires optimization, power consumption becomes more critical along with other physical considerations like thermal hotspots. Despite these technological and architectural difficulties, the design time is always expected to be shortened due to the time-to-market pressure. Reusing simple bus architectures for the communication does not satisfy the mentioned requirements [6].

Traditionally, SoC designs utilize topologies based on shared buses. However, crossbar and point-to-point interconnections usually scale efficiently only up to twenty cores [34], for larger number of IPs integrated in a single chip a more scalable and flexible solution is needed. When shared buses and custom point-to-point communication are no longer sufficient, more elaborate networks are the obvious choice. By turning from the current path of buses and custom communication designs for the higher levels of interconnection on the chip, it is possible to reach high performance with lower design and verification costs. The solution consists of an on-chip data-routing network consisting of communication links and routing nodes generally known as Network-on-Chip (NoC) architectures [5].

A generic NoC design divides a chip into a set of tiles, with each tile containing a processing element and an on-chip router. Each processing element is connected to the local router through a network interface controller that packetsizes/depacketizes the data into/from the underlying interconnection network. Each router is connected to other routers forming a packet-based on-chip network.

A NoC design is a very complex optimization problem and challenges lie in the huge design space. The NoC design space has several spacial dimensions [28]. The first one represented by the choice of the communication infrastructure, such as network topology, channel bandwidth, and buffer size. This dimension defines how routers are interconnected to each other and reflects the fundamental properties of the underlying network. Another dimension depends on the communication paradigm, which dictates the dynamics of transferring and routing messages on the network. The application task mapping is the last dimension: it decides how the different application tasks are mapped to the network nodes.

NoCs have been widely studied and reported in several special issues in journals, numerous special sessions in conferences, and books [26][7][17]. However, the aforementioned sources do not present many implementation examples about the current NoC proposals. This chapter deals with an in-depth review of the state-of-the-art of the existing implementations for NoC. Here, a vast set of studies are gathered from literature and analyzed. No optimal NoC exists in the general case, but we want to point out that the main goal of this review is only to show major architectural and technological trends.

The chapter is structured as follows. Sections 3.2 and 3.3 discuss main advantages and challenges introduced by the NoC paradigm. Section 3.4 presents the design principles of NoC, while Section 3.5 describes its fundamental building blocks. Section 3.6 presents and discusses the available NoC implementations in literature and on the market.

3.2 Advantages of the NoC Paradigm

NoCs are packet-switching networks, brought to the on-chip level. Even if on-chip networks can leverage ideas from off-chip ones (like multichassis interconnection networks used in supercomputers, clusters of workstations, or Internet routers), the NoC design requirements differ in number and type. However, by moving on-chip, the main bottlenecks that faced prior off-chip networks are alleviated or in some cases avoided. This is mainly due to the abundance of on-chip wiring, which can supply bandwidth that is orders of magnitude higher than off-chip I/Os while obviating the inherent delay overheads associated with off-chip I/O transmission.

NoCs have the potential to bring a large number of advantages to on-chip communication. One of those is the virtually unlimited architectural scalability. Clearly, it is easy to comply with higher bandwidth requirements by larger numbers of cores simply by deploying more switches and links.

NoCs also have a much better electrical performance since all connections are point-to-point. The length of interswitch links is a design parameter that can be adjusted. The wire parallelism in links can be controlled at will, since packet transmission can be serialized. All these factors imply faster propagation times and total control over crosstalk issues.

In on-chip networks, routing concerns are greatly alleviated due to the possibility of having narrower links than in buses. Wiring overhead is also dramatically reduced leading to higher wire utilization and efficiency. Moreover, physical design improvements make NoCs more predictable than buses, enabling faster and easier design closure achievement.

NoCs exhibit better performance under load than buses or custom communication architectures since their operating frequency can be higher, the data width is parameterizable, and communication flows can be handled in parallel with suitable NoC topology design. Virtually any bandwidth load can be tackled.

When dealing with system design embedding a NoC, IP cores are attached in point-to-point fashion to on-chip networks via dedicated network interfaces. Network interfaces can be specialized for any interface that may be needed and potentially any core may be seamlessly attached to a NoC given the proper network interface. Computation and communication concerns are clearly decoupled at the network interface level enabling a more modular and plug&play-oriented approach to system assembly.

Very often hierarchical buses are assembled by hands and therefore manual intervention must be taken into account to tune and validate the overall design. On the contrary NoCs can be designed, optimized, and verified by automated means, leading to large savings in design times, and getting a solution closer to optimality. Moreover, NoCs can be tuned in a variety of parameters (topology, buffering, data widths, arbitrations, routing choices, etc.), leading to higher chances of optimally matching design requirements. Being distributed, modular structures, NoCs can also accommodate differently tuned regions. For example, some portions of a NoC could be tuned statically for lower resource usage and lower performance, or could dynamically adjust their mode of operation.

3.3   Challenges of the NoC Paradigm

Network-on-chip is a communication network targeted for on chip data transfer. A number of stringent technology constraints present challenges for on-chip network designs. Specifically, on-chip networks must supply high bandwidth at low latencies, with tight power consumption constraint and area budget. In order for enabling an increasing adoption of on-chip networks, the communication latency of a NoC implementation must be competitive with crossbars. Moreover, they need to be carefully designed as on-chip network power consumption can be high.

In SoC designs, caches and interconnects compete with the cores for the same chip real estate, so integration a such a large number of components poses a significant challenge for architects to create a balance between them. On-chip network are facing a completely different set of constraints compared to off-chip ones. While in the latter environment a switch is implemented with at least one dedicated chip, in a NoC the switch must occupy a tiny fraction of the chip real estate. This means that some of the principles acquired in wide area networking have to be revisited.

Considering NoC, the tradeoffs among network features, area and power budgets have to be studied from scratch. Policies that are widely accepted in general networking (e.g., dynamic packet routing) must be reassessed to evaluate their impact on silicon area.

Performance requirements are very different in the on-chip domain, also due to the completely different properties of on-chip wiring. Bandwidth milestones are much easier to achieve, since information transfer across on-chip wires is much faster than across long cables. Conversely, latency bounds are much stricter; while milliseconds or even hundreds of milliseconds are acceptable for wide area networks, IP cores on a chip normally require response times of a few nanoseconds.

Contrary to wide area networks, where nodes may often be dynamically connected to and disconnected from the network, in NoCs the set of attached IP cores is obviously fixed. In many applications, it is also relatively easy to statically characterize the traffic profiles of such IP cores. This opens up the possibility of thoroughly customizing NoCs for specific workloads. How to achieve this goal is, however, less clear.

Design tools for NoCs can be developed, but, as above, how exactly is an open question. The customizability of NoCs, while an asset, is also an issue when it comes to devising tools capable of pruning the design space in search of the optimal solutions. The problem is compounded by the need to take into account both architectural and physical properties; by the need to guarantee design closure; and by the need to validate that the outcome is fully functional, e.g., deadlock-free and compliant with performance objectives.

NoCs are a recent technology, and as such, they are in need of the development of thorough infrastructure. In addition to design tools, this includes simulators, emulation platforms, and back-end flows for the implementation on both FPGAs and ASICS.

3.4 Principles of NoC Architecture

The design of an on-chip network can be broken down into its various building principles: the topology, the routing and the flow control. This section gives an overview on these architectural aspects.

3.4.1 Topology

The on-chip network topology determines the physical layout and connections between nodes and channels in the network. Many different topologies exist, from the simplest crossbar to very complex hierarchical cubes and beyond [14].

The effect of a topology on overall network cost-performance is profound. The implementation complexity cost of a topology depends on two factors: the number of links at each node and the ease of laying out a topology on a chip. A topology determines the number of hops a message must traverse as well as the interconnect lengths between hops, thus influencing network latency significantly. As traversing routers and links incurs energy, a topology's effect on hop count also directly affects network energy consumption. Furthermore, the topology dictates the total number of alternate paths between nodes, affecting how well the network can spread out traffic and hence support bandwidth requirements.

A network topology can be classified as either direct or indirect. With a direct topology, each terminal node (e.g., a processor core or cache in a chip multiprocessor) is associated with a router, so all routers are sources and destinations of traffic. In a direct topology, nodes can source and sink traffic, as well as switch through traffic from other nodes. In an indirect topology, routers are distinct from terminal nodes; only terminal nodes are sources and destinations of traffic, intermediate nodes simply switch traffic to and from terminal nodes. In a direct network, packets are forwarded directly between terminal nodes. With an indirect network, packets are switched indirectly through a series of intermediate switch nodes between the source and the destination. To date, most designs of on-chip networks have used direct networks. Co-locating switches with terminal nodes is often most suitable in area-constrained environments such as on-chip networks.

The basic regular network topologies are discussed below.

A **mesh-shaped network** consists of m columns and n rows (see Figure 3.1). The routers are situated in the intersections of two wires and the computational resources are near routers. Addresses of routers and resources can be easily defined as x-y coordinates in mesh.

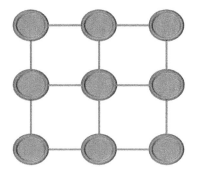

FIGURE 3.1
A Mesh Topology.

A **torus** network is an improved version of basic mesh network (see Figure 3.2). A simple torus network is a mesh in which the heads of the columns are connected to the tails of the columns and the left sides of the rows are connected to the right sides of the rows. Torus network has better path diversity than mesh network, and it also has more minimal routes.

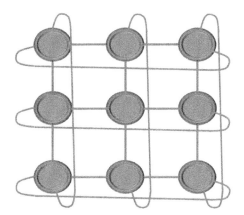

FIGURE 3.2
A Torus Topology.

In a **tree** topology nodes are routers and leaves are computational resources. The routers above a leaf are called the leaf's ancestors and correspondingly the leafs below the ancestor are its children. In a *fat tree topology* each node has replicated ancestors, which means that there are many alternative routes between nodes (see Figure 3.3).

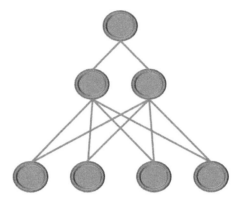

FIGURE 3.3
A Fat Tree Topology.

A **butterfly** network is uni- or bidirectional and butterfly-shaped network typically uses a deterministic routing (see Figure 3.4).

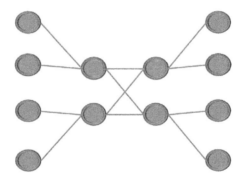

FIGURE 3.4
A Butterfly Topology.

The simplest **polygon** network is a circular network where packets travel in a loop from router to other. A network becomes more diverse when chords are added to the circle. When there are chords only between opposite routers, the topology is called **spidergon** (see Figure 3.5).

A **star** network consists of a central router in the middle of the star, and computational resources or subnetworks in the spikes of the star (see Figure 3.6). The capacity requirements of the central router are quite large, because all the traffic between the spikes goes through the central router. That causes a remarkable possibility of congestion in the middle of the star.

The **ring** topology is very easy to implement but is subject to the same fundamental limitations as a time-division bus since the only available resource

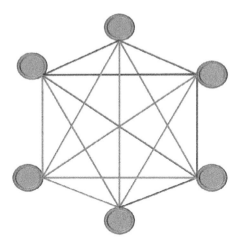

FIGURE 3.5
A Spidergon Topology.

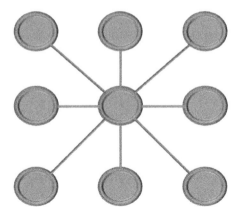

FIGURE 3.6
A Star Topology.

for transmission at a node will be occupied whenever a transmission wants to pass that node (see Figure 3.7).

On the other hand, the more powerful topologies like fat trees are very complex when it comes to wiring. The two-dimensional mesh and torus are very suitable for on-chip networks. The main advantages of these are the good performance-to-resource ratio, the ease of routing, and that the topologies are very easily mapped onto a chip. The torus have somewhat better properties for random traffic but the added resources may be unnecessary for many systems. The better performance of the torus is due to the uniform connectivity of that topology. The uniform connectivity also removes any boundary effects from

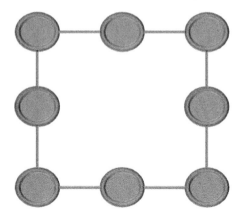

FIGURE 3.7
A Ring Topology.

the torus, which will appear totally homogeneous. The average logical distance between routers will become shorter in the torus, but at the cost of possibly longer physical distances.

Early works on NoC topology design assumed that using regular topologies, such as meshes, would lead to regular and predictable layouts [22][7]. While this may be true for designs with homogeneous processing cores and memories, it is not true for most MPSoCs as they are typically composed of heterogeneous cores in terms of area and communication requirements. A regular, tile-based floorplan, as in standard topologies, would result in poor performance, with large power and area overheads. Moreover, for most state-of-the-art MPSoCs the system is designed with static (or semistatic) mapping of tasks to processors and hardware cores, and hence the communication traffic characteristics of the MPSoC can be obtained statically. The clear advantage with allowing implementations with **custom** topology is the possibility to adapt the network to the application domain at hand. This can give significant savings for parts of the network where little traffic is handled while still being able to support significantly more traffic in other parts.

Since the first decision designers have to make when building an on-chip network is the choice of the topology, it is useful to have a means for quick comparisons of different topologies before the other aspects of a network are even determined. Bisection bandwidth is a metric that is often used in the discussion of the cost of off-chip networks. Bisection bandwidth is the bandwidth across a cut down the middle of the network. Bisection bandwidth can be used as a proxy for cost since it represents the amount of global wiring that will be necessary to implement the network. The degree of a topology instead refers to the number of links at each node. Degree is useful as an abstract metric of cost of the network, as a higher degree requires more ports at routers, which increases implementation complexity. The number of hops a message takes from source to destination, or the number of links it traverses, defines the hop count. This is a very simple and useful means for measuring network latency, since every node and link incurs some propagation delay, even when there is no contention.

The first aspect to take into account when selecting which topology to use for a network is the patterns of traffic that will go through the network. So, in order to determine the most appropriate topology for a system an investigation of the advantages and drawbacks of a number of common topologies with respect to the application at hand must be done during the early design stages.

The most common topologies in NoC designs are 2-D mesh and torus, which constitute over 60% of cases.

3.4.2 Routing

NoC architectures are based on packet-switched networks. Routers can implement various functionalities—from simple switching to intelligent routing. Routing on NoC is quite similar to routing on any network. A routing algorithm determines how the data is routed from sender to receiver.

The routing algorithm is used to decide what path a message will take through the network to reach its destination. The goal of the routing algorithm is to distribute traffic evenly among the paths supplied by the network topology, so as to avoid hotspots and minimize contention, thus improving network latency and throughput. All of these performance goals must be achieved while adhering to tight constraints on implementation complexity: routing circuitry can stretch critical path delay and add to a router's area footprint. While energy overhead of routing circuitry is typically low, the specific route chosen affects hop count directly, and thus substantially affects energy consumption. Since embedded systems are constrained in area and power consumption, but still need high data rates, routers must be designed with hardware usage in mind.

Routing algorithms can be classified in various ways. For on-chip communication, unicast routing strategies (i.e., the packets have a single destination) seem to be a practical approach due to the presence of point-to-point communication links among various components inside a chip. Based on the routing decision, unicast routing can be further classified into four classes: centralized routing, source routing, distributed routing, and multiphase routing.

In centralized routing, a centralized controller controls the data flow in a system. In case of source routing, the routing decisions are taken at the point of data generation, while in distributed routing, the routing decisions are determined as the packets/flits flow through the network. The hybrid of the two schemes, source and destination routing, is called *multiphase routing*.

Routing algorithms can also be defined based on their implementation: lookup table and Finite State Machine (FSM). Lookup table routing algorithms are more popular in implementation. They are implemented in software, where a lookup table is stored in every node. We can change the routing algorithm by replacing the entries of the lookup table. FSM-based routing algorithms may be implemented either in software or in hardware.

These routing algorithms may further be classified based on their adaptability. **Oblivious algorithms** route packets without any information about traffic amounts and conditions of the network.

Deterministic routing the path is determined by packet source and destination.

Adaptive routing the routing path is decided on a per-hop basis. Adaptive schemes involve dynamic arbitration mechanisms, for example, based on local link congestion. This results in more complex node implementations but offers benefits like dynamic load balancing.

Deterministic routing is preferred as it is easier to implement; adaptive routing, on the other hand, tends to concentrate traffic at the center of the network resulting in increased congestion there [31][32]. Adaptive routing algorithms need more information about the network to avoid congested paths in the network. These routing algorithms are obviously more complex to implement, thus, are more expensive in area, cost, and power consumption. Therefore, we must consider a right QoS (Quality-of-Service) metric before employing these algorithms. Deterministic routing always follows a deterministic path on the network. Problems in oblivious routing typically arise when the network starts to block traffic. The only solution to these problems is to wait for traffic amount to reduce and try again. Deadlock, livelock, and starvation are potential problems on both oblivious and adaptive routing.

Various routing algorithms have been proposed for the NoC. Most researchers suggested static routing algorithms and performed communication analysis based on the static behavior of NoC processes, thus, determining the static routing for NoC.

Packet-switched networks mostly utilize deterministic routing (about 70%) but adaptivity or some means for reprogramming the routing policy is necessary for fault-tolerance. Although deadlock is generally avoided, out-of-order packet delivery is problematic with adaptive routing.

3.4.2.1   Oblivious Routing Algorithms

Oblivious routing algorithms have no information about conditions of the network, like traffic amounts or congestions. A router makes routing decisions on the grounds of some algorithm. The simplest oblivious routing algorithm is a minimal turn routing. It routes packets using as few turns as possible.

Dimension order routing is a typical minimal turn algorithm. The algorithm determines to what direction packets are routed during every stage of the routing.

XY routing is a dimension order routing that routes packets first in x- or horizontal direction to the correct column and then in y- or vertical direction to the receiver (see Figure 3.8). XY routing suits well on a network using mesh or torus topology. Addresses of the routers are their xy-coordinates. XY routing never runs into deadlock or livelock. There are some problems in the traditional XY routing. The traffic does not extend regularly over the whole network because the algorithm causes the biggest load in the middle of the network. There is a need for algorithms that equalize the traffic load over the whole network.

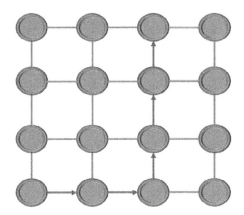

FIGURE 3.8
XY routing from router A to router B.

Pseudoadaptive XY routing works in deterministic or adaptive mode depending on the state of the network. The algorithm works in deterministic mode when the network is not or only slightly congested. When the network becomes blocked, the algorithm switches to the adaptive mode and starts to search routes that are not congested. While a traditional XY routing causes network loads more in the middle of the network than to lateral areas, the pseudoadaptive algorithm divides the traffic more equally over the whole network. On the contrary, it requires additional hardware resources (i.e., buffers) to handle congestion signaling and handshaking. *Surrounding XY* routing has three different routing modes. *Normal XY* mode works just like the basic XY routing. It routes packets first along the x-axis and then along the y-axis. Routing stays on The Normal XY mode as long as the network is not blocked and routing does not meet inactive routers. *Surround horizontal XY* mode is used when the router's left or right neighbor is deactivated (see Figure 3.9).

Correspondingly, the third mode *Surround vertical XY* is used when the upper or lower neighbor of the router is inactive (see Figure 3.10).

The surround horizontal XY mode routes packets to the correct column on the grounds of coordinates of the destination. The algorithm bypasses packets around the inactive routers along the shortest possible path. The situation is a little bit different in the surround vertical XY mode because the packets are already in the right column. Packets can be routed to the left or right. The routers in the surround horizontal XY and surround vertical XY modes add a small identifier to the packets that tells to other routers that these packets are

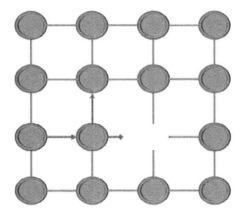

FIGURE 3.9
Surround horizontal XY.

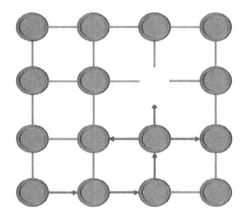

FIGURE 3.10
Surround vertical XY. There are 2 optional directions.

routed using surround horizontal XY or surround vertical XY mode. Thus, the other routers do not send the packets backwards.

Turn model algorithms determine a turn or turns that are not allowed while routing packets through a network. Turn models are livelock-free. A *west-first* routing algorithm prevents all turns to the west (see Figure 3.11). So the packets going to the west must be first transmitted as far to the west as necessary. Routing packets to the west is not possible later.

Turns away from the north are not possible in a *north-last* routing algorithm (see Figure 3.12). Thus, the packets that need to be routed to the north, must be transferred there at last.

The negative-first routing algorithm allows all other turns except turns

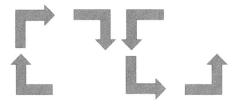

FIGURE 3.11
Allowed turns in west-first.

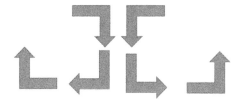

FIGURE 3.12
Allowed turns in north-last.

from a positive direction to a negative direction (see Figure 3.13). Packet routings to negative directions must be done before anything else.

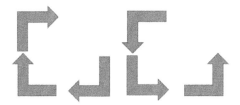

FIGURE 3.13
Allowed turns in negative-first.

3.4.2.2 Deterministic Routing Algorithms

Deterministic routing algorithms route packets every time from two certain points along a fixed path. In congestion-free networks deterministic algorithms are reliable and have low latency. They suit well on real-time systems because packets always reach the destination in correct order and so a reordering is not necessary. In the simplest case each router has a routing table that includes routes to all other routers in the network. When network structure changes, every router has to be updated.

A *shortest path routing* is the simplest deterministic routing algorithm. Packets are always routed along the shortest possible path. A distance vector routing and a link state routing are shortest path routing algorithms. In *Distance Vector Routing*, each router has a routing table that contains information about neighbor routers and all recipients. Routers exchange routing table information with each other and this way keep their own tables up to date. Routers route packets by counting the shortest path on the basis of their routing tables and then send packets forward. Distance vector routing is a simple method because each router does not have to know the structure of the whole network. *Link state routing* is a modification of distance vector routing. The basic idea is the same as in distance vector routing, but in link state routing each router shares its routing table with every other router in the network. Link state routing in a Network-on-Chip systems is a customized version of the traditional one. The routing tables covering the whole network are stored in a router's memory already during the production stage. Routers use their routing table updating mechanisms only if there are remarkable changes in the network's structure or if some faults appear.

In a *source routing* a sender makes all decisions about the routing path of a packet. The whole route is stored in the header of the packet before sending, and routers along the path do the routing just like the sender has determined it. A *vector routing* works basically like the source routing. In the vector routing the routing path is represented as a chain of unit vectors. Each unit vector corresponds to one hop between two routers. Routing paths do not have to be the shortest possible. *Arbitration lookahead scheme* (ALOAS) is a faster version of source routing. The information from the routing path has been supplied to routers along the path before the packets are even sent. Route information moves along a special channel that is reserved only for this purpose.

A *contention-free* routing is an algorithm based on routing tables and time division multiplexing. Each router has a routing table that involves correct output ports and time slots to every potential sender-receiver pairs.

A *destination-tag* routing is a bit like an inversed version of the source routing. The sender stores the address of the receiver, also known as a destination-tag, to the header of the packet in the beginning of the routing. Every router makes a routing decision independently on the basis of the address of the receiver. The destination-tag routing is also known as *floating vector routing*.

Deterministic routing algorithms can be improved by adding some adaptive features to them. A *topology adaptive* routing algorithm is slightly adaptive. The algorithm works like a basic deterministic algorithm but it has one feature that makes it suitable to dynamic networks. A systems administrator can update the routing tables of the routers if necessary. A corresponding algorithm is also known as *online oblivious routing*. The cost and latency of the topology adaptive routing algorithm are near to costs and latencies of basic deterministic algorithms. A facility of topology adaptiveness is its suitability to irregular and dynamic networks.

Routing with *stochastic routing* algorithms is based on coincidence and an assumption that every packet sooner or later reaches its destination. Stochastic algorithms are typically simple and fault-tolerant. Throughput of data is especially good but as a drawback, stochastic algorithms are quite slow and they use plenty of network resources. Stochastic routing algorithms determine a packet's time to live, which is how long a packet is allowed to move around in the network. After the determined time has been reached, the packet will be removed from the network. The most common stochastic algorithm type is the *flooding algorithm*. Here are three different versions of flooding. The costs of each of the three algorithms are equivalent. The simplest stochastic routing algorithm is the *probabilistic flooding* algorithm. Routers send a copy of an incoming packet to all possible directions without any information about the location of packet's destination. The packet's copies diffuse over the whole network like a flood. Finally, at least one of the copies will arrive to its receiver and the redundant copies will be removed. A *directed flood* routing algorithm is an improved version of the probabilistic flood. It directs packets approximately to the direction where their destination exists. The directed flood is more fault-tolerant than the probabilistic flood and uses less network resources. A *random walk* algorithm sends a predetermined amount of the packet's copies to the network. Every router along the routing path sends incoming packets forward through some of its output ports. The packets are directed in the same way as in the directed flood algorithm. The random walk is as fault-tolerant as the directed flood but consumes less energy and bandwidth.

3.4.2.3   Adaptive Routing Algorithms

The minimal adaptive routing algorithm always routes packets along the shortest path. The algorithm is effective when more than one minimal, or as short as possible, routes between sender and receiver exist. The algorithm uses the route that is least congested.

A *fully adaptive routing* algorithm always uses a route that is not congested. The algorithm does not care that the route is not the shortest path between sender and receiver. Typically, an adaptive routing algorithm sets alternative congestion-free routes in order of superiority. The shortest route is the best one.

A *congestion look-ahead* algorithm gets information about blocks from other routers. On the basis of this information, the routing algorithm can direct packets to bypass the congestion.

Turnaround routing is a routing algorithm for butterfly and fat-tree networks (see Figure 3.14). Senders and receivers of packets are all on the same side of the network. Packets are first routed from the sender to some random intermediate node on the other side of the network. In this node the packets are turned around and then routed to the destination on the same side of the network, where the whole routing started. The routing from the intermediate node to the definite receiver is done with the destination-tag routing. Routers in turnaround routing are bidirectional, which means that packets can flow through the router in both forward and backward directions. The algorithm is deadlock-free because packets only turn around once from a forward channel to a backward channel.

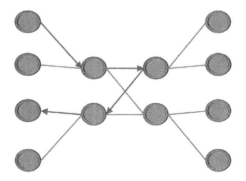

FIGURE 3.14
Turnaround routing in a butterfly network.

Turn-back-when-possible is an algorithm for routing on tree networks. It is a slightly improved version of the turnaround routing. When turn-back channels are busy, the algorithm looks for a free routing path on a higher switch level. A turn-back channel is a channel between a forward and a backward channel. It is used to change the routing direction in the network.

IVAL (Improved VALiant's randomized routing) is a bit similar to turn-around routing. In the algorithm's first stage, packets are routed to randomly chosen point between the sender and the receiver by using an oblivious dimension order routing. The second stage of the algorithm works almost equally, but this time the dimensions of the network are gone through in reversed order. Deadlocks are avoided in IVAL routing by dividing the router's channels to virtual channels. Full deadlock avoidance requires a total of four virtual channels per one physical channel.

The 2TURN algorithm itself does not have an algorithmic description. Only the algorithm's possible routing paths are determined in a closed form. Routing from sender to receiver with the 2TURN algorithm always consists of 2 turns that will not be U-turns or changes of direction within dimensions. Just as in the IVAL routing, a 2TURN router can avoid deadlock if all the router's physical channels are divided to four virtual channels. Locality is a routing algorithm metric that is expressed as the distance a packet travels on average. This metric largely determines the end-to-end delay of packets at low load.

An *odd-even* routing is an adaptive algorithm used in a dynamically adaptive and deterministic (DyAD) Network-on-Chip system. The odd-even routing is a deadlock-free turn model that prohibits turns from east to north and from east to south at tiles located in even columns and turns from north to west and south to west at tiles located in odd columns. The DyAD system uses the minimal odd-even routing, which reduces energy consumption and also removes the possibility of livelock.

A *hot-potato* routing algorithm routes packets without temporarily storing them in routers' buffer memory. Packets are moving all the time without stopping before they reach their destination. When one packet arrives to a router, the router forwards it right away towards a packet's receiver but if there are two packets going in the same direction simultaneously, the router directs one of the packets to some other direction. This other packet can flow away from its destination. This occasion is called *misrouting*. In the worst case, packets can be misrouted far away from their destination and misrouted packets can interfere with other packets. The risk of misrouting can be decreased by waiting a little random time before sending each packet. Manufacturing costs of the hot-potato routing are quite low because the routers do not need any buffer memory to store packets during routing.

3.4.2.4   Problems on Routing

Deadlock, livelock, and starvation are potential problems on both oblivious and adaptive routing.

Routing is in **deadlock** when two packets are waiting for each other to be routed forward. Both of the packets reserve some resources and both are waiting on each other to release the resources. Routers do not release the resources before they get the new resources and so the routing is locked.

Livelock occurs when a packet keeps going around its destination without ever reaching it. There are mainly two solutions to avoid in livelock. Time to live counter counts how long a packet has travelled in the network. When the counter reaches some predetermined value, the packet will be removed from the network. The other solution is to give packets a priority, which is based on the "age" of the packet. The oldest packet always finally gets the highest priority and will be routed forward.

Using different priorities can cause a situation where some packets with lower priorities never reach their destinations. This occurs when the packets with higher priorities reserve the resources all the time. **Starvation** can be avoided by using a fair routing algorithm or reserving some bandwidth only for low-priority packets.

3.4.3   Flow Control

Flow control determines how network resources, such as channel bandwidth, buffer capacity, and control state, are allocated to a packet traversing the network. It governs and determines when buffers and links are assigned to messages, the granularity at which they are allocated, and how these resources are shared among the many messages using the network. A well-designed flow control protocol decreased the latency in delivering messages at low loads without imposing high overhead in resource allocation, and increased network throughput by enabling effective sharing of buffers and links across messages. Flow control is instrumental in determining network energy and power consumption, since it determines the rate at which packets access buffers and traverse links. Its implementation complexity influences the complexity of the router microarchitecture as well as the wiring overhead that is required for communicating resource information between routers.

Usually when dealing with NoC, the message injected into the network is first segmented into packets. A message is the logical unit of communication above the network, and a packet is the physical unit that makes sense to the network. Packets are further divided into fixed-length flits. The packet consists of a head flit that contains the destination address, body flits, and a tail flit that indicates the end of a head, body, and packet. Flits can be further broken down into *phits*, which are physical units and correspond to the tail flits physical channel width. A packet contains destination information while a flit may not, thus all flits of a packet must take the same route.

On-chip networks present an abundance of wiring resources, so messages are likely to consist of a single or a few packets. In off-chip networks, channel widths are limited by pin bandwidth so that flits need to be broken down into smaller chunks called phits. In on-chip networks, flits are composed of a single phit and are the smallest subdivision of a message due to wide on-chip channels.

Flow control techniques are classified by the granularity at which resource allocation is handled: based on message, packet, or flit.

3.4.3.1 Message-Based

Circuit-switching is a technique that operates at the message-level, which is the coarsest granularity, and then refines these techniques to finer granularities. It preallocates resources (links) across multiple hops to the entire message.

A small setup message is sent into the network and reserves the links needed to transmit the entire message from the source to the destination. Once the setup message reaches the destination and has successfully allocated the necessary links, an acknowledgment message will be transmitted back to the source. When the source receives the acknowledgment message, it will release the message that can then travel quickly through the network. Once the message has completed its traversal, the resources are deallocated. After the setup phase, per-hop latency to acquire resources is avoided.

With sufficiently large messages, this latency reduction can amortize the cost of the original setup phase. In addition to possible latency benefits, circuit switching is also bufferless. As links are pre-reserved, buffers are not needed at each hop to hold packets that are waiting for allocation, thus saving on power. While latency can be reduced, circuit switching suffers from poor bandwidth utilization. The links are idle between setup and the actual message transfer and other messages seeking to use those resources are blocked.

3.4.3.2   Packet-Based

Packet-based flow control techniques first break down messages into packets, then interleave these packets on the links, thus improving link utilization. Unlike Message-Based, the remaining techniques will require per-node buffering to store in-flight packets. There are two main choices how packets are forwarded and stored: store-and-forward and cut-through.

The basic mode for packet transport is *store-and-forward* where a packet will be received at a router in its entirety before forwarding is done to the output. Clearly, the store-and-forward method waits for the whole packet before making routing decisions The node stores the complete packet and forwards it based on the information within its header. The packet may stall if the router does not have sufficient buffer space. The drawback is that store-and-forward is fairly inefficient for smaller, dedicated networks. Latency as well as the requirements on buffer memory size will be unnecessarily high.

Cut-through forwards the packet already when the header information is available. Cut-through works like the wormhole routing but before forwarding a packet the node waits for a guarantee that the next node in the path will accept the entire packet. The main forwarding technique used in NoCs is wormhole because of the low latency and the small realization area as no buffering is required. Most often connectionless routing is employed for best effort while connection-oriented routing is preferable for guarantee throughput needed when applications have QoS requirements. Once the destination node is known, in order to determine to which of the switch's output ports the message should be forwarded, static or dynamic techniques can be used.

Both store-and-forward and cut-through methods need buffering capacity for one full packet at the minimum. Wormhole switching is the most popular and well suited on chip. It splits the packets into several flits (flow control digits). Routing is done as soon as possible, similarly to cut-through, but the buffer space can be smaller (only one flit at the smallest). Therefore, the packet may be spread into many consecutive routers and links like a worm.

3.4.3.3   Flit-Based

To reduce the buffering requirements of packet-based techniques, flit-based flow control mechanisms exist. Low buffering requirements help routers meet tight area or power constraints on-chip.

In *wormhole*, the node looks at the header of the packet (stored in the first flit) to determine its next hop and immediately forwards it. The subsequent flits are forwarded as they arrive at the same destination node. As no buffering is done, wormhole routing attains a minimal packet latency. The main drawback is that a stalling packet can occupy all the links a worm spans. The general drawback with wormhole routing is the increased resource occupancy that can increase the deadlock problems in the network.

3.5 Basic Building Blocks of a NoC

3.5.1 Router

A router consists basically of a set of input and output buffers, an interconnection matrix, and some control logic. Buffers are exploited to enqueue transmission data allowing for the local storage of data that cannot be immediately routed, but unfortunately they have a high cost in terms of energy consumption. The interconnection matrix usually is realized by a single crossbar, but it can be also composed by the cascading of various stages. The control logic is represented by the circuitry handling additional functionalities such as arbitration, error detection and correction, control flow protocols, etc.

Routers must be designed to meet latency and throughput requirements under tight area and power constraints. The architecture of the router affects the NoC per-hop delay and overall network latency. Router microarchitecture also impacts network energy as it determines the circuit components in a router and their activity. The implementation of the routing, flow control, and the actual router pipeline will affect the efficiency at which buffers and links are used and thus overall network throughput. The area footprint of the router is clearly determined by the chosen router microarchitecture and underlying circuits.

The wide majority of NoCs are based on packet-switching networks, and packets are forwarded on a per-hop basis. While circuit-switching forms a path from source to destination prior to transfer by reserving the routers (switches) and links, packet-switching performs routing per-packet basis. Packet-switching is more common and it is utilized in about 80% of the available NoCs.

Circuit-switching is best suited for predictable transfers that are long enough to amortize the setup latency, and which require performance guarantees. Circuit-switching scheme also reduces the buffering needs at the routers. Packet-switching necessitates buffering and introduces unpredictable latency (jitter) but is more flexible, especially for small transfers. It is an open research problem to analyze the break-even point (predictability and duration of transfers) between the two schemes.

For embedded systems such as handheld devices, cost is a major driving force for the success of the product and therefore the underlying architecture as well. Along with being cost effective, handheld systems are required to be of small size and to consume significantly less power, relative to desktop systems. Under such considerations, there is a clear tradeoff in the design of a routing protocol. A complex routing protocol would further complicate the design of the router. This will consume more power and area without being cost effective. A simpler routing protocol will outperform in terms of cost and power consumption, but will be less effective in routing traffic across the system.

3.5.1.1 Virtual Channels

The throughput of interconnection networks is limited to a fraction (typically 20% to 50%) of the network capacity due to the coupled allocation of resources: buffers and channels. Typically, a single buffer is associated with each channel. However, a channel can be multiplexed in *Virtual Channels*. Virtual channels provide multiple buffers for each channel, increasing the resources allocation for each packet. A virtual channel is basically a separate queue in the router, and multiple virtual channels share the physical wires (physical link) between two routers. Virtual channels arbitrate for physical link bandwidth on a cycle-by-cycle basis. When a packet holding a virtual channel becomes blocked, other packets can still traverse the physical link through other virtual channels. Thus virtual channels increase the utilization of the physical links and extends overall network throughput.

A virtual channel splits a single channel into more channels, virtually providing two paths for the packets to be routed. The use of virtual channels reduces the network latency, but it costs in terms of area, power consumption, and production cost. However, there are various other added advantages offered by virtual channels. Virtual channels can be applied to all the above flow control techniques to alleviate head-of-line blocking. Virtual channels are also widely used to break deadlocks, both within the network, and for handling system-level or protocol-level deadlocks.

3.5.2 Network Interface

In addition to the router, a network interface is usually needed to handle the packetization, packet re-ordering, and for controlling the retransmissions. The network interface can be considered a protocol converter that maps the processing node I/O protocol into the protocol used within the NoC. It splits messages into packets and, depending on the underlying network, additional routing information are encoded into a packet header. Packets are further decomposed into flits (flow control units), which again are divided into phits (physical units), which are the minimum size datagram that can be transmitted in a single link transaction.

The network interface offers high-level services by abstracting the underlying network to an interface. It handles the end-to-end flow control, encapsulating the messages or transactions generated by the cores for the routing strategy of the network. Adaptive routing may deliver packets out-of-order. The network interface is responsible for reordering if processing the element requires that. Reordering typically needs large buffers and the receiver must send acknowledge at known intervals to avoid infinite buffers.

3.5.3 Link

The link represents the realization through wires of the physical connection between two nodes in a NoC. The transportation of data packets among various nodes in a NoC can be performed by using either a serial or a parallel link. Parallel links make use of a buffer-based architecture and can be operated at a relatively lower clock rate in order to reduce power dissipation. Unfortunately, parallel links usually incur a high silicon cost due to interwire spacing, shielding, and repeaters. This can be minimized up to a certain limit by employing multiple metal layers.

Serial links allow savings in wire area, reduction in signal interference and noise, and further eliminate the need for having buffers. However, serial links would need serializer and deserializer circuits to convert the data into the right format to be transported over the link and back to the cores. They offer the advantages of a simpler layout and simpler timing verification, but sometimes suffer from ISI (Intersymbol Interference) between successive signals while operating at high clock rates. Nevertheless, such drawbacks can be addressed by encoding and with asynchronous communication protocols.

3.6 Available NoC Implementations and Solutions

In this section we review and discuss some specific NoC implementations found in the open literature.

In the next subsections several concrete systems that employ NoC will be analyzed, as well as several NoC solutions developed by both academia and industry: a certain number of companies already exist that have started bringing the NoC technology to commercial products, namely Arteris [2], Silistix [36], and iNoCs [23], and more start-ups are likely to appear in the growing market of MPSoC on-chip interconnect designs.

3.6.1 IBM Cell

The *Cell* architecture [21] is a joint effort between IBM, Sony, and Toshiba. It represents a family of chips targeting game systems and high-performance computing, but are general enough for other domains as well (i.e., avionics, graphics, etc.). It is a 90 nm, 221 mm 2 chip running at frequencies above 4GHz.

It is composed of one IBM 64-bit Power Architecture core (PPE) and eight Synergistic Processing Elements (SPEs) (see Figure 3.15). The PPE is dedicated to the operating system and acts as the master of the system, while the eight synergistic processors are optimized for compute-intensive applications.

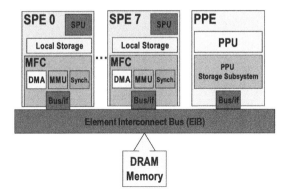

FIGURE 3.15

Cell Broadband Engine Hardware Architecture.

The PPE is a multithreaded core and has two levels of on-chip cache, however, the main computing power of the cell processor is provided by the eight SPEs.

The SPE is a compute-intensive coprocessor designed to accelerate media and streaming workloads. Each SPE consists of a synergistic processor unit (SPU) and a memory flow controller (MFC). The MFC includes a DMA controller, a memory management unit (MMU), a bus interface unit, and an atomic unit for synchronization with other SPUs and the PPE.

Efficient SPE software should heavily optimize memory usage, since the SPEs operate on a limited on-chip memory (only 256 KB local store) that stores both instructions and data required by the program. The local memory of the SPEs is not coherent with the PPE main memory, and data transfers to and from the SPE local memories must be explicitly managed by using asynchronous coherent DMA commands. Both PPE and SPEs can execute vector operations.

3.6.1.1 Element Interconnect Bus

The processing elements (i.e., PPE and SPEs), the memory controller, and the two I/O interfaces are interconnected with an on-chip Element Interconnect Bus (EIB), with a maximum bisection bandwidth of over 300 GBytes/s. The EIB consists of four unidirectional rings (see Figure 3.16), two in each direction, as each ring is 16 bytes wide, runs at 1.6GHz, and can support 3 concurrent transfers.

The rings are accessed in a bus-like manner, with a sending phase where the source element initiates a transaction (e.g., issues a DMA), a command phase through the address-and-command bus where the destination element is informed about this impending transaction, then the data phase where access to rings is arbitrated and if access is granted, data are actually sent from source to destination. Finally, the receiving phase moves data from the network interface (called the Bus Interface Controller (BIC)) to the actual

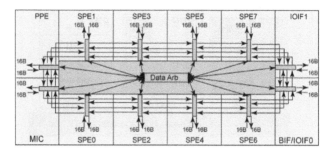

FIGURE 3.16
Element Interconnect Bus Architecture.

local or main memory or I/O. However, the bus access semantics and the ring topology can lead to a worst-case throughput of 50% with adversarial traffic patterns.

The access to rings (i.e., resource allocation) is controlled for the ring arbiter by priority policy. The highest priority is given to the memory controller so requestors will not be stalled on read data. Other elements on the EIB have equal priority and are served in a round-robin manner.

The routing in the EIB consists in the choice between left or right on each of the four unidirectional rings. The ring arbiter will prevent allocations of transfers that are going more than halfway around the ring, i.e., only the shortest path routes are permitted. As the Cell interfaces with the EIB through DMA bus transactions, the unit of communications is large DMA transfers in bursts, with flow control semantics of buses rather than packetized networks.

As the IBM Cell essentially mimics a bus using the four rings, it does not have switches or routers at intermediate nodes. Instead, bus interface units (BIU) arbitrate for dedicated usage of segments of the ring, and once granted, injects into the ring.

3.6.2   Intel TeraFLOPS

The TeraFLOPS processor [25] is an Intel research prototype chip. It is targeted at exploring future parallel processor designs with high core counts. It is a 65 nm, 275 mm^2 chip with 80 tiles running at a targeted frequency of 5 GHz. Each tile has a processing engine (PE) connected to an on-chip network router. The PE embeds two single-precision floating-point multiply-accumulator (FP-MAC) units; 3 KB of single-cycle instruction memory (IMEM), and 2 KB of data memory (DMEM).

The router interface block (RIB) interfaces the PE with the router, and any PE can send or receive instructions and data packets to or from any other PE. Two FPMACs in each PE providing 20 GigaFLOPS of aggregate performance, partnered with a maximum bisection bandwidth of 320 GBytes/s in the on-chip network enable the chip to realize a sustained performance of one TeraFLOPS while dissipating less than 100 W.

3.6.2.1 TeraFLOPS Network

The topology of the TeraFLOPS network is an 8x10 mesh. Each channel consists of two 38-bit unidirectional links. The network runs at a clock rate of 5 GHz on a 65 nm process. The network provides a bisection bandwidth of 380 GB/s or 320 GB/s of actual data bisection bandwidth, since 32 out of the 38 bits of a flit are data bits; the remaining 6 bits are used for sideband. Depending on the traffic, routing and flow control, the real throughput will be a fraction of that.

The network routing is source-based. Each hop is encoded as a 3-bit field corresponding to the 5 possible output ports a packet can take at each router. The hops supported by the packet format are up to ten. However, a chained header bit can be set, indicating that routing information for more hops can be found in the packet data. The TeraFLOPS can tailor routes to specific applications: source routing can enable the use of many possible adaptive or deterministic routing algorithms.

The TeraFLOPS minimum packet size is of two flits (38-bit flits with 6 bits of control data and 32 bits of data), while the router architecture does not impose any limits on the maximum packet size. The flow control is wormhole with two virtual channels. However, the virtual channels are used only to avoid system-level deadlock, and not for flow control. This simplifies the router design since no virtual channel allocation needs to be done at each hop. Buffer backpressure is maintained using on/off signaling, with software programmable thresholds.

The router architecture takes advantage of a 5-stage pipeline: buffer write, route computation, two separable stages of switch allocation, and switch traversal. Single hop delay is just 1 ns. Each port has two input queues, one for each virtual channel.

The TeraFLOPS network supports message passing, through send and receive instructions in the ISA. Any tile can send/receive to any other tile, and send/receive instructions have latencies of 2 cycles, and 5 cycles of router pipeline along with a cycle of link propagation for each hop. The 2-cycle latency of send/receive instructions is the same as that of local load/store instructions. In addition, there are sleep/wakeup instructions to allow software to put entire router ports to sleep for power management. These instructions trigger sleep/wakeup bits in the packet header to be set, which can turn on/off 10 sleep regions in each router as they traverse the network.

3.6.3 RAW Processor

The RAW microprocessor is a general purpose multicore CPU designed at MIT [37]. The main focus is to exploit instruction level parallelism (ILP) across several core tiles whose functional units are connected through a NoC. The first prototype (realized in 1997) had 16 individual tiles, running at a clock speed of 225 MHz.

Each tile contains a general-purpose processor, which is connected to its neighbors by both the static router and the dynamic router. The processor is an eight-stage single-issue MIPS-style pipeline. It has a four-stage pipelined FPU, a 32 KByte two-way associative SRAM data cache, and 32 KBytes of instruction SRAM. The tiles are connected through four 32-bit NoCs with a length of wires that is no greater than the width of a tile, allowing high clock frequencies. Two of the networks are static and managed by a single static router (which is optimized for lowlatency), while the remaining two are dynamic. The networks are integrated directly into the pipeline of the processors, enabling an ALU-to-network latency of 4 clock cycles (for a 8-stage pipeline tile).

The static router is a 5-stage pipeline that controls 2 routing crossbars and thus 2 physical networks. Routing is performed by programming the static routers in a per-clock-basis. These instructions are generated by the compiler and as the traffic pattern is extracted from the application at compile time, router preparation can be pipelined allowing data words to be forwarded towards the correct port upon arrival.

The dynamic network is based on packet-switching and the wormhole routing protocol is used. The packet header contains the destination tile, a user field, and the length of the message. Two dynamic networks are implemented to handle deadlocks. The memory network has a restricted usage model that uses deadlock avoidance. The general network usage is instead unrestricted and when deadlocks happen the memory network is used to restore the correct functionality.

The success of the RAW architecture is demonstrated by the Tilera company, founded by former MIT members, which distributes CPUs based on the RAW design. The current top microprocessor is the Tile-GX, which packs into a single chip 100 tiles at a clock speed of 1.5 Ghz [39].

3.6.4 Tilera Architectures

Tilera produces the TILE64, TILE64Pro, and Tilera Gx architectures [38]. These platforms have been designed to target high-performance embedded applications such as networking and real-time video processing. TILE64 and TILE64Pro support a shared memory space across the 64 tiles (see Figure 3.17).

Each tile consists of a 3-issue VLIW processor core, L1 and L2 caches. TILE64Pro has additional support for cache coherence, called *Dynamic Distributed Cache*. The TILE64 chip at 90 nm, 750 MHz, has 5 MB of on-chip cache and on-chip networks that provide a maximum bisection bandwidth of 2Tb/s with each tile dissipating less than 300 mW.

There is no floating-point (FP) hardware in Tilera TILE64 family. With

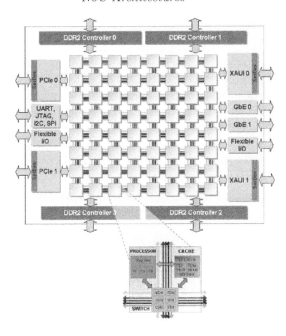

FIGURE 3.17
TILE64 Processors Family Architecture.

the Gx series chips, there is some FP hardware to catch the odd instruction without a huge speed hit. The Tilera Gx can host up to 100 cores and can provide 50 GigaFLOPS of FP (see Figure 3.18). The TILE64 cores are a proprietary 32-bit ISA, and in the Gx it is extended to 64-bit.

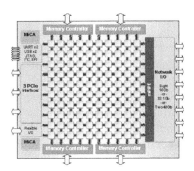

FIGURE 3.18
TILE-Gx Processors Family Architecture.

3.6.4.1 iMesh

Tilera's processors are based on a mesh networks, called *iMesh*. The iMesh consists of five up to 8x8 (10x10 in Gx) meshes. In the TILE64 generations of Tilera chips, all of these networks were 32 bits wide, but on the Gx, the widths vary to give each one more or less bandwidth depending on their functions. Traffic is statically distributed among the five meshes: each mesh handles a different type, namely user-level messaging traffic (UDN), I/O traffic (IDN), memory traffic (MDN), intertile traffic (TDN), and compiler-scheduled traffic (STN). The chip frequency is of 1 GHz and iMesh can provide a bisection bandwidth of 320GB/s.

The user-level messaging is supported by UDN (user dynamic network): threads can communicate through message passing in addition to the cache coherent shared memory. Upon message arrivals, user-level interrupts are issued for fast notification. Message queues can be virtualized onto off-chip DRAM in case of buffer overflows in the network interface. IDN is instead I/O Dynamic Network, and passes data on and off the chip. The MDN (memory dynamic network) and TDN (tile dynamic network) connect the caches and memory controllers, with intertile cache transfers going through the TDN and responses going through the MDN. The usage of two separate physical networks thus provides system-level deadlock freedom.

The four dynamic networks (UDN, IDN, MDN, and TDN) use the dimension-ordered routing algorithm, with the destination address encoded in X-Y ordinates in the header. The static network (STN) allows the routing decision to be preset. This is achieved through circuit switching: a setup packet first reserves a specific route, the subsequent message then follows this route to the destination.

The iMesh's four dynamic networks use simple wormhole flow control without virtual channels to lower the complexity of the routers, trading off the lower bandwidth of wormhole flow control by spreading traffic over multiple networks. Credit-based flow control is used. The static network uses circuit switching to enable the software to preset arbitrary routes while enabling fast delivery for the subsequent data transfer; the setup delay is amortized over long messages.

The iMesh' wormhole networks have a single-stage router pipeline during straight portions of the route, and an additional route calculation stage when turning. Only a single buffer queue is needed at each of the 5 router ports, since no virtual channels are used. Only 3 flit buffers are used per port, just sufficient to cover the buffer turnaround time. This emphasis on simple routers results in a low-area overhead of just 5.5% of the tile footprint.

3.6.5 Intel Single-Chip Cloud Computer

Intel has recently presented the Single-chip Cloud Computer (SCC), which integrates 48 Pentium class IA-32 cores [24] on a 6×4 2D-mesh network of tiled core clusters with high-speed I/Os on the periphery. Each core has a private 256 KB L2 cache (12 MB total on-die) and is optimized to support a message-passing-programming model whereby cores communicate through shared memory. A 16 KB message-passing buffer (MPB) is present in every tile, giving a total of 384 KB on-die shared memory, for increased performance. Memory accesses are distributed over four on-die DDR3 controllers for an aggregate peak memory bandwidth of 21 GB/s. The die area is 567 mm^2, implemented in 45 nm.

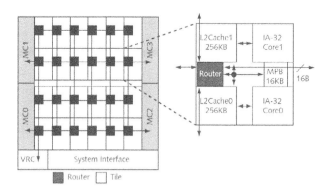

FIGURE 3.19
The Single-Chip Cloud Computer Architecture.

The design is organized in a 6×4 2D-array of tiles (see Figure 3.19). Each tile is a cluster of two enhanced IA-32 cores sharing a router for intertile communication. A new message-passing memory type (MPMT) is introduced as an architectural enhancement to optimize data sharing. A single bit in a core's TLB designates MPMT cache lines. The MPMT retains all the performance benefits of a conventional cache line, but distinguishes itself by addressing noncoherent shared memory. All MPMT cache lines are invalidated before reads/writes to the shared memory to prevent a core from working on stale data.

The 5-port virtual cut-through router used to create the 2D-mesh network employs a credit-based flow-control protocol. Router ports are packet-switched, have 16-byte data links, and can operate at 2 GHz at 1.1 V. Each input port has five 24-entry queues, a route precomputation unit, and a virtual-channel allocator. Route precomputation for the outport of the next router is done on queued packets.

An XY-dimension ordered routing algorithm is strictly followed. Deadlock free routing is maintained by allocating 8 virtual channels between 2 message classes on all outgoing packets.

Input port and output port arbitrations are done concurrently using a wrapped wavefront arbiter. Crossbar switch allocation is done in a single clock cycle on a packet's granularity. No-load router latency is 4 clock cycles, including link traversal. Individual routers offer 64 GB/s interconnect bandwidth, enabling the total network to support 256 GB/s of bisection bandwidth.

3.6.6 ST Microelectronics STNoC

The STNoC [33] by ST Microelectronics is a prototype architecture and methodology thought to replace the widely used STBus in MPSoCs. It is a flexible and scalable packet-based on-chip micronetwork designed for interconnecting many IP blocks that make up SoC devices. Typically, these include one or more general purpose processor cores as well as complex, dedicated IP blocks such as audio/video codecs, a wide range of connectivity IPs (USB, Ethernet, serial ATA, DVB-H, HDMI, etc.), and memories.

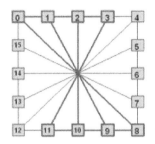 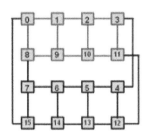

FIGURE 3.20
The ST Microelectronics STNoC Spidergon.

The STNoC proposes a pseudoregular topology, called *Spidergon* (see Figure 3.20), that can be customized depending on the actual application traffic characteristics. In the general Spidergon topology, all of the IP blocks are arranged in a ring. In addition, each IP block is also connected directly to its diagonal counterpart in the network, which allows the routing algorithm to minimize the number of nodes that a data packet has to traverse before reaching its destination [13].

Thanks to the ring-like spidergon topology, the STNoC can be routed using regular routing algorithms that are identical at each node. For instance, the *Across-First* routing algorithm sends packets along the shortest paths, using the long across links that connect nonadjacent nodes in STNoC only when that gives the shortest paths, and only as the first hop. Despite a low hop count, long link traversals cycles may increase the packet latency for Across-First routing. The Across-First routing algorithm is not deadlock-free, relying on the flow control protocol to ensure deadlock freedom instead. STNoC routing is implemented through source routing, encoding just the across link turn and the destination ejection, since local links between adjacent rings are default routes. The across-first algorithm can be implemented within the network interface controller either using routing tables or combinational logic.

STNoC uses wormhole flow control, supporting flit sizes ranging from 16 to

512 bits depending on the bandwidth requirements of the application. Virtual channels are used to break deadlocks. Actual flit size, number of buffers, and virtual channels are determined for each application-specific design through design space exploration.

Since the STNoC targets MPSoCs, it supports a range of router pipelines and microarchitectures. As MPSoCs do not require GHz frequencies, STNoC supports up to 1 GHz in 65 nm ST technology. Since the degree is three, the routers are four-ported. Buffers are placed at input ports, statically allocated to each virtual channel, with the option of also adding output buffer queues at each output port to further relieve head-of-line blocking. The crossbar switch is fully synthesized.

3.6.7 Xpipes

Xpipes was developed by the University of Bologna and Stanford University [8]. Xpipes consists of a library if soft macros of switches and links that can be turned into instance-specific network components at instantiation time. Xpipes library components are fully synthesizeable and can be parameterized in many respects, such as buffer depth, data width, arbitration policies, etc. Components can be assembled together allowing users to explore several NoC designs (e.g., different topologies) to better fit the specific application needs.

The Xpipes NoC library also provides a set of link design methodologies and flow control mechanisms to tolerate any wiring parasitics, as well as network interfaces that can be directly plugged to existing IP cores, thanks to the usage of the standard OCP interface. It promotes the idea of pipelined links with a flexible number of stages to increase throughput. Important attention is given to reliability as distributed error detection techniques are implemented at link level.

Xpipes is fully synchronous, however facilities to support multiple frequencies are provided in the network interfaces but only by supporting integer frequency dividers. Routing is static and determined in the network interfaces (source routing). Xpipes adopts wormhole switching as the only method to deliver packets to their destinations. Xpipes supports both input and/or output buffering, depending on circumstances and designer choices. In fact, since Xpipes supports multiple flow controls, the choice of the flow control protocol is intertwined with the selection of a buffering strategy. Xpipes does not leverage virtual channels. However, parallel links can be deployed among any two switches to fully resolve bandwidth issues. Deadlock resolution is demanded to the topology design phase.

One of the main advantages of Xpipes over other NoC libraries is the provided tool set. The XpipesCompiler is a tool to automatically instantiate an application-specific custom communication infrastructure using Xpipes components. It can tune flit size, degree of redundancy of the CRC error detection, address space of cores, number of bits used for packet sequence count, maximum number of hops between any two network nodes, number of flit sizes, etc. In a top-down design methodology, once the SoC floorplan is decided, the required network architecture is fed into the XpipesCompiler. The output of the XpipesCompiler is a SystemC description that can be fed to a back-end RTL synthesis tool for silicon implementation.

3.6.8 Aethereal

The Aethereal is a NoC developed by Phillips that aims at achieving composability and predictability in system design [19]. It also targets eliminating uncertainties in interconnects, by providing guaranteed throughput and latency services.

The Aethereal NoC has an instance of a 6-port router with an area of 0,175 mm^2 after layout, and a network interface with four IP ports having a synthesized area of 0,172 mm^2. All the queues are 32-bits wide and 8-words deep. With regard to buffering, input queuing is implemented using custom-made hardware FIFOs to keep the area costs down. Both the router and the network interface are implemented in 0.13 μm technology, and run at 500 MHz. The network interface is able to deliver the bandwidth of 16 Gbits/sec to all the routers in the respective directions.

Aethereal is a topology-independent NOC and mainly consists of two components: the network interface and the router, with multiple links between them. The Aethereal router provides best-effort (BE) and guaranteed-throughput (GT) service levels. Aethereal uses wormhole routing with input queuing to route the flits and the router exploits source routing. The architecture of the combined GT-BE router is depicted in Figure 3.21.

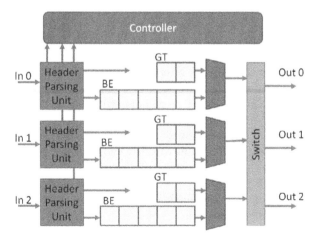

FIGURE 3.21
Aethereal Router Architecture.

The Aethereal uses virtual channels and shares the channels for different connections by using a time-division multiplexing. In the beginning of the routing the whole routing path is stored on the header of the packet's first flit. When the flits arrive to a router a header parsing unit extracts the first hop from the header of the first flit, moves the flits to a GT or BE FIFO and notifies the controller that there is a packet. The controller schedules flits for the next cycle. After scheduling the GT-flits, the remaining destination ports can serve the BE-flits.

A time-division multiplexed circuit switching approach with contention-free routing has been employed for guaranteed throughput. All routers in the network have a common sense of time, and the routers forward traffic based on slot allocation. Thus, a sequence of slots implement a virtual circuit. The allocation of slots can be setup statically, during an initialization phase, or dynamically, during runtime. Best-effort traffic makes use of non-reserved slots and of any slots reserved but not used. Best-effort packets are used to program the guaranteed-throughput slots of the routers.

The Aethereal implements the network interface in two parts: the kernel and the shell. The kernel communicates with the shell via ports.

3.6.9 SPIN

The Scalable Programmable Integrated Network-on-chip (SPIN) is a packet-switching on-chip micronetwork, which is based on a fat-tree topology [1]. It is composed of two types of components: initiators and targets. The initiator components are traffic generators, which send requests to the target components. The target component sends a response as soon as it receives a request. The system can have different numbers of cores for each type, and all the components composing the system are designed to be VCI (Virtual Socket Interface) compliant.

SPIN uses wormhole switching, adaptive routing, and credit-based flow control. The packet routing is realized as follows. First, a packet flows up the tree along any one of the available paths. When the packet reaches a router, which is a common ancestor with the destination terminal, the packet is turned around and routed to its destination along the only possible path. Links are bidirectional and full-duplex, with two unidirectional channels. The channel's width is 36 bits wide, with 32 data bits and 4 tag bits used for packet framing, parity, and error signaling. Additionally, there are two flow control signals used to regulate the traffic on the channel.

SPIN's packets are defined as sequences of data words of 32 bits, with the header fitting in the first word. An 8-bit field in the header is used to identify the destination terminal, allowing the network to scale up to 256 terminals. The payload has an unlimited length as defined by two framing bits (Begin Packet / End of Packet). The input buffers have a depth of 4 words, which results in cheaper routers.

The basic building block of the SPIN network is the RSPIN router, which includes eight ports having a pair of input and output channels compliant with the SPIN link.

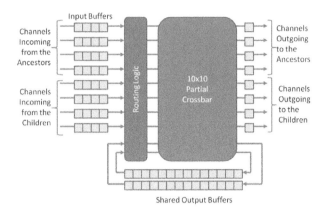

FIGURE 3.22
RSPIN Router Architecure used in SPIN Systems.

The architecture of the RSPIN router is represented in Figure 3.22. The RSPIN router includes a 4 words buffer at each input channel and two 18 words output buffers, shared by the output channels. The output buffers have greater priority to use the output channels than input buffers. This reduces contention. RSPIN contains a partial 10x10 crossbar, which implements only the connections allowed by the routing scheme: all the packets flowing down the tree can be forwarded to children and only such packets can use the output buffers when the required output channel is busy. Nevertheless, only the packets incoming from children can flow up the tree and be forwarded to the fathers.

In 2003, a 32-port SPIN network was implemented in a 0.13 μm CMOS process, the total area was 4.6 mm^2 (0.144 mm^2 per port), for an accumulated bandwidth of about 100 Gbits/s.

3.6.10   MANGO

MANGO (Message-Passing Asynchronous Network-on-Chip providing Guaranteed services through OCP interfaces) is a clockless Network-on-Chip system [9, 11, 10]. It uses wormhole network flow control with virtual channels and provides both guaranteed throughput and best-effort routing. Because the network is clockless, the time-division multiplexing cannot be used in sharing the virtual channels. Therefore, some virtual channels are dedicated to best-effort traffic and others to guaranteed-throughput traffic. The benefits of the clockless system are maximum possible speed and zero idle power.

The MANGO router architecture (depicted in Figure 3.23) consists of separated guaranteed throughput and best-effort router elements, input and output ports connected to neighboring routers, and local ports connected to the local IP core through network adapters that synchronize the clockless network

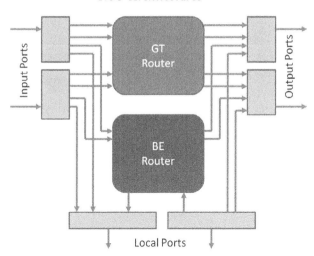

FIGURE 3.23
MANGO Router Architecture.

and clocked IP core. The output port elements include output buffers and link arbiters.

The BE router routes packets using basic source routing where the routing path is stored in the header of the packet. The paths are shaped like in the XY routing. The GT connections are designed for data streams and the routing acts like a circuit switched network. In the beginning of GT routing, the GT connection is set up by programming it into the GT router via the BE router.

3.6.11 Proteo

The Proteo network consists of several sub-networks that are connected to each other with bridges [35]. The main subnetwork in the middle of the system is a ring but the topologies of the other subnetworks can be selected freely.

The layered structure of the Proteo router is depicted in Figure 3.24. Each layer has one input and one output port so a router with one layer is one-directional and suits only on subnetworks with simple ring topology. In more complex networks, more than one layer have to be connected together.

The Proteo system has two different kinds of routers, initiators and targets. The initiator routers can generate requests to the target routers while targets can only respond to these requests. The only difference between initiator and target routers is a structure of the interface. The task of the interface is to create and extract packets.

The routing on the Proteo system is destination-tag routing, where the destination address of the packet is stored on the packet's header. When a packet arrives to the input port the greeting block detects the packet's destination

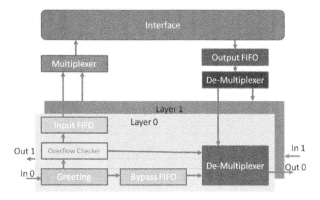

FIGURE 3.24
Proteo Router Architecture.

address and compares it to the address of the local core. If the addresses are
equal the greeting block writes the packet to the input FIFO through the overflow checker, otherwise the packet is written to the bypass FIFO. Finally, the
distributor block sends packets forward from the output and bypass FIFOs.

3.6.12 XGFT

XGFT (eXtended Generalized Fat Tree) Network-on-Chip is a fault-tolerant
system that is able to locate the faults and reconfigure the routers so that
the packets can be routed correctly [27]. The network is a fat tree and the
wormhole network flow control is used. Besides the traditional wormhole mechanism, there is a variant called *pipelined circuit switching*. If the packet's first
flit is blocked, it is routed one stage backwards and routed again along some
alternative path.

When there are no faults in the network, the packets are routed using
adaptive turn-around routing. However, when faults are detected, the routing
path is determined to be deterministic using source routing so that packets
are routed around faulty routers. To detect the faults, there has to be some
system that diagnoses the network.

3.6.13 Other NoCs

There have been a sizeable number of proposals/implementations of NoCs in
the literature.

3.6.13.1 Nostrum

The Nostrum NoC is the work of researchers at KTH in Stockholm and the implementation of guaranteed services has been the main focus point. The Nostrum network adopts a mesh-based approach, and guaranteed services are provided by so-called looped containers. These are implemented by virtual circuits, using an explicit time-division multiplexing mechanism that they call *Temporally Disjoint Networks (TDN)*.

The Nostrum uses a deflective routing algorithm aimed at keeping its area small and its power consumption low due to the absence of internal buffer queues.

More detailed information on Nostrum can be found in [30].

3.6.13.2 QNoC

The architecture of QNoC is based on a regular mesh topology. It makes use of wormhole packet routing and packets are forwarded using the static X-Y coordinate-based routing.

QNoC does not provide any support for error correction logic and all links and data transfers are assumed to be reliable. Packets are forwarded based on the number of credits remaining in the next router.

QNoC aims at providing different levels of quality of service for the end users. QNoC has identified four different service levels based on the on-chip communication requirements. These service levels include Signaling, Real-Time, Read/Write (RD/WR) and Block Transfer, Signaling being the top priority and Block transfer being the least in the order as listed.

More detailed information on QNoC can be found in [12, 16, 15].

3.6.13.3 Chain

The CHAIN network (CHip Area INterconnect) has been developed at the University of Manchester. CHAIN is implemented entirely using asynchronous, or clockless, circuit techniques.

CHAIN is targeted for heterogeneous low-power systems in which the network is system specific.

More detailed information on CHAIN can be found in [3, 4].

3.7 Bibliography

[1] A. Adriahantenaina, H. Charlery, A. Greiner, L. Mortiez, and C. Albenes Zeferino. Spin: A scalable, packet switched, on-chip micro-network. In *DATE '03: Proceedings of the Conference on Design, Automation and*

Test in Europe, 70–73, Washington, DC, USA, 2003. IEEE Computer Society.

[2] Arteris. Company Website: http://www.arteris.com/.

[3] John Bainbridge and Steve Furber. Chain: A delay-insensitive chip area interconnect. *IEEE Micro*, 22:16–23, 2002.

[4] W. J. Bainbridge, L. A. Plana, and S. B. Furber. The design and test of a smartcard chip using a chain self-timed network-on-chip. In *DATE '04: Proceedings of the Conference on Design, Automation and Test in Europe*, 274–279, Washington, DC, USA, 2004. IEEE Computer Society.

[5] L. Benini. Networks on chip: A new paradigm for systems on chip design. In *In Proceedings of the Conference on Design, Automation and Test in Europe*, 418–419. IEEE Computer Society, 2002.

[6] L. Benini and G. De Micheli. Networks on chips: A new SoC paradigm. *Computer*, 35:70–78, 2002.

[7] L. Benini and G. De Micheli, editors. *Networks on Chip, Technology and Tools*. Morgan Kaufmann, 2006.

[8] D. Bertozzi and L. Benini. Xpipes: A network-on-chip architecture for gigascale systems-on-chip. *IEEE Circuits and Systems Magazine*, 4(2):18–31, 2004.

[9] T. Bjerregaard. *The MANGO Clockless Network-on-Chip: Concepts and Implementation*. PhD thesis, Informatics and Mathematical Modelling, Technical University of Denmark, DTU, Richard Petersens Plads, Building 321, DK-2800 Kgs. Lyngby, 2005.

[10] T. Bjerregaard, S. Mahadevan, R. G. Olsen, and J. Sparsø. An OCP compliant network adapter for GALS-based SoC design using the MANGO network-on-chip. In *ISSOC'05 Proceedings of the 2005 International Symposium on System-on-Chip*, 171–174, 2005.

[11] T. Bjerregaard and J. Sparsø. A router architecture for connection-oriented service guarantees in the MANGO clockless network-on-chip. In *DATE '05 2005. Proceedings of the Conference on Design, Automation and Test in Europe*, 1226–1231, 2005.

[12] E. Bolotin, I. Cidon, R. Ginosar, and A. Kolodny. QNoC: QoS architecture and design process for network on chip. *Journal of Systems Architecture*, 50:105–128, 2004.

[13] L. Bononi and N. Concer. Simulation and analysis of network on chip architectures: Ring, spidergon, and 2D mesh. In *DATE '06: Proceedings of the Conference on Design, Automation and Test in Europe*, 154–159, 3001 Leuven, Belgium, Belgium, 2006. European Design and Automation Association.

[14] W. Dally and B. Towles. *Principles and Practices of Interconnection Networks*. Morgan Kaufmann Publishers Inc., San Francisco, CA, USA, 2003.

[15] R. Dobkin, R. Ginosar, and I. Cidon. QNoC asynchronous router with dynamic virtual channel allocation. In *NOCS'07 First International Symposium on Networks-on-Chip*, 218–218, May 2007.

[16] R. Dobkin, R. Ginosar, and A. Kolodny. QNoC asynchronous router. *Integration, the VLSI Journal*, 42(2):103–115, March 2009.

[17] F. Gebali, H. Elmiligi, and M. W. El-Kharashi. *Networks-on-Chips: Theory and Practice*. CRC Press, Inc., Boca Raton, FL, USA, 2009.

[18] P. Gepner and M. F. Kowalik. Multi-core processors: New way to achieve high system performance. In *Proceedings of the International Symposium on Parallel Computing in Electrical Engineering*, 9–13, 2006.

[19] K. Goossens, J. Dielissen, and A. Radulescu. Aethereal network on chip: Concepts, architectures, and implementations. *IEEE Des. Test*, 22(5):414–421, 2005.

[20] R. Ho, K. W. Mai, Student Member, and M. A. Horowitz. The future of wires. In *Proceedings of the IEEE*, 490–504, 2001.

[21] H. P. Hofstee. Power efficient processor architecture and the cell processor. *High-Performance Computer Architecture, International Symposium on*, 258–262, 2005.

[22] J. Hu and R. Marculescu. Exploiting the routing flexibility for energy/performance aware mapping of regular noc architectures, 2002.

[23] iNOCs. Company Website: http://www.iNOCs.com/.

[24] Intel Single-chip Cloud Computer. Project Website: http://techresearch.intel.com/articles/Tera-Scale/1826.htm.

[25] Intel TeraFLOPS. Project Website: http://techresearch.intel.com/articles/Tera-Scale/1449.htm.

[26] A. Jantsch and H. Tenhunen, editors. *Networks on Chip*. Kluwer Academic Publishers, Hingham, MA, USA, 2003.

[27] K. Kariniemi and J. Nurmi. Fault tolerant XGFT network on chip for multi-processor system on chip circuits. In *International Conference on Field Programmable Logic and Applications*, 203–210, 2005.

[28] R. Marculescu, U. Y. Ogras, L.-S. Peh, N. E. Jerger, and Y. Hoskote. Outstanding research problems in NoC design: System, microarchitecture, and circuit perspectives. *Trans. Comp.-Aided Des. Integ. Cir. Sys.*, 28(1):3–21, 2009.

[29] S. Medardoni, M. Ruggiero, D. Bertozzi, L. Benini, G. Strano, and C. Pistritto. Interactive presentation: Capturing the interaction of the communication, memory and I/O subsystems in memory-centric industrial mpsoc platforms. In *DATE '07: Proceedings of the Conference on Design, Automation and Test in Europe*, 660–665, San Jose, CA, USA, 2007. EDA Consortium.

[30] M. Millberg, E. Nilsson, R. Thid, and A. Jantsch. Guaranteed bandwidth using looped containers in temporally disjoint networks within the nostrum network on chip. In *DATE '04: Proceedings of the Conference on Design, Automation and Test in Europe*, 890–895, Washington, DC, USA, 2004. IEEE Computer Society.

[31] S. Murali, D. Atienza, L. Benini, and G. De Micheli. A method for routing packets across multiple paths in NoCs with in-order delivery and fault-tolerance gaurantees. *VLSI Design*, 2007:11, 2007.

[32] J. L. Nunez-Yanez, D. Edwards, and A. M. Coppola. Adaptive routing strategies for fault-tolerant on-chip networks in dynamically reconfigurable systems. *IET Computers & Digital Techniques*, 2(3):184–198, May 2008.

[33] G. Palermo, G. Mariani, C. Silvano, R. Locatelli, and M. Coppola. A topology design customization approach for STNoC. In *Nano-Net '07: Proceedings of the 2nd International Conference on Nano-Networks*, pages 1–5, ICST, Brussels, Belgium, 2007. ICST (Institute for Computer Sciences, Social-Informatics and Telecommunications Engineering).

[34] M. Ruggiero, F. Angiolini, F. Poletti, D. Bertozzi, L. Benini, and R. Zafalon. Scalability analysis of evolving soc interconnect protocols. In *Int. Symp. on Systems-on-Chip*, 169–172, 2004.

[35] I. Saastamoinen, M. Alho, J. Pirttimaki, and J. Nurmi. Proteo interconnect IPS for networks-on-chip. In *In Proc. IP Based SoC Design*, Grenoble, France, 2002.

[36] silistix. Company Website: http://www.silistix.com/.

[37] M. Taylor, J. Kim, J. Miller, F. Ghodrat, B. Greenwald, P. Johnson, W. Lee, A. Ma, N. Shnidman, D. Wentzlaff, M. Frank, S. Amarasinghe, and A. Agarwal. The raw processor: A composeable 32-bit fabric for embedded and general purpose computing. In *In Proceedings of HotChips 13*, 2001.

[38] Tilera. Company Website: http://www.tilera.com.

[39] Tilera Tile-GX. Project Website: http://www.tilera.com/products/ TILE-Gx.php.

3.8 Glossary

Flit: A flit is the smallest flow control unit handled by the network. The first flit of a packet is the head flit and the last flit is the tail.

Flow Control: In computer networking, flow control is the process of managing the rate of data transmission between two nodes to prevent a fast sender from outrunning a slow receiver.

Network Interface: A network interface (NI) is a hardware device that handles an interface to the network and allows a network-capable device to access that network.

NoC: Network-on-Chip (NoC) is a new approach to design the communication subsystem of System-on-a-Chip (SoC). NoC brings networking theories and systematic networking methods to on-chip communication and brings notable improvements over conventional bus systems. NoC greatly improve the scalability of SoCs, and shows higher power efficiency in complex SoCs compared to buses.

Packet: In information technology, a packet is a formatted unit of data carried by a packet mode computer network.

Router: A router is a networking device customized to the tasks of routing and forwarding information.

Routing: Routing is the process of selecting paths in a network along which to send network traffic.

Topology: In computer networking, topology refers to the layout of connected devices. Network topology is defined as the interconnection of the various elements (links, nodes, etc.) of a computer network.

4

Quality-of-Service in NoCs

Federico Angiolini

iNoCs SaRL, 1007 Lausanne, Switzerland

Srinivasan Murali

iNoCs SaRL, 1007 Lausanne, Switzerland

CONTENTS

4.1 Introduction

Network-on-Chip (NoCs) are being envisioned for, and adopted in, extremely complex SoCs, featuring tens of processing elements, rich software stacks, and many operating modes. Consequently, the on-chip traffic is very varied in nature and requirements. For example, consider the following:

- A processor running into a cache miss needs to transfer few bytes, but as urgently as possible, else it cannot resume execution.

- A Direct Memory Access (DMA) controller is programmed to transfer large, contiguous chunks of data. Sometimes the transfers may be particularly urgent, sometimes they may be background activities. Whenever the DMA starts accessing the network, it is going to flood it with traffic. As soon as it is done, it may then remain idle for extended periods.

- An H.264 decoder block, when in operation, generates streams of data at constant bandwidth. These streams may persist for hours, depending on end-user demands. If the network introduces jitter or bandwidth drops, the user experience is affected.

- An on-chip Universal Serial Bus (USB) controller can transfer high-bandwidth streams of data if given the opportunity, but in most applications, lower-bandwidth transfers may still be acceptable to the end user if this allows more critical tasks (e.g., a phone call) to complete successfully.

- A graphics accelerator used for 3D gaming typically needs to maximize frames per second, and its traffic is high priority. Still, other simultaneous activities—such as processing keystrokes and timely generating audio—take precedence.

These examples show that the problem of getting satisfactory system performance is multifaceted. The NoC, with its finite set of resources, is subject to contrasting demands by multiple cores trying to simultaneously access the interconnect, and should allocate performance optimally.

In the NoC context, the **Quality-of-Service (QoS)** perceived by a given traffic flow is defined as how well the NoC is fulfilling its needs for a certain "service." Usually, some "services" of the NoC are taken for granted, even contrary to some assumptions for wide-area networks, such as reliable and in-order delivery of messages. Other "services" are instead subject to resource availability, for example the performance-related ones, such as the availability of at least a certain average or peak bandwidth threshold, the delivery of messages within a deadline, or the amount of jitter (deviation in or displacement of some aspect of the pulses in a high-frequency digital signal) in delivery times. Yet other "services" are more functional in nature, for example the capability to broadcast messages or to ensure cache coherence. In this chapter, the focus will be on performance-related QoS metrics, as they are one of the most common NoC design issues.

To discuss the QoS offered by a Network-on-Chip, it is first of all necessary to understand the requirements of the traffic flows traversing it. The traffic flows can be clustered in a number of **traffic classes**. The number of classes depends on the design. In the simplest classification [15], all traffic is split as either **Best-Effort (BE)** or **Guaranteed Service (GS)**. The latter represents all urgent communication, and must be prioritized; all remaining traffic is delivered on a best-effort basis, i.e., when resources permit. The classification of traffic can be much more detailed, with three [10], four [3] or more [24] classes. Each class may have a completely different set of requirements, as in the example in Table 4.1.

It is important to note that it may not always be easy to map specific transactions to traffic class. For instance, considering communication between a processor and a memory, the transactions may belong to different applications (due to multitasking) and thus belong to different classes, both over time and on a target-address basis. In the extreme case of symmetric multi-core designs, the traffic classes do not depend at all on the type of source and destination cores, but rather on the type of application running on each core at any time. Thus, the traffic must be constantly tracked and reassigned to new classes dynamically.

Once the traffic classes have been identified, *service contracts* can be established to guarantee a certain QoS level. The guarantees can be expressed in many forms [4], such as absolute (worst-case), average, or on a percentile basis. As in the real-time domain, it is very common to define **hard QoS guarantee** a guarantee that applies in worst-case conditions, and **soft QoS guarantee** a solution that attempts to satisfy the constraints but may fail in some conditions. Naturally, softly guaranteeing QoS is often much cheaper, while hard QoS guarantees generally incur higher overdesign penalties.

This chapter discusses methods to guarantee QoS levels on NoCs. Section 4.2 introduces the possible alternative architectural underpinnings that QoS-aware NoCs must feature, while Section 4.2.6 describes how QoS can be guaranteed by either analytic or synthetic methods with design-time tools.

4.2 Architectures for QoS

The exact architectural implementation of a NoC has a large impact on the predictability of its performance and its suitability to QoS provisioning. Major NoC architectural choices, such as routing mechanisms and policies, switching mechanisms, and buffering, all affect the capability to reliably prioritize traffic streams with respect to each other.

For example, consider the well-known effect called *head-of-line blocking*. Under the most natural (and common) implementation approach, NoC buffers

TABLE 4.1
Traffic classes

Traffic Type	Bandwidth	Latency	Jitter
Control traffic	Low	Low	Low
Cache refills	Medium	Low	Tolerant
Cache prefetches	High	Tolerant	Tolerant
Hard real-time video	High	Tolerant	Low
Soft real-time video	High	Tolerant	Tolerant
Audio and MPEG2 bitstreams	Medium	Tolerant	Low
Graphics	Tolerant	Tolerant	Tolerant

are handled as First In First Out queues (FIFOs). Whenever a packet does not immediately find a way to leave a FIFO (e.g., due to congestion), all packets enqueued behind it in the same FIFO are also unable to make any progress. This queuing effect can propagate backwards to upstream switches as more packets queue up, potentially stalling large parts of the NoC until the first packet eventually frees its resources. The phenomenon is also known as "saturation tree" since it resembles a tree, with the root in the congestion point and branches propagating outwards. If a packet ever finds itself in such a queue, it may be severely delayed, disrupting the QoS. Notably, best-effort, low-priority packets are more likely to incur stalling and head-of-line blocking due to the QoS mechanisms themselves. If a higher-priority packet ever finds itself on the same route as lower-priority packets, it may be unable to proceed, despite its priority level, due to stalling ahead on its path, as it cannot overtake the lower-priority packets ahead. In principle, high-priority packets could then become unexpectedly stalled due to contention among low-priority flows originated at the opposite side of the NoC. Figure 4.1 illustrates this condition.

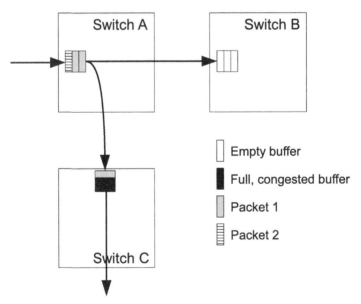

FIGURE 4.1
Head-of-line blocking. Switch C is congested, while Switch B is free. Packet 1 enters Switch A from West, requests to exit from South, and remains blocked along the way. Packet 2 enters the same port of Switch A and tries to go East. Although Switch B is completely free, Packet 2 cannot make progress as it must wait for Packet 1 to move ahead first. Congestion can build up further behind Packet 2, leading to a "saturation tree," until Packet 1 finally resumes.

As this example hints, in general, a large part of the problem of guaranteeing some degree of QoS in a NoC can be cast into a problem of segregating traffic classes from each other. This can be achieved in space (different traffic classes are routed so as to never share resources), in time (different traffic classes are allotted different time slots for propagation), or in a mix of the two (for example by allowing classes to share resources, but dynamically varying priorities or resource allocations).

Traffic segregation, however, incurs overheads. If it is done in space, more physical resources need to be devoted to the NoC, with a corresponding area and power penalty. If it is done in time, performance is affected, as bandwidths are intrinsically capped; to recover the required performance level, additional physical resources may again become necessary. Moreover, the architectural support to enable traffic segregation (e.g., priority-based allocators, priority encoders and decoders, and all other logic required to implement each specific method) has an additional cost. A variety of solutions have been studied to offer the best tradeoffs among the comprehensiveness of the guarantees and the implementation cost.

Another general problem is that of allocation fairness among flows that belong to the same traffic class—for example, two high-bandwidth transfers among video accelerators. Traffic class segregation, by itself, does not prevent the occurrence of network contention among two such flows. Thus, any traffic, even high priority, may still experience collisions with other streams belonging to the same class. If such contention disrupts the intended QoS provisioning, two general solutions are available. One is to adopt a "worst-case" position by simply assigning every flow to a new service class. While this enables designs with hard QoS guarantees, the number of required physical resources may become impractically large—up to one dedicated channel per QoS-provisioned flow. Alternatively, the intraclass contention may be considered as accepted behavior, but architectural extensions should be adopted to provide fairness among all affected flows.

It is very simple to guarantee local fairness at one switch in the NoC, e.g., by round-robin arbitration. However, networks are distributed entities, and the fairness issue does not have equally trivial solutions. A famous [6] analogy, that of the parking lot exit, is depicted in Figure 4.2. A number of cars, each leaving a parking lot row, must access the exit road before they can leave. Each intersection admits cars in round-robin fashion, i.e., in a locally fair way. At the first time event, cars A_1, B_1, C_1, etc., will access the exit road from their row. Since arbitration at intersection A is fair, A_1 will leave the lot first, followed immediately by B_1. At this point, A_2 and C_1 will contend access to the exit; local fairness will admit A_2 before C_1, which however is unfair in a global sense. Car C_1 will only be able to leave in the 4th time slot, car D_1 in the 8th, car E_1 in the 16th, and so on. In terms of rate of cars leaving the parking, or bandwidth in the NoC realm, row A will exploit half of the total capacity of the exit, row B one fourth, row C one eighth, and so

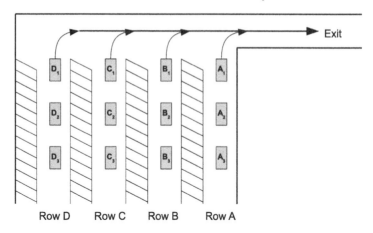

FIGURE 4.2
Allocation fairness. Cars attempting to leave a parking lot through a single exit. If fair local arbitration occurs at each intersection, the cars in the leftmost rows must wait much longer than the cars in the rightmost rows to exit.

on. Hence, it can be understood that, in addition to traffic class segregation, fairness within a given class is another requirement to QoS provisioning.

4.2.1 Traffic Class Segregation in Space

A natural way to guarantee that essential traffic is delivered according to the required QoS is to make sure that high-priority traffic never contends with lower-priority traffic for any physical resource. This can be done by choosing disjoint routes, i.e., by instantiating **multiple physical channels** and switches, and avoiding any collision among different-priority flows, a method known as *noninterfering networks*. A different formulation of the same basic approach envisions multiple disjoint NoCs in the same system; each NoC is tailored to a different type of traffic. For example, this can separate dataflow traffic (large bandwidth, with typically loose latency constraints) from control traffic (smaller bandwidth, but latency-critical). Dedicated networks have also been proposed for "sideband" activities, such as monitoring, testing, or debugging.

Alternatively, it is possible to deploy **virtual channels (VCs)** [7]. Virtual channels permit different traffic classes to share a single physical link while keeping buffers disjoint, so that high-priority packets are not affected by the queuing of low-priority flows. A virtual channel implementation provides multiple buffers in parallel at each link endpoint (Figure 4.3). For example, in a switch, the incoming packets from a link are processed based on their identity (e.g., with a packet tag, with sideband wires, etc.) and stored in one of the virtual channel buffers for that link. The switch arbitration occurs, in princi-

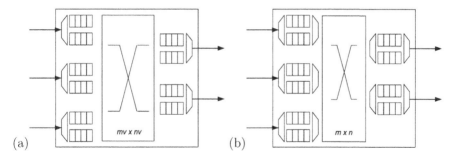

FIGURE 4.3
Virtual channels. Virtual channels implementations in a $m \times n$ NoC router ($m = 3, n = 2$). $v = 2$ virtual channels are shown. Variant (a): full crossbar implementation for maximum efficiency; variant (b): hierarchical implementation for smaller area and higher frequency.

ple, among all virtual channel buffers of all input ports for all virtual channel buffers of all output ports; as this impacts the area and latency of the switch arbiter, hierarchical arbitration has been proposed. In particular, the switch arbitration will take into account the QoS requirements, e.g., by prioritizing some virtual channels (traffic classes) [3, 10]. Once the packet has been arbitrated towards an output port, it is enqueued into one of the virtual channel buffers of that output port. The buffers of the output port will take turns in sending traffic on the link, in multiplexed fashion, again paying attention to the QoS policy.

The choice of using VCs has advantages and disadvantages compared to the full decoupling ensured by multiple physical links [14]. In large-area networks, where cabling is very expensive compared to on-router buffers, a key benefit of virtual channels is the multiplexing of traffic streams onto a single cable. However, this is a less compelling argument in NoCs; on-chip wires are cheap, while a tighter constraint is the power consumption of the datapath, which VCs do not significantly modify. Instead, virtual channels can provide advantages over physical channels in terms of flexibility and reconfigurability at runtime. Virtual channels however provide less total bandwidth than a solution with the same number of physical channels, since the channels are multiplexed. Further, if the links are to be pipelined due to timing constraints, the pipeline stages must also be made VC-aware (i.e., have multiple buffers in parallel), with the corresponding overhead, else the architecture will not any longer be able to offer traffic class separation.

Whether using physical or virtual channels, additional pitfalls may be encountered. A constraint to keep in mind is that the NoC can only guarantee the requested Quality-of-Service if the cores attached to it are designed to guarantee a suitable lower bound on ejection rates. Buffering in the Network

Interfaces can alleviate this concern. Moreover, the spatial separation of traffic classes requires a preliminary step to concretely decide how many channels are needed, and how to allocate flows and classes to channels. Automatic tools have been designed to tackle this challenge, as will be seen in Section 4.2.6.

4.2.2　Traffic Class Segregation in Time

To provide a certain level of Quality-of-Service, traffic classes can also be segregated in time. The simplest way is to just multiplex all classes of traffic onto the same links and switches, but adding **priority mechanisms** to the architecture. Packets can be tagged at the source with a class field, modifying their priority upon arbitration in switches. As a variant, packets can be tagged with a **Time-to-Live (TTL)**-like field. In this idea, derived from wide area networks, the TTL field acts as a progressive priority boost over time, as switches observe that the deadline for the delivery of packets is approaching. Even more simply, the priority tables in NoC switch arbiters can be tuned at design time so that particular routes are privileged; in this case, however, all packets along that route (including low-priority ones) receive the same boost. In general, plain priority schemes are a simple and very-low-overhead way of meeting QoS requirements. However, in the absence of other architectural provisions, they can only provide soft QoS guarantees. For example, head-of-line blocking effects are not solved by prioritized arbitration.

An approach to time segregation of flows leverages **Time-Division Multiplexed Access (TDMA)** techniques [15]. First, a route table and an injection schedule are built at design time. This schedule specifies which traffic can be injected in each clock cycle (or other time period); for example, a given master may be able to inject once in five time slots. The schedule is built so as to ensure that the packets, once injected in the NoC, will be able to traverse it without collisions. In this way, the latency and bandwidth of the NoC can be analytically evaluated in a fine-grained fashion, with hard QoS guarantees and by-construction ensured fairness. Some time slots can be left available for best-effort traffic transmission. An added benefit is that, for completely contention-free schedules, the NoC does not any longer require internal buffering, except just the flip-flops imposed by the timing convergence.

The main challenge of TDMA systems is how to find optimal schedules, able to maximize performance and resource usage. Moreover, precise traffic regulation is needed at the injection points. This can be counterproductive in some cases. For example, once a schedule has been set, a core may not be allowed to inject traffic ahead of time. While this ensures that subsequent communication will be fast, in some cases it may artificially delay the injection of packets even when the network would have enough free resources. For some latency-sensitive traffic flows (e.g., processor cache refills), such a missed opportunity can be expensive. More in general, when the chip operates at less than full schedule utilization, it may be complex to reclaim the unused schedule slots to boost the performance of the remaining traffic. Thus, average-case performance is sacrificed in return for predictability. Further, traffic regulation may re-introduce the need for buffering, this time at the **Network Interfaces (NI)**. It is also difficult to support low-latency and low-bandwidth flows, because the injection latency is inversely proportional to the allotted bandwidth.

QoS guarantees can also be strictly enforced with **preemption** mechanisms. A preemptive NoC is able to cast aside lower-priority packets whenever a higher-priority packet needs to be transmitted. This can be done with spare buffering resources, analogous to ambulances and road traffic: some packets can be "parked aside" while others, more urgent, "overtake" them. If the additional buffering cannot be provided, lower-priority packets could even be dropped and their retransmission demanded. However, while dropping packets is a relatively common occurrence and often necessity in wide-area networks, and those network stacks include software to handle retransmissions, it is complex to support it on-chip, where communication is supposed to be reliable and the whole network stack is mapped into NoC hardware. Dropped packets can still be transparently recovered with hardware retransmission facilities, but this again entails the need for large buffering resources, typically at the source Network Interface.

Many authors have proposed **virtual circuits** to guarantee QoS [18, 2]. As in wide area networks, the switching mechanism of NoCs does not necessarily need to be based on packets; circuit switching [40], or more commonly a hybrid of packet and circuit switching, is also possible. In a circuit switching regime, all resources can be reserved end-to-end among cores attached to the NoC. This is usually just done on top of a packet-switched architecture, by locking all switch arbiters along a route, hence the name of "virtual" circuits. The circuit connections are temporary, so to reuse resources over time. A circuit therefore needs to be set up and torn down (Figure 4.4). This can be done either with dedicated signals, e.g., sideband wires traversing the NoC, or by crafting special packets that are routed as usual, along with normal packet-switched traffic. Once a circuit is set up, all switches along that end-to-end route will not admit any other interfering traffic. Thus, the establishment of virtual circuits can comprehensively guarantee the highest QoS in many respects—bandwidth, latency, jitter. Once the need for arbitration is removed from a route, proposals have been formulated to even bypass arbitration and buffering stages along the way, reducing the number of clock cycles to traverse the circuit. The main obvious downside is that virtual circuits essentially prevent any other traffic to traverse a region of the NoC, thus offering a very coarse-grained approach to prioritizing streams. Another challenge is in the circuit set-up phase, whose latency may be hard to control, especially if the circuit is opened by means of regular best-effort packets opening the way (Scouting Routing [11] may help). Thus, the technique may be best used for streams that transfer large blocks of data in a burst, but without particular initial latency bounds. The set-up process may in fact even fail, for example when two circuits are opened simultaneously along intersecting routes. Care must also be taken not to trigger deadlocks during set-up, either by centralizing the set-up decision process in a dedicated block, or by proper architectural techniques [15].

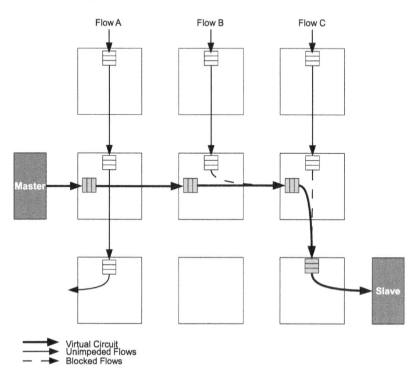

FIGURE 4.4
Virtual circuit. Virtual circuit set up among a Master and a Slave, across four switches. During a set-up phase, the shaded buffers are reserved for the circuit. Hence, during operation, Flow B and Flow C cannot make progress. When communication along the circuit is over, the circuit is torn down and the other flows can resume.

Even without resorting to virtual circuit switching, the switching mechanism can still play a role in the ability to provision QoS of the network. Wormhole packet switching is the most common choice for NoCs, since it minimizes the buffering requirements: switches and NIs can have, in the leanest configuration, input/output buffers as shallow as just a single flit. Unfortunately, wormhole switching does not simplify the QoS provisioning. Since a single packet can be simultaneously in transit across multiple switches (as the name of the technique suggests), the effects of local congestion can quickly spread across the whole NoC. This exacerbates head-of-line blocking phenomena and dramatically increases the worst-case latency bounds of congested NoCs. If hard QoS guarantees are demanded, a wormhole-switched NoC must achieve them by dedicated means, such as a circuit reservation overlay, as mentioned above. **Store-and-forward switching** takes the opposite approach, and demands relatively deep buffering (at least enough locations to hold a whole packet) in return for simple, predictable packet forwarding patterns. Store-and-forward is very amenable to characterization and simplifies the task of predicting and bounding switching latencies. A store-and-forward NoC with simple priority mechanisms can already provide latency and bandwidth guarantees. Of course, this type of switching incurs a large area cost, and is not optimized for average-case performance.

4.2.3 Other Methods for Traffic Class Segregation

The two techniques for traffic class segregation discussed above are not necessarily used in isolation. The clearest example is **dynamic routing**, whereby packets do not necessarily always follow the same route to get to their destination; instead, routes change over time. This can be exploited to ensure QoS, for example by deflecting low-priority traffic away and by directing a high-priority flow through shorter, or less congested, routes.

Wide-area networks make extensive use of dynamic routing, both as a way to balance the communication load and for fault tolerance purposes. Several works have implemented the same principles in NoC, allowing NoC switches to route packets differently, e.g., depending on instantaneous congestion metrics. Unfortunately, although benefits can be had, a NoC implementation of classical dynamic routing encounters major challenges. First, dynamic routing can in principle trigger deadlocks, demanding either deadlock-free approaches (e.g., with deep buffering) or deadlock-removal mechanisms (e.g., by dropping packets). In a NoC, it may be impractical to provide either of these, as already discussed above; thus, few alternative routing options may be available, restricting the usefulness of the scheme. Second, dynamic routing can result in out-of-order packet delivery among packets of the same flow. The buffering and logic required for reordering can be implemented in the NoC, but at an overhead.

At a coarser level, dynamic routing can also be implemented in a simpler way by dynamically explicitly reprogramming the routing tables of the NoC [38]. This step, as part of NoC reconfiguration, is common practice for instance in NoCs on Field Programmable Gate Arrays (FPGAs), whereby the NoC can be morphed according to the current needs. A similar approach can be taken to guarantee a certain QoS level.

Other authors, drawing inspiration from cellular networks, have proposed wireless NoCs [39]. While intra-chip wireless communication has not been broadly adopted yet, it is interesting to note that in this case traffic can be transmitted with **Code-Division Multiplexed Access (CDMA)**. CDMA offers a mechanism to segregate traffic classes based neither on space, nor time. A similar principle applies to optical NoCs [12], where multiple packets can be simultaneously multiplexed on a single waveguide by proper modulation— **Wavelength Division Multiplexing (WDM)**. In this case too, high-priority traffic classes can be devoted to a configurable portion of the total available bandwidth, without risk of collisions.

4.2.4   Fairness of Traffic Delivery

As the example of Figure 4.2 shows, it is very complex to ensure global delivery fairness among traffic of the same class if only local information is known. One possibility, already mentioned above, is that packets can carry a timestamp or deadline field. The switches can also gather from the packets information on their sources or destinations, thus becoming able to evaluate whether the packets are progressing satisfactorily across the network. Still, it is hard to avoid that two packets, both nearing their deadline, collide, causing one of them to miss it. Another challenge is that older packets may end up queued behind newer ones, a condition known as *priority inversion*. Thus, only soft guarantees can be provided.

An alternate approach relies on **traffic regulation** [34]. Under this approach, the flows injected into the NoC are first subject to appropriate *traffic shaping*. A simple logic block, for example a *rho, sigma* regulator, is interposed between cores and the NoC, for example at the core's NI. The regulator is composed of just a few gates and counters, and enforces an average (*rho*) and peak (*sigma*) admission bandwidth allotted to the core, delaying its transmissions if necessary. (The peak bandwidth can be equivalently described as burstiness of the traffic). By suitably restricting the injection rates of cores, it is possible to control congestion and ensure fairness. This equates to a looser version of TDMA approaches, providing softer guarantees. As in the case of TDMA, analytical methods however are needed to beneficially parameterize the regulation (Section 4.2.6). Further, since regulators are actually slowing down the local injection, they may also reduce average performance in return for fairness.

Fairness can also be improved by **end-to-end flow control** [37]. Instead of using flow control to regulate the transmission across a point-to-point link, it can be used to negotiate transmission among NoC endpoints. For example, credits can be assigned by a memory controller sharing them among multiple system masters. This ensures that traffic can be fairly serviced by its recipient, without clogging the NoC. End-to-end flow control can be implemented efficiently, for example, by piggybacking [15] credits to normal packets to be injected into the NoC. A disadvantage of end-to-end flow control is that it does not guarantee the availability of resources along the route; thus, even if the destination has available buffers, the packets may still encounter congestion elsewhere in the NoC. Further, as network size scales up, end-to-end messages require an increasing number of clock cycles and it becomes more complex to provide credits to a plurality of cores. Thus, the practical implementation may become challenging.

4.2.5 Monitoring and Feedback

The architectural approaches described above all attempt to improve the QoS perceived by a given set of flows in transit on the NoC. However, most of them can become ineffective or pessimistic if unforeseen traffic conditions occur on-chip. In modern SoCs, this can be a frequent occurrence. For instance, the end users of a SoC-powered device may download new application software, which may trigger traffic patterns at the hardware level that were unexpected by the chip designers. Further, with many tens of cores (processors, accelerators, controllers, memories) integrated in a SoC, even usage scenarios that can be predicted upfront may be hard to characterize and optimize for.

To tackle this challenge, a feedback mechanism can be used. The NoC can be extended with performance monitoring hardware [5], such as probes to measure congestion, buffer occupancy, and transmission latencies. Such probes can be queried in real time by the operating system running on the chip, providing a detailed and accurate snapshot of the currently available performance level. The operating system can react either by modifying the workload or, rather, by tuning NoC configuration parameters exposed to the software stack. These may include priority levels, VC mappings, routing tables, etc. An alternative to the operating system, a firmware layer, a hardware core [33], or even an off-chip controller could perform the same function.

Real-time performance monitoring does not, in itself, provide any means to enforce QoS guarantees. However, if the necessary dedicated architectural features are present, it provides additional opportunities to exploit them better. Further, a suitable control layer can adaptively adjust the QoS to unknown operating conditions.

4.2.6 Memory Controllers

Increasing interest has been devoted towards the particular issue of memory controllers paired with NoCs. In fact, in many designs, the memory controller is the main bottleneck of the chip; if Quality-of-Service is desired for on-chip communication, it is at least equally crucial to also tackle the challenge of off-chip memory access [37].

The main problem is that real-world memory controllers [9, 36] are very complex devices. To maximize bandwidth utilization, they process many memory requests at once, then schedule them to best reuse the open banks, rows and columns of **Dynamic Random Access Memory (DRAMs)** with minimum command overhead. Unfortunately, this causes the response time of the memory controller to be hardly predictable from the NoC point of view, affecting the possibility of supplying QoS guarantees. Predictable memory controllers [1] have been conceived, but not widely utilized. The dual problem also exists; NoCs are often not designed to present memory controllers with optimized traffic patterns. Hence, the memory bottleneck may be tackled suboptimally.

Memory schedulers have been developed by NoC vendors. These schedulers do not implement all functionality of a real memory controller, and they still rely on having one in the system. However, they mediate the memory accesses by sitting between the NoC and the memory controller. Their role is to expose guaranteed-QoS interfaces on the NoC side, and feeding a prescheduled stream of requests for consumption by the memory subsystem. For example, Sonics provides MemMax [35], which exposes three traffic classes: Priority, Controlled Bandwidth, and Best Effort. This scheduler is based on multithreaded communication; flows can be tagged with a thread identifier and processed in parallel, possibly out-of-order, while maintaining the capability to subsequently reorder them. Crucially, the prioritization of traffic is done with awareness not only of memory locality concerns like in traditional controllers, but also leveraging the QoS classes defined in the NoC. This permits lower-latency and more predictable operation of the overall system.

Other architectural approaches have been undertaken specifically to handle the memory controller hotspot, for instance end-to-end admission control [37] so that a temporary overload does not overflow into the NoC. The memory controller can also be given additional information about the congestion status of the network [21], so as to selectively prioritize which responses to issue first. The NIs and switches of the NoC can also be extended to optimize memory-directed traffic [22, 17, 8] in various ways, for example by coalescing threads and reordering transactions based on the target bank, row, and column.

Looking ahead, a more scalable approach to memory communication would probably demand the presence of multiple memory channels in parallel. This is impractical with off-chip DRAMs, but may become commonplace with multiple on-chip integrated memory banks [20]. Vertical stacking technologies have been proposed to enable on-chip memory integration.

Design methods for supporting QoS can be classified into two categories: analysis and synthesis. The methods used for analysis takes as inputs the NoC topology, routing function, application communication pattern and computes worst-case or average-case network metrics (usually bandwidth and latency) for the different flows. The synthesis methods builds topologies that meet specific QoS constraints on traffic flows. In this chapter, we will show how both of these methods can be applied to architectures that have specific features to support QoS as well as on generic best-effort based NoC architectures.

4.2.7   Methods for QoS Analysis

The methods for analysing a NoC topology to calculate the QoS metrics can be further classified into two categories: (i) worst-case analysis, (ii) average-case analysis. The worst-case methods guarantee an upper bound on latency and lower bound on supported bandwidth that is never violated under any network conditions. The average-case methods are based on queuing theory and probabilistic analysis or network calculus. The latter methods can only provide a *soft* QoS guarantee for latency or bandwidth, for example, a latency of less than a bound for a certain percentage of packets, while individual packets can incur a higher latency. The bounds obtained by the worst-case and average-case methods can be further optimized by considering application traffic that is regulated, where each core injects packets at a specific rate and burstiness. It is interesting to note that most works for average case analysis have an inherent assumption of regulated traffic at the cores, either to apply suitable queuing theory models (invariably the assumption of Poisson arrival pattern for packets) or to apply network calculus.

4.2.7.1 Worst-Case Analysis

For architectures that have specific hardware support for hard QoS, computing the worst case bounds is fairly simple. Let us consider the architecture from [32], where TDMA slots are allocated to traffic flows, thereby guaranteeing a contention-free network. Once a packet gets a free time slot to enter into the network, its latency to reach the destination is just the zero-load latency in the network, which can be easily computed. The allocation of time slots to traffic streams already guarantees a minimum bandwidth for each stream.

For a general best effort NoC, computing the worst-case bounds is a challenging problem. In [23], methods for calculating the worst-case latency and bandwidth bounds for traditional multiprocessor networks are presented. In [31], the authors present methods for NoCs, offering better bounds and applicable to varying burstiness of traffic.

The underlying principle of these works is the following: for a particular traffic flow, trace the path from the destination to the source. At each switch, consider a worst-case scenario, where the packet loses arbitration with packets from all other flows at the inputs and outputs. Since each packet's latency at a switch depends on the congestion at the downstream links, the latency calculations are done in a recursive manner from last to the first switch of a path. In order to achieve a system free from starvation and to have bounded worst-case latencies, a fair arbitration scheme such as round-robin needs to be used at the switches.

We explain briefly the latency calculation models from [31]. Let us consider an input-queued switch architecture using credit-based wormhole flow control. We assume the topology of the design, the routing function, and the communication traffic pattern of the application to be inputs to the method. We consider a case where a core (source node) can inject packets with arbitrary burstiness. When the network gets congested, the use of the credit-based flow control will send back-pressure information, ultimately throttling the traffic injected from the source nodes. We assume perfect source and destination end-nodes, i.e., the source has infinite buffering and destination can eject packets instantly. Note that this assumption is only to facilitate the illustration of the models. In reality, if a source node's buffer is full, the node (processor or memory core) can be stopped. An end-to-end flow control mechanism can be used to send packets to a destination only if enough buffering is available at the receiving network interface.

The worst-case behavior is obtained when all the buffers in the switches are full and when a packet of a flow loses arbitration with all other flows that can contend with it. Let u_i^j be the worst-case delay at switch j for a packet of flow i, which needs to be computed. Let ts_1 and ts_2 be the injection and ejection times at the end points of the network for flow i. Let h_i be the total number of hops in the path for flow i. Then, the worst-case latency for flow is given by:

$$UB_i = ts_1 + ts_2 + \sum_{\forall j} u_i^j \quad with \ j = 0 \ldots h_i \tag{4.1}$$

A source node can have many flows, each to a different destination. In practice, the cores should have support for multiple outstanding transactions. In such a case, a packet generated by a core can contend with packets generated by the same core to other destinations. We call this *source contention*. We can model this contention by using a virtual switch as the first hop of the path.

Let $I(x)$ returns the index of a flow from the pool of flows that contend with flow i at switch j. Let the number of flows that contend with flow i at switch j and use the output port c is denoted by $z_c(i,j)$. The source-contention latency can be modeled by the following equation:

$$u_i^0 = MAX(U_i^0, U_{I(x)}^0) + \sum_{\forall x} U_{I(x)}^0 \\ with \ x = 0 \ldots z_0(i,0) \tag{4.2}$$

Similarly, the delay for a packet at other switches can be calculated using the following equation:

$$u_i^j = MAX(U_i^j, U_{I(x)}^j) + \sum_{\forall x} U_{I(x)}^j \\ with \ x = 1 \ldots z_c(i,j), \ \ 1 \le j \le h_i \tag{4.3}$$

The value of U_i^j at a current switch depends on the delays on the next switch:

$$U_i^j = MAX(U_i^{j+1}, U_{I(x)}^{j+1}) + \sum_{\forall x} U_{I(x)}^{j+1} \\ with \ x = 1 \ldots z_c(i, j+1), \ \ 0 \le j \le h_i - 1 \tag{4.4}$$

To calculate the upper bound delay, the Equations 4.2, 4.3, and 4.4 have to be calculated in a recursive manner. The recursive formulation is guaranteed to complete because the delay of any flow at the last switch in the path is fixed. The termination conditions are given by Equation 4.5:

$$U_i^{h_i} = L_i, \quad U_{l(x)}^{h_{l(x)}} = L_{l(x)} \tag{4.5}$$

The ejection time of a packet at the last switch in cycles for flow i is denoted as L_i.

The time needed for a source i to inject the next packet is the time to

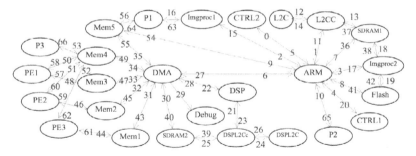

FIGURE 4.5
Multimedia benchmark used for analysis.

create it and the time needed for the current packet to move to the input buffer of the first switch, given by:

$$MI_i = ts_1 + u_i^0 \tag{4.6}$$

Let fw be the flit width, L be the length of a packet (number of flits in a packet) and $freq$ be the frequency of operation of the NoC. The minimum injectable bandwidth for the flow i is given by:

$$mbw_i = L * fw/MI_i * freq \tag{4.7}$$

By applying the equations to all the flows and switches, the worst-case latencies and bandwidths can be calculated. In [31], the authors show the comparisons with earlier works and show the practicality of the methods for critical flows.

When the traffic injected by the cores can be regulated such that a subsequent packet is injected only after a precomputed time interval after a packet, the latency bounds can be made much tighter. This interval to be computed will also provide a minimum bound on the available bandwidth value for the flow. Let us consider overlapping flows, flows that contend for the same output port at a switch and which also share the same input port. When a flow F_i contends with multiple overlapping flows at a switch, it is possible to locally coalesce all such overlapping flows into a single one. This is because the arbitration cannot be lost to many of those flows, as they cannot physically produce a contending packet simultaneously given that they enter the switch through the same input port. If there exist, e.g., two overlapping contending flows at hop j having delay parameters U_{i1}^j and U_{i2}^j, then it is possible to consider $max(U_{i1}^j, U_{i2}^j)$, as their representative delay instead of their sum. By applying this optimization across all the switches on a path, significant reduction in the maximum latency bound can be obtained. The minimum interval MI_i should be respected to inject the next packet by any source to a particular destination.

Here we reproduce the result from [31], showing the bandwidth and latency

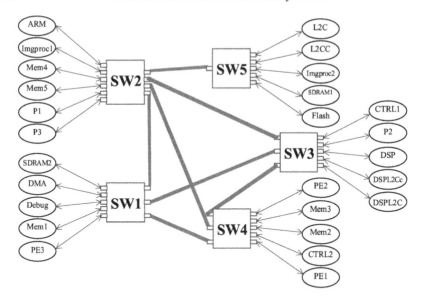

FIGURE 4.6
A five-switch NoC topology for the benchmark.

bounds obtained on a multimedia benchmark. We consider three methods: Wormhole Channel Feasibility Checking (WCFC), the method from [23], Real Time Bound for High-Bandwidth (RTB-HB), which is the one presented in Equations 4.1–4.5 and Real Time Bound for Low-Latency (RTB-LL), where the traffic from the cores are regulated and the above optimization is applied. The multimedia benchmark has 26 cores and 67 communications flows. The average bandwidth of communication across the flows is shown in Figure 4.5. A 5-switch NoC is designed to connect the cores and support the application characteristics using existing tools [27]. The NoC topology is shown in Figure 4.6.

In Figure 4.7, we show the latency bounds computed for all the flows using the three methods. The RTB-LL model always provides the tightest bounds. Compared to WCFC, the largely improved tightness (more than 50% on average) is due to the analysis of overlapping flows, but without any impact on the accuracy of the bounds, which are still under worst-case assumptions. RTB-HB naturally returns higher worst-case latency, due to the assumption that no hardware traffic injection regulation facilities are available. In fact, due to the different calculation approach, the bounds are on average still 30% lower than in WCFC, despite the less restrictive assumptions. In Figures 4.8, and 4.9 we show the interval for injecting subsequent packets and the minimum bandwidth guaranteed computed for the different traffic flows.

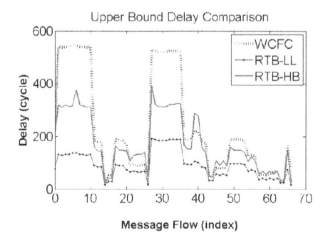

FIGURE 4.7
Worst-case latency values for the flows of the benchmark.

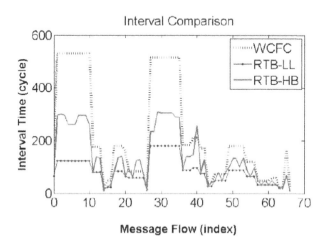

FIGURE 4.8
Minimum interval between packets for the flows of the benchmark.

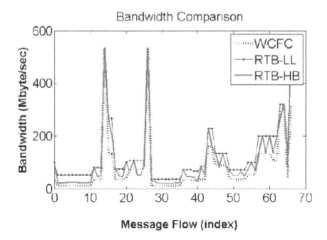

FIGURE 4.9
Guaranteed bandwidth for the flows of the benchmark.

4.2.7.2   Average-Case Analysis

In this subsection, we outline some of the works on average-case analysis. The methods for average-case analysis can give soft QoS guarantees on latency and bandwidth. Most average-case analysis are based on application of queuing theory for NoCs. In [19], the authors present a Markov model-based approach for latency estimation. They also assume a Poisson arrival process. They consider a uniform traffic pattern with the use of virtual channels in the switches. In [28], the authors present a performance model for a NoC router, which is a generalization of the delay modes for single queues. The method works for wormhole networks with arbitrary message lengths and finite buffering for specific traffic patterns of applications. The method also computes other important network metrics, such as buffer utilization and latency of a flow at each router.

An issue in using a single queue model is that the effect of back-pressure and congestion from the downstream buffers are not reflected accurately in the delay calculations. In [13], the authors present a method for considering this when computing the average-case bounds. They consider arbitrary NoC topologies and routing functions with known traffic patterns, like most of the works on QoS analysis. They also have the assumption of Poisson distribution for the packets from each source. For computing the latency of a particular packet, termed as the *tagged packet*, the mean latency at each switch of the path is computed. In order to do this, the probabilities and delays of contention between the tagged packet with other contending packets arriving at the switch is computed. Then, the average latencies of the switches on path are summed up to get the overall latency for the packet. This method's essence of recursively computing the delays is similar to the method for computing worst-case bounds in the previous subsection. The authors present a recursive algorithm to consider the reciprocal impact of all incoming flows to a switch on the others and the values are computed following a reverse order of the dependencies.

Several works have applied network calculus for performance analysis of Asynchronous Transfer Mode (ATM) traffic in the Internet with different services and several other networks. In [30], the authors present worst-case delay bounds for best-effort traffic using network calculus. They assume that the application traffic can be characterized in terms of *arrival curves* at each core. The router service model is abstracted by a *service curve*. The objective of the work is to derive the equivalent service curve the tandem of routers on a path provides to a flow. The contention of a flow on a routing path is considered by analyzing flow-interference patterns by building a contention tree model. Based on the sharing of the flows at each router, the equivalent service curve is constructed. Once this is computed, the delay bounds are computed using network calculus.

4.2.8 Synthesis Methods for Supporting QoS

In this section, we show outline methods to synthesize NoCs to meet average and worst-case QoS constraints. We first show methods for meeting the average-case constraints and then present the worst-case methods.

4.2.9 Meeting Average-Case Constraints

Several works have been presented to synthesize NoC topologies to meet average bandwidth constraints of the traffic flows and zero-load latency constraints [27]–[29]. In [26], a method to map cores onto standard NoC topologies, such, as mesh and Tours is presented. The work uses several different routing functions and ensures that the average bandwidth constraints are met for the traffic flows.

In [27], a method to build topologies, find deadlock-free paths for packets and set architectural parameters is presented. We explain briefly the synthesis method here. The application traffic characteristics, such as the bandwidth and latency constraints and the objective for synthesis (power or performance optimization) are obtained as inputs to the synthesis process. An input floorplan of the design with the size and placement of the cores is taken as an optional input. The area and power models of the network components are computed for the target technology and also taken as an input. The models are obtained by synthesizing the Register Transfer Level (RTL) design of each component and performing place and route using standard CAD tool flows. With these inputs, the NoC architecture that optimizes the user objectives and satisfies the design constraints is automatically synthesized. The different steps in this phase are presented in Figure 4.10. In the outer iterations, the key NoC architectural parameters (NoC frequency of operation and linkwidth) are

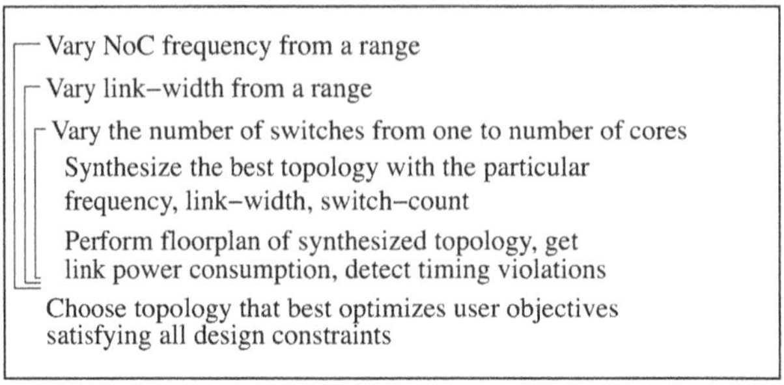

Vary NoC frequency from a range

Vary link–width from a range

Vary the number of switches from one to number of cores

Synthesize the best topology with the particular
frequency, link–width, switch–count

Perform floorplan of synthesized topology, get
link power consumption, detect timing violations

Choose topology that best optimizes user objectives
satisfying all design constraints

FIGURE 4.10

NoC architecture synthesis steps.

varied in a set of suitable values. The bandwidth available on each NoC link is the product of the NoC frequency and the link width. During the topology synthesis, the algorithm ensures that the traffic on each link is less than or equal to its available bandwidth value.

The synthesis step is performed once for each set of the architectural parameters. In this step, several topologies with different number of switches are explored, starting from a topology where all the cores are connected to one switch, to one where each core is connected to a separate switch. The synthesis of each topology includes finding the size of the switches, establishing the connectivity between the switches and connectivity with the cores, and finding deadlock-free routes for the different traffic flows.

In the next step, to have an accurate estimate of the design area and wire lengths, the floorplanning of each synthesized topology is automatically performed. The floorplanning process finds the 2D position of the cores and network components used in the design. Based on the frequency point and the obtained wire lengths, the timing violations on the wires are detected and the power consumption on the links is obtained. In the last step, from the set of all synthesized topologies and architectural parameter design points, the topology and the architectural configuration that best optimizes the user's objectives, satisfying all the design constraints is chosen. Thus, the output is a set of application-specific NoC topologies that meet the input constraints.

The process of meeting the QoS constraints is performed during the synthesis step. For a particular switch count, the cores are assigned to the different switches, such that cores that have high bandwidth and low-latency traffic between them are mapped onto the same switch. When computing the paths for a particular traffic flow, all available paths from the source to the destination that support the bandwidth requirement of the flow are checked and the least cost path is chosen. At the beginning, the cost of a path is computed only based on the power consumption of the traffic flows on that path. If no existing path can support the bandwidth, then new physical links are opened between one or more switches to route the flow. Once a path is found, it is checked to see whether the latency constraints are met. If not, then the cost of a path is gradually changed to the length of the path, rather than just the power consumption of the flows. This helps in achieving the zero-load latency constraint of the flow. The process is repeated for all the flows of the application.

4.2.9.1   Meeting Worst-Case QoS Constraints

In [16]–[25], the authors present a method to synthesize a NoC that has, TDMA scheme to support hard QoS constraints. Many of the synthesis issues, such as finding the switch count and finding paths are similar to the case of designing a best effort NoC. An important step that needs to be performed for the TDMA scheme is to design slot tables and allocate time slots for the different traffic flows. For example, in [25], the slot table size is increased until a valid topology is synthesized. Having a large slot table implies a finer breakdown of the time among different streams, thereby leading to a better allocation of network bandwidth to different streams. On the other hand, increased slot table size increases complexity of the hardware mechanism (especially the table to signify which slot belongs to which stream). For a particular traffic stream, the slots themselves could be assigned in many ways. One possible scenario is a set of contiguous slots allocated to a particular traffic. On the other extreme, the slots for a particular flow could be distributed uniformly in time across all the slots from a particular source. For a given slot table size and the assignment strategy, it is a challenging problem to allocate the NoC resources along the path to the stream, so that once a packet enters the network it has consecutive slots on the downstream ones and can reach the destination without contention. In [16], a slot optimization method is presented.

Based on the worst-case models from [31], the authors have modified the method in Figure 4.10 to synthesize NoCs to meet hard QoS constraints. For illustrative purposes, let us term the former synthesis approach as *ORIG* and the proposed modifications to support hard real-time constraints as *RT*. Apart from the average-case constraints, the worst-case bandwidth and latency constraints for some (or all) the flows are obtained as inputs. During path computation process, the worst-case latency models are applied to calculate the worst-case latency for all flows that have paths. If the current flow that is being routed creates a timing violation, the function will try again iteratively giving more importance to worst-case latency when computing the costs. If this also fails to meet the worst-case latency constraints, then the core to switch assignment is also changed to connect cores with tight worst-case latency constraints to the same switch.

In this set of experiments, we show how the *RT* method performs in meeting the hard latency and bandwidth constraints when compared to *ORIG*. Let us consider the 26-core multimedia benchmark presented earlier in Figure 4.5. The *RT* algorithm is applied on this benchmark and the smallest latency

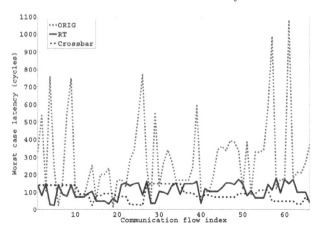

FIGURE 4.11
Worst-case latency on each flow.

bound for which valid topologies could be synthesized is obtained. While in many real applications only a subset of flows have real-time constraints, in this study, we show an illustrative case to highlight the overhead involved in the method. Thus, we put real-time constraints on all flows. From running the *RT* algorithm, the tightest constraint for the upper-bound delay for which feasible topologies could be build is computed to be 180 cycles. To find out how tight the constraint was, the worst-case latencies of the flows only due to source and destination contentions were calculated separately. To perform this, all the cores were connected through a single crossbar switch and the worst-case latencies computed. On the crossbar, the average worst-case latency for this benchmark was found 92 cycles and the maximum value across all flows was 148 cycles. This shows that the constraint imposed for the *RT* algorithm (of 180 cycles) is quite tight, as it is only 1.25× the maximum value of the flows from the ideal case.

Solutions with several different switch counts were synthesized by the *RT* method and we chose a 14-switch NoC solution for further analysis. The *ORIG* method was also applied to the benchmark and the 14-switch solution was generated without any hard latency constraints.

In Figure 4.11, we show the worst-case latencies for the flows of the benchmark for 3 cases: *ORIG*, *RT*, and a full crossbar. As can be seen, for the topology designed with the *RT* algorithm, the worst-case latency of the flows is in the same range as the worst-case latency for the flows mapped on the crossbar. On the topology designed with the original algorithm, most flows have worst-case latency values much higher than those of the crossbar. Another less intuitive effect that is visible in the plot is that the *RT* algorithm provides lower worst-case latency than the crossbar. This is because, in a crossbar, each flow will have to contend with all the other flows to the same

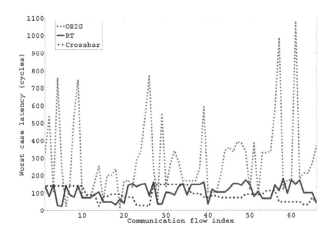

FIGURE 4.12
Minimum guaranteed bandwidth for each flow.

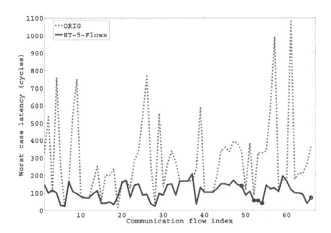

FIGURE 4.13
Worst-case latency when only 5 flows are constrained.

destination. Whereas, in a multiswitch case, this may not happen, for example, if there are 3 flows to the same destination. In the multiswitch case, two of them may share a path until a point where they contend with the third flow. The third flow only has to wait for one of them (with the maximum delay) to go through. Whereas, in a full crossbar, the third flow will have to wait for both the flows, in the worst case. Thus, we can see that, when only few flows require real-time guarantees a multiswitch topology can give better bounds and it is really difficult to come with the best topology directly using designer's intuition. In Figure 4.12, we show the calculated minimum guaranteed bandwidth for the communication flows for the 14-switch topology.

So far we showed what happened to the worst-case latency when a constraint is set to all the flows. In Figure 4.13, we show the behavior of the *RT* synthesis algorithm when only 5 flows have worst-case latency constraints. The flows that had constraints are marked with bubbles on the figure. The latency constraints were added to flows going to and from peripherals. This is a realistic case, as many peripherals have small buffers and data has to be read at a constant rate, so that it would not be overwritten. In this case, the bounds on those 5 flows could be tightened further (two flows at 160 cycles and three flows at 60 cycles). Putting these constraints also leads to a reduction in the worst-case latency of other flows as well. Due to the tight constraints, the *RT* algorithm maps the *RT* flows first. Then, the unconstrained flows also have to be mapped with more care so that they do not interfere with the previously mapped ones.

4.3   Glossary

CDMA: Code-Division Multiplexed Access, a mechanism to share the utilization of a transmission medium based on the use of orthogonal codes to differentiate simultaneously transmitting channels.

DMA Controller: Direct Memory Access Controller, a programmable core that can transfer chunks of data across the chip without the need for other cores to directly supervise the process.

DRAMS: Dynamic Random Access Memory, the most common technology to implement off-chip fast and volatile memories.

FIFO: First-In First-Out, a type of buffer where the first input data must be the first to be output.

Jitter: The variation in message delivery latency across different messages.

NI: Network Interface, a NoC component in charge of packetizing the communication requests of cores for transmission across the NoC, and of the inverse process at the receiving core.

NoC: Network-on-Chip, an on-chip interconnect design style based on packet switching.

QoS: Quality-of-Service, the fulfillment of a NoC service objective, such as the offered bandwidth or the transmission latency of a packet.

TDMA: Time-Division Multiplexed Access, a mechanism to share the utilization of a resource based on the allocation of time slots to requestors.

TTL: Time-to-Live, a field inserted in packets, and periodically decremented, which bounds the maximum time the packet should be in flight before delivery or dropping.

VC: Virtual Channel, a mechanism to multiplex two or more channels onto a single physical NoC link, based on dedicated buffers for each channel.

WDM: Wavelength-Division Multiplexing, a mechanism to share the utilization of a transmission medium based on the use of different wavelengths to differentiate simultaneously transmitting channels.

4.4 Bibliography

[1] B. Akesson, K. Goossens, and M. Ringhofer. Predator: A predictable SDRAM memory controller. In *International Conf. on Hardware/Software Codesign and System Synthesis (CODES+ISSS)*, 251–256. ACM, October 2007.

[2] T. Bjerregaard and J. Sparsø. Scheduling discipline for latency and bandwidth guarantees in asynchronous network-on-chip. In *Proceedings of the 11th IEEE International Symposium on Asynchronous Circuits and Systems (ASYNC)*, 34–43, 2005.

[3] E. Bolotin, I. Cidon, R. Ginosar, and A. Kolodny. QNoC: QoS architecture and design process for network on chip. In *J. Syst. Archit.*. Elsevier, North Holland, New York, 2004.

[4] A. Campbell, C. Aurrecoechea, and L. Hauw. A review of QoS architectures. *Mult. Syst.*, 6:138–151, 1996.

[5] C. Ciordas, A. Hansson, K. Goossens, and T. Basten. A monitoring-aware network-on-chip design flow. *J. Syst. Archit.*, 54(3-4):397–410, 2008.

[6] W. Dally and B. Towles. *Principles and Practices of Interconnection Networks*. Morgan Kaufmann Publishers Inc., San Francisco, CA, USA, 2003.

[7] W. J. Dally. Virtual-channel flow control. In *ISCA '90: Proceedings of the 17th Annual International Symposium on Computer Architecture*, 60–68, 1990.

[8] M. Daneshtalab, M. Ebrahimi, P. Liljeberg, J. Plosila, and H. Tenhunen. A Low-Latency and Memory-Efficient On-chip Network. In *Proceedings of the 4th ACM/IEEE International Symposium on Networks-on-Chip (NOCS10)*, May 2010.

[9] Denali. Databahn DDR memory controller IP, 2010. http://www.denali.com.

[10] J. Diemer and R. Ernst. Back Suction: Service Guarantees for Latency-Sensitive On-Chip Networks. In *Proceedings of the 4th ACM/IEEE International Symposium on Networks-on-Chip (NOCS10)*, May 2010.

[11] J. Duato, S. Yalamanchili, and N. Lionel. *Interconnection Networks: An Engineering Approach*. Morgan Kaufmann Publishers Inc., San Francisco, CA, USA, 2002.

[12] M. Kobrinsky et al. On-chip optical interconnects. *Intel Tech. J.*, 8(2):129–142, 2004.

[13] S. Faroutan et al. An analytical method for evaluating network-on-chip performance. In *DATE '10: Proceedings of the Conference on Design, Automation and Test in Europe*, 2010.

[14] F. Gilabert, M. E. Gómez, S. Medardoni, and D. Bertozzi. Improved Utilization of NoC Channel Bandwidth by Switch Replication for Cost-Effective Multi-Processor Systems-on-Chip. In *Proceedings of the 4th ACM/IEEE International Symposium on Networks-on-Chip (NOCS10)*, May 2010.

[15] K. Goossens, J. Dielissen, and A. Radulescu. Aethereal network on chip: Concepts, architectures, and implementations. *Design Test of Computers, IEEE*, 22(5):414–421, Sept.-Oct. 2005.

[16] A. Hansson, M. Coenen, and K. Goossens. Channel trees: Reducing latency by sharing time slots in time-multiplexed networks on chip. In *CODES+ISSS '07: Proceedings of the 5th IEEE/ACM International Conference on Hardware/Software Codesign and System Synthesis*, 149–154, 2007.

[17] W. Jang and D. Z. Pan. An SDRAM-aware router for networks-on-chip. In *Proceedings of the 46th Annual Design Automation Conference*, 800–805, 2009.

[18] F. Karim, A. Nguyen, and S. Dey. An interconnect architecture for networking systems on chips. *Micro, IEEE*, 22(5):36–45, Sep./Oct. 2002.

[19] A. E. Kiasari, D. Rahmati, H. Sarbazi-Azad, and S. Hessabi. A Markovian performance model for networks-on-chip. In *PDP '08: Proceedings of the 16th Euromicro Conference on Parallel, Distributed and Network-Based Processing*, 157–164, 2008.

[20] D. Kim, K. Kim, J.-Y. Kim, S. Lee, and H.-J. Yoo. Implementation of memory-centric NoC for 81.6 GOPS object recognition processor. In *Proceedings of Solid-State Circuits Conference*, 47–50, 2007.

[21] D. Kim, S. Yoo, and S. Lee. A Network Congestion-Aware Memory Controller. In *Proceedings of the 4th ACM/IEEE International Symposium on Networks-on-Chip (NOCS10)*, May 2010.

[22] D. Lee, S. Yoo, and K. Choi. Entry control in network-on-chip for memory power reduction. In *ISLPED '08: Proceeding of the 13th International Symposium on Low Power Electronics and Design*, 171–176, New York, NY, 2008. ACM.

[23] S. Lee. Real-time wormhole channels. *J. Parallel Distrib. Comput.*, 63(3):299–311, 2003.

[24] D. E. McDysan and D. L. Spohn. *ATM Theory and Application / David E. McDysan, Darren L. Spohn*. McGraw-Hill, New York, 1994.

[25] S. Murali, M. Coenen, A. Radulescu, K. Goossens, and G. De Micheli. A methodology for mapping multiple use-cases onto networks on chips. In *DATE '06: Proceedings of the Conference on Design, Automation and Test in Europe*, 118–123, 2006.

[26] S. Murali and G. De Micheli. Sunmap: A tool for automatic topology selection and generation for NoCs. In *DAC '04: Proceedings of the 41st Annual Design Automation Conference*, 914–919, 2004.

[27] S. Murali, P. Meloni, F. Angiolini, D. Atienza, S. Carta, L. Benini, G. De Micheli, and L. Raffo. Designing application-specific networks on chips with floorplan information. In *ICCAD '06: Proceedings of the 2006 IEEE/ACM International Conference on Computer-Aided Design*, 355–362, 2006.

[28] U. Y. Ogras and R. Marculescu. Analytical router modeling for networks-on-chip performance analysis. In *DATE '07: Proceedings of the Conference on Design, Automation and Test in Europe*, 1096–1101, 2007.

[29] A. Pinto, L. P. Carloni, and A. L. Sangiovanni-Vincentelli. Efficient synthesis of networks on chip. In *ICCD '03: Proceedings of the 21st International Conference on Computer Design*, 146, 2003.

[30] Y. Qian, Z. Lu, and W. Dou. Analysis of worst-case delay bounds for best-effort communication in wormhole networks on chip. In *NOCS '09: Proceedings of the 2009 3rd ACM/IEEE International Symposium on Networks-on-Chip*, 44–53, 2009.

[31] D. Rahmati, S. Murali, L. Benini, F. Angiolini, G. De Micheli, and H. Sarbazi-Azad. A method for calculating hard QoS guarantees for networks-on-chip. In *ICCAD '09: Proceedings of the 2009 International Conference on Computer-Aided Design*, 579–586, 2009.

[32] E. Rijpkema, K. G. W. Goossens, A. Radulescu, J. Dielissen, J. van Meerbergen, P. Wielage, and E. Waterlander. Trade offs in the design of a router with both guaranteed and best-effort services for networks on chip. In *DATE '03: Proceedings of the Conference on Design, Automation and Test in Europe*, 350–355, 2003.

[33] A. Sharifi, H. Zhao, and M. Kandemir. Feedback control for providing QoS in NoC based multicores. In *Proceedings of the Design, Automation, & Test in Europe Conference*, 1384–1389, 2010.

[34] D. A. Sigüenza-Tortosa and J. Nurmi. Packet scheduling in proteo network-on-chip. In *Proceedings of the IASTED International Conference on Parallel and Distributed Computing and Networks*, Innsbruck, Austria, February 17–19, 2004, 116–121.

[35] Sonics. MemMax 2.0, 2010. http://www.sonicsinc.com.

[36] Synopsys. DesignWare DDRn memory interface IP, 2010. `http://www.synopsys.com`.

[37] I. Walter, I. Cidon, R. Ginosar, and A. Kolodny. Access regulation to hot-modules in wormhole NoCs. In *Networks-on-Chip, 2007. NOCS 2007. First International Symposium on Networks-on-Chip (NOCS)*, 137 –148, 7-9 2007.

[38] L. Wang, H. Song, Y. Jiang, and L. Zhang. A routing-table-based adaptive and minimal routing scheme on network-on-chip architectures. *Computers & Electrical Engineering*, 35(6):846–855, 2009. High Performance Computing Architectures–HPCA.

[39] X. Wang, T. Ahonen, and J. Nurmi. Applying CDMA technique to network-on-chip. *IEEE Trans. Very Large Scale Integr. Syst.*, 15(10):1091–1100, 2007.

[40] D. Wiklund and D. Liu. SoCBUS: Switched network on chip for hard real time embedded systems. In *Proceedings of the 17th International Symposium on Parallel and Distributed Processing Symposium*, 781–789, 2003.

5

Emerging Interconnect Technologies

Davide Sacchetto

Ecole Polytechnique Fédérale de Lausanne, Switzerland

Mohamed Haykel Ben-Jamaa

Commissariat a l'Energie Atomique(CEA/LETI), Grenoble, France

Bobba Shashi Kanth

Ecole Polytechnique Fédérale de Lausanne, Switzerland

Fengda Sun

Ecole Polytechnique Fédérale de Lausanne, Switzerland

CONTENTS

5.1 Introduction

The recent years have seen an exponential increase of devices per unit area. As the transistors' size shrinks following Moore's Law, the increased delay of wires began to outweigh the increased performance of small transistors. Decreasing the wire delay, e.g., by using the low-dielectric (low-κ) isolating materials releases the problem in 90 nm; but for 65 nm and beyond, an ultralow-κ material is required. Although this is technologically feasible, it may increase the fabrication costs [59]. Industry has been following the scaling dictate and technology innovation was pushed by the design requirements. Along with the miniaturization coming with new technology nodes, profound modifications to routing have been carried out. An increasing number of metal layers at ever smaller lithography pitch became necessary for the increasing number of devices to interconnect. In the last decade the scaling of the interconnect has slowed down due to technological barriers such as wire resistivity increase as well as high capacitive coupling between adjacent wires. All this requires more effort on the technological research side for less resistive materials and low-κ insulators with ever lower dielectric constant.

In the International Roadmap for Semiconductors [74], the future of interconnects may lead to completely new concepts (see Table 5.1), either exploiting a totally different physics, such as wireless or optical/plasmonics signaling or a radical re-adaptation of the actual dual damascene Cu process into another one that employs innovative conductors, such as carbon nanotubes, nanowires, or graphene. These novel approaches leverage specific properties of the underlying technologies. For instance the optical signalling has the advantages of high bandwidth and no cross-talk; the radio frequency (RF) wireless invests on the possibility of interconnecting different components without the need of routing. In the case of nanotubes/nanowires/graphene, the intrinsic force as interconnect is a higher conductivity than Cu when properly engineered. Moreover, since those innovative materials are also investigated for their promising properties as field effect transistors for future technology nodes, there is the likelihood that these two branches of research, the one being on interconnects, and the other one being on devices, will one day lad to a unique platform entirely based on nanowires/nanotubes technology.

Another solution is 3D integration. This technology was proposed first in the 1970s. The main goal of 3D circuit processing is creating additional semiconducting layers of silicon, germanium, gallium arsenide, or other materials on top of an existing device layer on a semiconducting substrate. There are several possible fabrication technologies to form these layers. The most promis-

TABLE 5.1

Alternative interconnect technologies [74]

Technology	Advantages	Disadvantages
Optical	High Bandwidth	Low Density
Plasmonics	High Density	Low Distance
Wireless	Scaling Compatible	Process Integration
Nanowires/Nanotubes	High Density	CMOS Compatibility

TABLE 5.2

Electrical interconnects issues [74]

Problem	Potential Solution
Electromigration	Blech Length Via
Stress Migration	Via-to-Line Geometric Design
Dielectric Breakdown	N/A
Line Edge Roughness	N/A
κ increase in low-κ dielectrics due to CMP	N/A

ing near-term techniques are wafer bonding [61], silicon epitaxial growth [77], and recrystallization of polysilicon [39].

5.1.1 Traditional Interconnects

The ultimate goal in interconnect technology is the transmission of a signal without cost and power consumption. Since the advent of electronics, the traditional way of signal transmission has been of an electrical type that is still in use in today's integrated circuit technology. The increase of device density pushed the technology to stack more interconnect layers (see Figure 5.1), pushing the miniaturization of interconnect in the last decades towards its limits (see Table 5.2). While the ultimate performance would target zero resistance, zero capacitive coupling and zero inductance, scaling is working against these achievements so that technologists have to fight even more to match the requirements for future technology nodes. For instance, low resistivity copper wires have been introduced in the mid-90s as a solution for the problems of electromigration and failure rates of small pitch aluminum wires. This step could only be achieved by the introduction of highly reliable processes for Cu diffusion barriers, solving the well-known issue of Cu contamination in integrated circuit technology. Similarly, capacitive and inductive coupling is being minimized by the use of low-κ intermetal dielectrics, and values of the dielectric constant as low as 2.7–3.2 are already in use today. The final limit is to reach the dielectric constant of the vacuum, and porous materials are currently under investigation. However, all these new materials cannot be introduced in the actual dual damascene processing scheme, due to the reduced mechanical strength. Moreover, an increase of the dielectric constant has been observed after chemical mechanical polishing, which makes these low-κ materials difficult to be adopted.

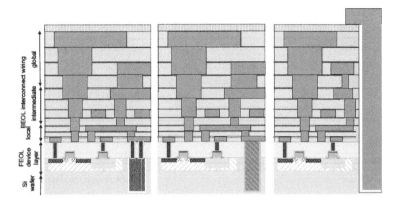

FIGURE 5.1
SoC interconnect layers. Lateral cross-section of a System-on-Chip showing
the Back End of Line with the different interconnect layers [74].

5.1.2   General Organization of the Chapter

The chapter is organized as a literature survey for the different emerging technologies candidates for future interconnects. Section 5.2 discusses the opportunities and challenges of optical interconnects based on photonic devices. Then
Section 5.3 presents an evolutionary approach for optical signaling based on
plasmon-polaritons waveguiding. Either photonic and plasmonic devices can
be efficiently implemented in 1D structures, such as silicon nanowires and
carbon nanotubes, which are also independent fields of research for emerging interconnects. Thus, a natural prosecution of the chapter are Section 5.4
and Section 5.5, which survey several approaches for the fabrication of Si
nanowires and carbon nanotubes, respectively. The 3D through silicon vias
(TSVs) approach based on traditional electrical interconnects compared with
implementations based on Carbon Nanotube (CNT) TSVs and optical TSVs
is discussed in Section 5.6. Finally, general conclusions summarizing advantages and opportunities of the different technologies are drawn in Section 5.7,
Summary and Conclusions.

5.2   Optical Interconnects

Because of the underlying physics, photonic technologies require a new set of
devices to enable the good signal link between components in a integrated circuit [13]. The different components for an on-chip optical interconnect system
are depicted in Figure 5.2(a). As it can be seen in Figure 5.2(b), optical data
link on 1 cm long optical waveguides with respect to Cu wires becomes ad-
vantageous when used in combination with wavelength division multiplexing

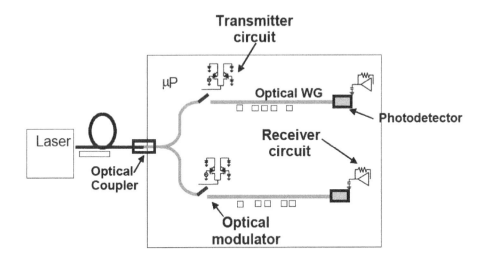

(a) Optical system for On-Chip signaling

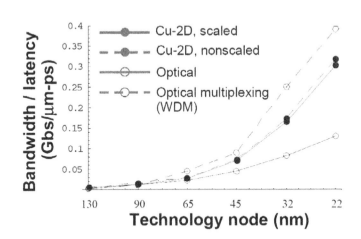

(b) Optical Bandwidth vs. traditional Cu interconnects

FIGURE 5.2
Optical Interconnects [13]: (a) System for on-chip optical interconnects. (b) Bandwidth/latency ratio for 1 cm long optical, nonscaled Cu, WDM optical and scaled Cu interconnect.

TABLE 5.3

Optical vs. electrical interconnects

Optical	Electrical
Speed-of-Ligth	Speed Requirements Matched
High Bandwidth	Bandwdith Requirements Matched
No Crosstalk	Crosstalk Constraint
Low Density	High Density
Global Interconnects	All Interconnect Levels

(WDM), which is a unique property of optical waveguides. Moreover, as the device scaling progresses, global interconnects using WDM systems become more and more competitive with respect to Cu. Advantages and disadvantages of optical physics for interconnects are summarized in Table 5.3. One of the major difficulties is the hybrid integration of photonic devices with state-of-the-art electronic devices, whose scaling properties have been the base of their successful implementation during the last 50 years.

Successful implementation of optical devices has been achieved in telecom applications, where the advantage of speed-of-light communication together with the high bandwidth replaced other competitive technologies. In the case of microelectronic circuits, more work has to be done in the miniaturization side, in order to integrate photonic devices at lower cost than standard Very Large Scale of Integration (VLSI) processing [50]. Thus, light sources, detectors, modulators, waveguides and couplers need to be integrated in a Complementary Metal Oxide Semiconductor (CMOS)-compatible way in order to make photonic interconnects appealing. Si, Ge, or SiGe materials are suitable candidates for all of these components, although every device would give the best performance when using a different material combination, such as III–V or II–VI semiconductors. For instance, compact lasers for light generation can be made with CdS nanowires (NWs) Fabry-Perot cavities (Figure 5.3(a)). Moreover, nanowire lasers have been demonstrated capable of light emission at different wavelengths, enabling the generation of distinct optical signals on-chip, as required for WDM data links.

It is worth noticing that even though these materials may sound exotic, III–V or II–VI compound semiconductors and nanowire structures are actually considered for future technology nodes in high-performance CMOS devices, so that the CMOS compatibility in this case will be extended to a broader group of materials. For instance, III–V quantum well modulators have been successfully co-integrated with CMOS. In any case, Si process technology seems to be the easiest way for CMOS compatible photonics at low cost. Typical wavelengths that can be used with Si are 850 nm, 1310 nm, and 1550 nm. Thus, cost-effective CMOS-compatible optical interconnects would, most likely, be composed of waveguides made of Si, Si_3N_4, or SiO_xN_y core embedded in a SiO_2 cladding. Depending on the refractive index contrast between core and cladding, high turns radii can be done at the cost of lower signal speed. The lowest turn radii of a few μms are achievable with Si/SiO_2 waveguides, which is also representative of the integration limit for this type of optical interconnect. As the other type of components would require to be significantly scaled, optical interconnects will most likely be used for top-level global signaling or for clock-tree distribution. Another constraint is that the waveguide pitch that cannot be reduced more than 300 nm–400 nm lateral dimensions, although the WDM technique can be useful in packing more signals per waveguide.

A different type of photonic device that requires to be miniaturized is the light modulator, the equivalent of the switch for optical signals. Among many alternatives, one possible implementation is the gate-all-around construction over a silicon nanowire waveguide [52]. This solution efficiently combines the waveguiding of Si/SiO_2 nanowire together with high-speed and high-efficiency modulation due to the capacitive system. Applying a voltage bias to the gate modifies the free-carrier concentration at the Si/SiO_2 interface where the maximum of the electric field is present. This variation in carrier concentration can be obtained in either accumulation or inversion mode, and both modes modify the effective refractive index of the modulator. Thus, Si nanowires embedded in SiO_2 cladding can be efficiently used for implementing both waveguides and light modulators.

5.3 Plasmonic Interconnects

An extension of the photonic devices would use a particular way of waveguiding light, which exploits the interaction of photons with the oscillations of the free electron gas, called *plasmons* [26, 48]. Plasmons can strongly interact with light, forming a quasi-particle (plasmon-polariton) that is bound at the surface of the material where the plasmon forms. Depending on the frequency of the photons impinging a given material, and depending on the frequency of plasmon oscillations, absorption or transmission of light can occur. By properly engineering the structure of the material, plasmonic antennae can be fabricated and used for different applications, such as waveguiding or high-efficiency light detection [1, 21]. Moreover, new types of lasers based on plasmonics can be built. For instance, high-gain CdS nanowire waveguides can be combined with metallic thin-films plasmonic antennae [58]. The interaction between the light confined into the CdS nanowire and a 5 nm-distant Ag thin-film layer gives spontaneous emission for a specific wavelength (see Figure 5.3(b)). One of the advantages of plasmonic devices compared with standard photonics, is that plasmons enable waveguiding of light at higher density and larger bandwidth [23]. A typical plasmonic waveguide can achieve turn radii below 0.5 μm. However, light cannot be guided for distances longer than some hundreds of μms [57], so plasmonic waveguides would fulfill very well the task for interconnecting between global wires and intermediate inter-

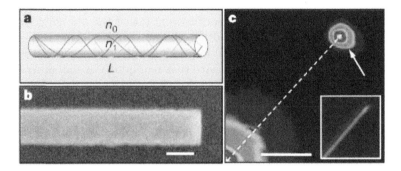

(a) Single nanowire lasing principle

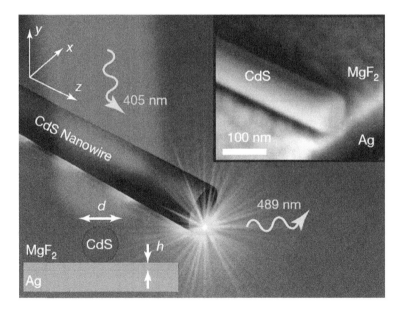

(b) Plasmonic laser

FIGURE 5.3
Nanowire lasing: (a) Single nanowire lasing principle [20]. (b) Plasmonic lasing based on a compound II–VI semiconducting nanowire on top of a MgF_2 gain layer [58].

connect levels. Plasmonic waveguide design includes different types, such as metal or semiconductors nanowires, grooves, nanoparticles, or metal stripes. Thus, nanowires or particles with subwavelength dimensions can effectively be used for light waveguiding. For instance, 200 nm diameter semiconductor wires have been demonstrated to transmit optical signals over 40 μm – 150 μm distance maintaining a high level of optical confinement at low loss [57].

5.4   Silicon Nanowires

Nanowires can be used for a broad range of applications, from ultimate high-performance field effect transistors, to high-sensitivity sensors, resonators, or nanomechanical devices. This large range of properties are coming from the one dimensional confinement of nanowires, which brought about quantum physics effects into play.

There are different approaches for the fabrication of nanowires, but all can be separated into two different techniques: bottom-up and top-down. The bottom-up approach consists in the growth of nanowires and it ends with the placement of grown nanowires on specific locations. Conversely, top-down approach starts by defining the positions of nanowires inside a device frame and ends with the formation, generally obtained by etching, of the nanowires. The different techniques have advantages and disadvantages that are summarized in Table 5.4. A large part of this section surveys these techniques and has been published in [8]. The authors would like to acknowledge those who contributed to this work.

Silicon nanowires have been proven excellent candidates for state-of-the-art field effect transistors, nonvolatile memories, biosensors, and nanoelectro-mechanical devices. Moreover, Si nanowires can be also exploited as interconnects in novel type of circuit constructions, such as crossbars. In addition, CMOS compatibility is a big advantage compared to other types of nanowires, and hybrid integration with CMOS is possible. Different approaches to build NW crossbars achieved i) metallic arrays, which do not have any semiconducting part that can be used as an access transistor, or ii) silicon-based crossbars with fluidic assembly, which have a larger pitch on average than the photolithography limit. Table 5.5 surveys the reported realized crossbars and

TABLE 5.4
Top-down vs. bottom-up approach

	Top-Down	**Bottom-Up**
Advantages	Alignment, Reliability	Dimension Control, Heterojunctions, Materials
Disadvantages	Variability	No Alignment, Metal Catalyst Contaminations

TABLE 5.5
Survey of reported nanowire crossbars. Functionalized arrays are those including molecular switches

Reference	[24]	[30]	[6]	[7]
NW material	Si/Ti	Ti/Pt	Si	poly-Si
NW width [nm]	16	30	20	54
NW pitch [nm]	33	60	> 1000	100
Crossbar density [cm^{-2}]	10^{11}	2.7×10^{10}	N/A	10^{10}
Technique	SNAP	NIL	Self-assembly	MSPT
Functionalized?	yes	yes	no	no

shows that the MSPT patterning technique has both advantages of yielding semiconducting NW and a high-crosspoint density of $\sim 10^{10}$ cm^{-2} while using conventional photolithographic processing steps.

5.4.1 Bottom-Up Techniques

5.4.1.1 Vapor-Liquid-Solid Growth

One of the widely used bottom-up techniques is the vapor-liquid-solid (VLS) process, in which the generally very slow adsorption of a silicon-containing gas phase onto a solid surface is accelerated by introducing a catalytic liquid alloy phase. The latter can rapidly adsorb vapor to a supersaturated level; then the crystal growth occurs from the nucleated catalytic seed at the metal-solid interface. Crystal growth with this technique was established in the 1960s [71] and silicon nanowire growth is today mastered with the same technique.

The VLS process allows for the control of the nanowire diameter and direction growth by optimizing the size and composition of the catalytic seeds and the growth conditions, including temperature, pressure, and gas composition in the chamber. In [31], defect-free silicon nanowires were grown in a solvent heated and pressurized above its critical point, using alkanethiol-coated gold monocrystals. Figure 5.4 depicts the growth process and transmission electron microscope (TEM) images of the grown nanowires. The nanowire diameters were ranging from 40 to 50 Å, their length was about several micrometers, and their crystal orientation was controlled with the reaction pressure.

5.4.1.2 Laser-Assisted Catalytic Growth

A related technique to VLS is the laser-assisted catalytic growth. High-powered, short laser pulses irradiate a substrate of the material to be used for the nanowire growth. The irradiated material either evaporates, sublimates, or converts into plasma. Then, the particles are transferred onto the substrate containing the catalyst, where they can nucleate and grow into nanowires. This technique is useful for nanowire materials that have a high melting point,

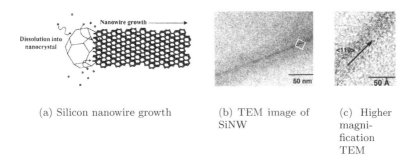

(a) Silicon nanowire growth   (b) TEM image of SiNW   (c) Higher magni- fication TEM

FIGURE 5.4
Vapor-liquid-solid growth of a silicon nanowire [31]: (a) Free Si atoms from silane dissolve in the Au seed until reaching the Si:Au supersaturation. Then Si is expelled as nanowire. (b) TEM image of SiNW synthesized at 500°C in hexane at 200 bar. (c) TEM of a part of SiNW inside the square in (b) shows high crystalline SiNWs.

since the laser pulses locally heat the substrate generating the particle for the nanowire growth. It is also suitable for multicomponent nanowires, including doped nanowires, and for nanowires with a high-quality crystalline structure [18].

5.4.1.3 Chemical Vapor Deposition

The chemical vapor deposition (CVD) method was shown to be an interesting technique used with materials that can be evaporated at moderate temperatures [28]. In a typical CVD process, the substrate is exposed to volatile precursors, which react on the substrate surface producing the desired nanowires. In [42], the CVD method was applied to fabricate nanowires based on different materials or combinations of materials, including Si, SiO_2 and Ge.

5.4.1.4 Opportunities and Challenges of Bottom-Up Approaches

The bottom-up techniques offer the ability of doping the as-grown nanowires in situ, i.e., during the growth process. In [18], the laser catalytic growth was used in order to control the boron and phosphorus doping during the vapor phase growth of silicon nanowires. The nanowire could be made heavily doped in order to approach a metallic regime, while insuring a structural and electronic uniformity. Another more advanced option offered by the bottom-up approaches consists in alternating the doping regions or the gown materials along the nanowire axis, as illustrated in Figure 5.5(a) [25, 76]. The growth of concentric shells with different materials around the nanowire axes was also demonstrated in [42], as illustrated in Figure 5.5(b).

The grown nanowires can either represent a random mesh laid out laterally

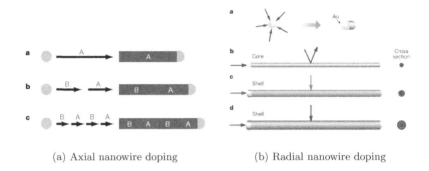

(a) Axial nanowire doping   (b) Radial nanowire doping

FIGURE 5.5
In situ axial and nanowire doping: (a) Doping along the nanowire axis (axial doping) [25]. (b) Doping around the nanowire axis (radial doping) [42].

over the substrate (Figure 5.6) [34], or they can stand vertically aligned with respect to the substrate (Figure 5.7) [29, 64]. The growth substrate is in general different from the functional substrate. Consequently, it is necessary to disperse the as-grown nanowires in a solution, and then to transfer them onto the functional substrate, making the process more complex. In [27], the nanowires were dispersed in ethanol; then the diluted nanowire suspension was used to flow-align the nanowires by using microfluidic channels. A similar technique was used in [35] in order to assemble arrays of nanowires through fluidic channel structures formed between a polydimethylsiloxane (PDMS) mold and a flat substrate. This technique yields parallel nanowires over long distances, as shown in Figure 5.8.

5.4.2 Top-Down Techniques

The top-down fabrication approaches have in common the utilization of CMOS steps or hybrid steps that can be integrated into a CMOS process, while keeping the process complexity low and the yield high enough. They also have in common the ability of defining the functional structures (nanowires) directly onto the functional substrate, with no need of dispersion and transfer of nanowires. Any top-down process uses patterning in a certain way: the patterning technique can be based on a mask, such as in standard photolithography or in other miscellaneous mask-based techniques, or it can be maskless, i.e., using a nanomold for instance.

5.4.2.1 Standard Photolithography Techniques

These techniques use standard photolithography to define the position of the nanowire. Then, by using smart processing techniques, including the accurate

(a)   (b)

FIGURE 5.6
Growth of meshed nanowires [34]: (a) Scanning electron microscope (SEM) image of gold-catalyzed growth of SiNWs on Si_3N_4/Si substrate. Image width $= 7$ μm. (b) High-magnification image of branched nanowires. Image width $= 0.7$ μm.

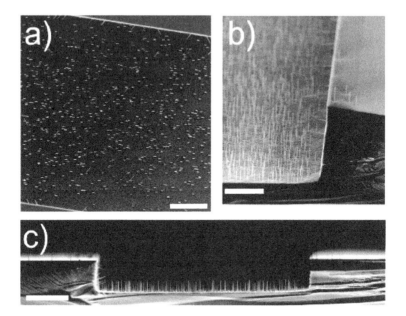

FIGURE 5.7
Growth of vertical nanowires [29]: (a) Conformal growth of nanowires to the substrate. (b) Tilted SEM image and (c) a cross-sectional SEM image of the structure. Scale bars are 10 μm.

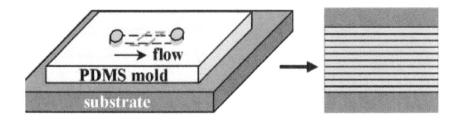

(a) Principle of the technique

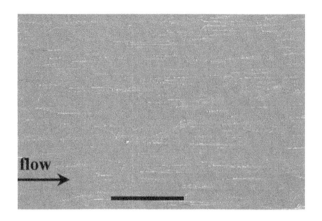

(b) SEM image

(c) Higher magnification SEM image

FIGURE 5.8
PDMS-mold-based assembly of InP nanowires [35]: (a) Schematic representation of the technique. (b) SEM image of the aligned nanowires (scale bar
= 50 μm). (c) Higher magnification SEM image of the aligned nanowires (scale
bar = 2 μm).

control of the etching, oxidation and deposition of materials, it is possible to scale the dimensions down far below the photolithographic limit.

In [51], silicon nanowires were defined on bulk substrates by using CMOS processing steps. First, a Si_3N_4 nitride rib was defined on the substrate. Then, the isotropic etch defined the nanowire underneath the rib. Well-controlled self-limited oxidation and subsequent etching steps resulted in silicon nanowires with different cross-section shapes and dimensions, and with a nanowire diameter down to 5 nm.

A related fabrication approach was presented in [43], whereby the nanowire dimensions were defined by an accurate control of the silicon oxidation and etch. The authors transferred the as-fabricated nanowires onto a different substrate in order to arrange them into parallel arrays, which makes this approach partly reminiscent of the bottom-up techniques explained previously.

Another approach was presented in [67], which uses epitaxial Si and Ge layers on a bulk substrate. A thin epitaxial Si layer was sandwiched between Si_3N_4 and SiGe layers. Then, the Si_3N_4 and SiGe were selectively etched, leading to a partial etch of the sandwiched Si layer. The remaining edges of the Si layer were thinned out and lead to 10 nm nanowire diameter.

5.4.2.2 Miscellaneous Mask-Based Techniques

Instead of using standard lithography and thinning out the devices by means of well-controlled oxidation and selective etching, an alternative approach is to use electron-beam lithography [40] that offers a higher resolution below 20 nm, and then eventually further reduce the nanowire diameter by stress-limited oxidation [38].

A higher resolution can be achieved by using extreme ultraviolet interference lithography (EUV-IL) [3]. Metallic nanowires with the width of 8 to 70 nm and a pitch of 50 to 100 nm could be achieved with this technique. However, this approach needs a highly sophisticated setup in order to provide the required EUV wavelength, which is not available in state-of-the-art semiconductor fabrication lines. And it has not be proven so far how this technique may be used in order to fabricate semiconducting nanowires.

The stencil lithography is another approach, which is inherently different from the previous ones, but it shares the same feature of using a mask while avoiding the classical paradigm of CMOS processing, which consists in patterning a photoresist through the mask and then patterning the active layer through the patterned photoresist. The stencil approach [68] is based on the definition of a mask that is fully open at the patterned locations. The mask is subsequently clamped onto the substrate, and the material to be deposited is evaporated or sputtered through the mask openings onto the substrate. Nanowires with a width of 70 nm could be achieved this way. Even though only metallic nanowires have been demonstrated, the technique can be extended to semiconducting nanowires as well.

5.4.2.3   Spacer Techniques

The spacer technique is based on the idea of transforming thin lateral dimensions, in the range of 10 to 100 nm, into vertical dimension by means of anisotropic etch of the deposited materials. In [33], spacers with a thickness of 40 nm were demonstrated with a line-width roughness of 4 nm and a low variation across the wafer.

In [16], spacers were defined by means of low-pressure chemical vapor deposition (LPCVD), then their number was duplicated by using the spacers themselves as sacrificial layers for the following spacer set. This technique, the iterative spacer technique (IST), yields silicon structures with sub-10 nm width and a narrower half-pitch than the photolithography limit.

In [14], the multispacer patterning technique (MSPT) was developed as in the previous approach, by iterating single spacer definition steps. The spacers were reminiscent to nanowires with a thickness down to 35 nm. The multispacer array was not used as a nanomold to define the nanowires, but it was rather used as the actual nanowire layer.

5.4.2.4   Nanomold-Based Techniques

Alternative techniques use the nanoimprint lithography (NIL), which is based on a mold with nanoscale features [73] that is pressed onto a resist-covered substrate in order to pattern it. The substrate surface is scanned by the nanomold in a stepper fashion. The as-patterned polymer resist is processed in a similar way to photolithographically patterned photoresist films. The advantage of this technique is its ability to use a single densely patterned nanomold to pattern a large number of wafers. The obtained density of features on the substrate depends on the density of features in the nanomold, i.e., it mainly depend on the technology used to fabricate the nanomold. A related technique to NIL, called the superlattice nanowire pattern transfer technique (SNAP), was presented in [37], in which the nanowires are directly defined on the mold; then, they are transferred onto the polymer resist. Nanowires with a pitch of 34 nm could be achieved using the SNAP technique. The superlattice was fabricated by defining 300 successive epitaxial $GaAs/Al_xGa_{1-x}As$ layers on a GaAs wafer. Then, the wafer was cleaved, and the GaAs layers were selectively etched, so that the edge of each layer became an initial nanowire template. Then, a metal was deposited onto the exposed $Al_xGa_{1-x}As$ ridges during a self-aligned shadow mask step. The self-aligned metal nanowires at the $Al_xGa_{1-x}As$ planes were subsequently transferred with the SNAP technique. The spacer technique was used in an iterative way in [65] in order to define a nanomold yielding sub-10 nm nanowires with a 20 nm pitch. The process, planar edge defined alternate layer (PEDAL), uses standard photolithography to define a sacrificial layer for the first spacer, then by iterating the spacer technique, a layer of dense spacers was defined. A shadow-mask deposited metal at the partially released spacer ridges formed thin nanowires, which were subsequently transferred onto the functional substrate. Since the spacer technique was used to define the nanomold and not the nanowires directly, this process is closer in nature to the nanomold-based techniques than to the spacer techniques.

5.4.2.5 Opportunities and Challenges of Top-Down Approaches

Top-down approaches are attractive because of their relatively easier required methods. The standard photolithography and spacer techniques can be integrated in a straightforward way into a CMOS process, whereas miscellaneous mask-based techniques may be more expensive and slower than the desired level for large production, and the maskless approaches may require the hybridization of the process with the nonconventional steps. A promising opportunity that is offered by the spacer and nanomold-based techniques is the definition of devices with a subphotolithographic pitch. In contrast to standard photolithography techniques, whose pitch is ultimately defined by the lithography limit, using spacer- and nanomold-based techniques represents an elegant way to circumvent the photolithographic limitations, and has a potential application field with regular architectures such as crossbar circuits. The alignment of different processing steps is straightforward with standard lithography and miscellaneous mask-based techniques; while the spacer techniques are self-aligned. Nevertheless, whenever a nanomold-based step is introduced, the alignment becomes a very challenging issue, making these techniques more likely to be used at the early processing stages. A general drawback of all these bottom-up techniques is that the obtained semiconducting nanowires are undifferentiated, meaning that the doping profile along the nanowires is generally the same and cannot be modified at later process stages after the nanowires are defined. In order to uniquely address every nanowire, it is highly desirable to associate a different doping profile to every nanowire in order to uniquely address them. Another challenge that is specific to the nanomold-based technique is the metallic nature of the most demonstrated nanowires. However, in order to fabricate the access devices to the nanowires, it is required to have semiconducting nanowires that can be field-effect-controlled. Fortunately, there are still many opportunities promising the fabrication with semiconducting nanowires with these techniques as well.

5.4.3 Focus on the Spacer Technique

The goal of this part of the chapter is to define a process flow for a nanowire crossbar framework using standard CMOS processing steps and the available micrometer scale lithography resolution. The proposed approach is based on the spacer patterning technique presented in Section 5.4.2.3. The iteration of the spacer steps has been shown to be an attractive and cost-efficient way to fabricate arrays of parallel stripes used for the definition of nanomolds [65] or directly as nanowire arrays [14]. The approach presented in this part of the chapter is based on the idea of MSPT demonstrated in [14] for a single nanowire layer. In this part of the chapter, the efforts are concentrated on related challenges: first, the demonstration of the ability of this technology to yield a crossbar structure; then the assessment of the limits of this technology in terms of nanowire dimensions and pitch; and finally, the characterization of access devices operating as single poly-Si nanowire field effect transistors (poly-SiNWFET). The main idea of the process is the iterative definition of thin spacers with alternating semiconducting and insulating materials, which result in semiconducting and insulating nanowires. The structures are defined inside a 1 μm high wet SiO_2 layer over the Si substrate (Figure 5.9(a)). This SiO_2 layer has two functions: on the one hand, it insures the isolation between the devices; on the other hand, it is used to define a 0.5 μm high-sacrificial layer on which the multispacer is defined. Then, a thin conformal layer of poly-Si with a thickness ranging from 40 to 90 nm is deposited by LPCVD in the *Centrotherm* tube 1-1 (Figure 5.9(b)). During the LPCVD process, silane (SiH_4) flows into the chamber and silicon is deposited onto the substrate.

The type of deposited silicon (amorphous or poly-crystalline) depends on the chamber temperature and pressure [69, 2, 70]. The deposition has been specifically optimized for the CMI facilities [60]. At the deposition temperature of 600°C, the LPCVD process yields poly-crystalline silicon. Thereafter, this layer is etched with the Reactive Ion Etching (RIE) etchant *STS Multiplex ICP* using a Cl_2 plasma, in order to remove the horizontal layer while keeping the sidewall as a spacer (Figure 5.9(c)). As the densification of deposited silicon improves the crystalline structure [54], the poly-Si spacer is densified at 700°C for 1 hour under N_2 flow in the *Centrotherm* tube 2-1. Then, a conformal insulating layer is deposited as a 40 to 80 nm thin Low-Temperature Oxide (LTO) layer obtained by LPCVD in the *Centrotherm* tube 3-1 following the reaction of SiH_4 with O_2 at 425°C (Figure 5.9(d)). The quality of the LTO can be improved through densification [9]. Thus, the deposited LTO is densified at 700°C for 45 minutes under N_2 flow. Then it is etched in the RIE etchant *Alcatel AMS 200 DSE* using C_4F_8 plasma in order to remove the horizontal layer and just keep the vertical spacer (Figure 5.9(e)). Alternatively, instead of depositing and etching the LTO, the previously defined poly-Si spacer can be partially oxidized in the *Centrotherm* tube 2-1 in order to directly form the following insulating spacer. These two operations (poly-Si and insulating spacer definition) are performed one to six times in order to obtain a multispacer with alternating poly-Si and SiO_2 nanowires (Figure 5.9(f)). Then, the batch is split into two parts: some of the wafers are dedicated to the definition of a second perpendicular layer of nanowires, some others are processed further with the gate stack and the back-end steps and are dedicated to perform electrical measurements.

In order to address the issue of characterizing a single access device (poly-SiNWFET), a single nanowire layer is used, on top of which a poly-Si gate stack is defined with an oxide thickness of 20 nm, obtained by dry oxidation of the poly-SiNW, and different gate lengths (Figure 5.9(g)). The drain and source regions of the undoped poly-SiNW are defined by the e-beam evapo-

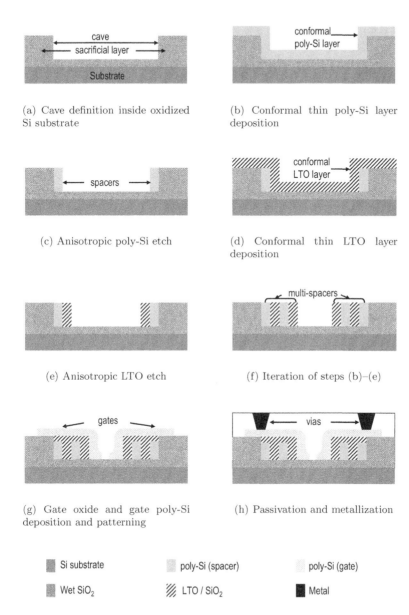

(a) Cave definition inside oxidized Si substrate

(b) Conformal thin poly-Si layer deposition

(c) Anisotropic poly-Si etch

(d) Conformal thin LTO layer deposition

(e) Anisotropic LTO etch

(f) Iteration of steps (b)–(e)

(g) Gate oxide and gate poly-Si deposition and patterning

(h) Passivation and metallization

Si substrate   poly-Si (spacer)   poly-Si (gate)

Wet SiO_2 LTO / SiO_2 Metal

FIGURE 5.9
MSPT process steps in (a)-(h).

ration and lift-off of 10 nm Cr and 50 nm nichrome ($Ni_{0.8}Cr_{0.2}$) with *Alcatel EVA 600* (Figure 5.9(h)). The Cr enhanced the adhesion and resistance of Ni to oxidation during the two-step annealing (including 5 minutes at 200°C, then 5 minutes at 400°C [12]) performed in the *Centrotherm* tube 3-4. Using $Cr/Ni_{0.8}Cr_{0.2}$ is a simple way to contact undoped nanowires, since Ni is used to form mid-gap silicides. If the nanowires are doped, then it is possible to use aluminum as contact metal, which is then evaporated and patterned after a passivation layer is deposited and vias are opened, as depicted in Figure 5.9(h).

In order to address the issue of realizing a crossbar framework, the bottom multispacer is fabricated as explained previously in Figure 5.9(a) to 5.9(f), then a 20 nm dry oxide layer is grown as an insulator between the top and bottom nanowire layers. The top sacrificial layer is defined with LTO perpendicular to the direction of the bottom sacrificial layer. Then a poly-Si spacer is defined at the edge of the top sacrificial layer in a similar way to the bottom poly-Si spacers. Therafter, the separation of dry oxide and both sacrificial layers are removed in a Buffered Hydrofluoridric Acid (BHF) solution in order to visualize the crossing poly-Si spacers realizing a small poly-Si nanowire crossbar.

5.4.4 Focus on the DRIE Technique

Another type of processing which enables top-down fabrication of Si nanowires at a pitch that is not limited by the lithographic resolution is based on the deep reactive ion etching (DRIE) technique. This approach is utilized to obtain arrays of vertically stacked Si nanowires. While the density of horizontal strands is limited by the lithographic pitch, each strand can be composed of several vertically stacked nanowires by adjusting the number of cycles in the DRIE process.

In [63], the process begins by defining a photoresist line on a p-type silicon bulk wafer (see Figure 5.10(a). This mask will be used as a protective layer for the successive DRIE technique. This technique, that alternates a plasma etching with a passivation step, has been optimized to produce a scalloped trench in silicon with high reproducibility. Etching time, passivation time, and plasma platen power can be changed in order to enhance the scalloping effect. The application of the DRIE technique gives a trench like the one depicted in Figure 5.10(b). The flexibility of the process allows us to change the number of scallops easily. After trench definition, a sacrificial oxidation step is carried out. The effect of oxidation results in the total Si consumption of the smaller portions of the trench. The wider parts of the trench leave vertically stacked Si nanowires embedded in the grown oxide (see Figure 5.10(c)). Then the cavities produced by the Bosch process are filled with photoresist (Figure 5.10(d)). After a combination of chemical mechanical polishing (CMP) and BHF dip, the wet oxide is removed around the nanowires (see Figure 5.10(e)). After removal of the resist, caves with stacks of several nanowires are freestanding on a layer of thick wet oxide, which is left to isolate the substrate from

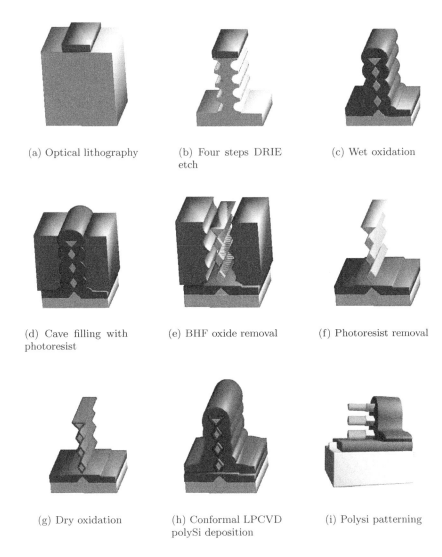

(a) Optical lithography

(b) Four steps DRIE etch

(c) Wet oxidation

(d) Cave filling with photoresist

(e) BHF oxide removal

(f) Photoresist removal

(g) Dry oxidation

(h) Conformal LPCVD polySi deposition

(i) Polysi patterning

FIGURE 5.10
Vertically stacked Si nanowire process steps in (a)-(i).

the successive processes (Figure 5.10(f)). Nanowires are oxidized in dry atmosphere, for a 10 – 20 nm higher quality oxide, as the dielectric for Field Effect Transistor (FET) devices (Figure 5.10(g)) as gate dielectric. Then between 200 nm and 500 nm of LPCVD polysilicon is deposited (Figure 5.10(h)). The LPCVD polySi layer allows conformal coverage of the 3D structure, enabling the formation of gate-all-around devices, such as FETs [62] or optical modulators [52]. The polysilicon gate is patterned by means of a combination of isotropic and anisotropic recipes (see Figure 5.10(i)). Depending then on the structure, implantation or metalization of the Si pillars can be carried out, so to produce Metal Oxide Semiconductor Field Effect Transistors (MOSFETs) or Schottky Barrier Field Effect Transistors (SBFETs), respectively.

Examples of fabricated structures demonstrating arrays having from 3 up to 12 vertically stacked Si nanowires are shown in Figure 5.11. The obtained nanowires can be used to build gate-all-around field effect transistors (see Figure 5.12) interconnected through Si pillars.

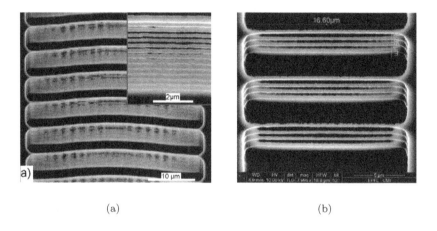

(a)   (b)

FIGURE 5.11
Arrays of vertically stacked Si nanowires [63]: (a) Silicon nanowire arrays with 12 vertical levels. (b) Silicon nanowire arrays with 3 vertical levels.

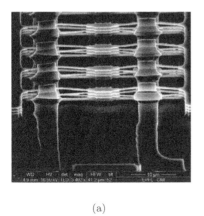

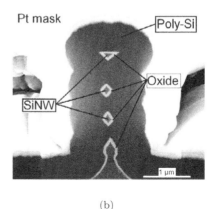

(a)   (b)

FIGURE 5.12
Vertically stacked Si nanowire transistors [63]: (a) Three horizontal Si nanowire strands with two parallel polysilicon gates. (b) Focused ion beam (FIB) cross-section showing triangular and rhombic nanowires embedded in a gate-all-around polysilicon gate.

5.5   Carbon Nanotubes

The resistance of the copper interconnects, for the future technology nodes, is increasing rapidly under the combined effects of enhanced grain boundary scattering, surface scattering, and the presence of the highly resistive diffusion barrier layer [66]. The quest for technologies with superior properties such as large electron mean free paths, mechanical strength, high thermal conductivity, and large current carrying capacity (Table 5.6), has showcased Carbon nanotubes (CNTs) as a possible replacement for copper interconnects [41, 74]. On-chip interconnects employing CNTs can improve the performance in terms of speed, power dissipation, and reliability.

TABLE 5.6
Cu vs. CNT

Properties	CNT	Copper
Mean Free Path [nm]	>1000 [49]	40
Max Current Density [A/cm^2]	>10^{10} [72]	10^6
Thermal Conductivity [W/mK]	5800 [32]	385

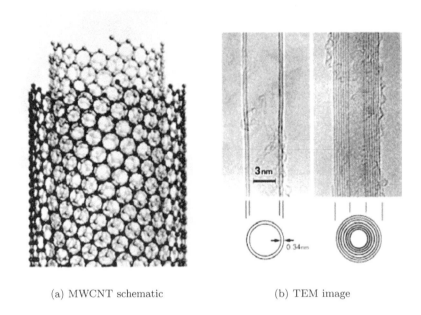

(a) MWCNT schematic   (b) TEM image

FIGURE 5.13
Multiwall carbon nanotube discovered by Sumio Iijima in 1991 [36]: (a) MWCNT schematic. (b) TEM image.

5.5.1 Physics of Carbon Nanotubes

Carbon nanotubes are graphene sheets rolled up into cylinders with diameters in the order of few nanometers. CNTs can be categorized into two groups: single-walled carbon nanotubes (SWCNTs) and multiwalled carbon nanotubes (MWCNTs). Figure 5.13 shows the TEM image of a MWCNT discovered by Sumio Iijima [36]. A SWCNT is made up of only one graphene shell with diameters ranging from 0.4 nm to 4 nm [19]. A MWCNT is composed of a number of SWCNTs nested inside one another with diameters ranging from a few to 100 nm [44].

The physical nature of SWNCTs has a huge impact on their electrical characteristics. Due to their cylindrical symmetry, there is a discrete set of directions in which a graphene sheet can be rolled to form a SWCNT. Hence, SWCNT can be characterized depending on their chiral vector (n, m) [10]. Figure 5.14 illustrates two main types of SWCNT, zigzag nanotube with m = 0 and armchair nanotube with m = n. A zigzag CNT behaves as a semiconductor whereas an armchair CNT is metallic in nature. On the other hand MWCNTs are not characterized in this way because they are composed of nanotubes with varying chirality. However, it should be noted that the band gap varies inversely with SWCNT diameter, approximately as $E_G = 0.84/D$ eV,

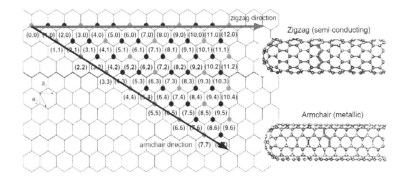

FIGURE 5.14
Chiral vectors of SWCNTs determining the type of CNTs: zigzag (semi-conducting CNTs) and armchair (metallic CNTs) [10].

where D is given in nm. Hence, the large semiconducting shells (D >5 nm) of the MWCNTs have bandgaps comparable to the thermal energy of electrons and act like conductors at room temperature [47]. This makes the MWCNTs mostly conducting.

5.5.2 Types of CNTs for Various Interconnects

For the implementation as future interconnects, CNTs have to fulfill the requirements of high current carrying capability and resistances comparable to or better than copper. Moreover, before taking CNTs for gigascale interconnect applications, there are few critical questions that have to be answered: first, the type of CNTs (bundle of SWCNTs or MWCNTs); second, the level of interconnect hierarchy at which CNTs can come into picture.

Metallic SWCNTs are vital for VLSI interconnect applications. However, the high resistance associated with an isolated CNT (with quantum resistance of 6.5 KΩ) [49] necessitates the use of a bundle of CNTs conducting current in parallel to form an interconnection. While dispersed SWCNTs can form dense regular arrays with constant 0.34 nm intertube spacing [46], in-place grown CNTs reported to date have been quite sparse. Table 5.7 [74] gives the minimum densities of metallic SWCNTs required to outperform minimum-size copper wires in terms of conductivity. This is due to the fact that at present, several fabrication challenges need to be overcome for densely packed metallic SWCNT bundles. At a fundamental level, the CNT synthesis technique developed to date cannot control chirality. Statistically, only one third of SWCNTs with random chirality are metallic. Improving the ratio of metallic to semiconductor tubes would proportionally increase the conductivity of SWCNT-bundles.

TABLE 5.7

Minimum density of metallic SWCNTs needed to exceed Cu wire conductivity [74]

Year	2010	2012	2014	2016	2018	2020	2024
Cu R_{eff} [μm-cm]	4.08	4.53	5.2	6.01	6.7	8.19	12.91
Min Density [nm^{-2}]	0.175	0.158	0.138	0.119	0.107	0.087	0.055

It is also important to investigate how SWCNT-bundle and MWCNT compare with each other [5]. If the SWCNTs are all metallic, SWCNT interconnect can outperform MWCNT at highly scaled technologies. On the other hand, if the SWCNTs have random chiralities, the MWCNT interconnect can achieve better performance. However, one of the major technology concern for MWCNTs is the connection between the contact and all the shells within a MWCNT. It has been proven experimentally that all shells within MWCNTs can conduct if proper connections are made to all of them [44, 47].

CNTs can potentially replace Cu/low-κ interconnects at most levels of interconnect hierarchy [53]. On-chip interconnects are categorized as local, semi-global, and global interconnects, depending on their length. A bundle of densely packed SWCNTs with high-quality contacts, with the electrodes, lowers the interconnect resistance and address the problem of size effects in copper wires. This integration option provides significant delay improvement for long (global) interconnects where the R(L)C delay is dominant. In the case of local interconnects, the capacitance has a bigger impact on delay than resistance. Hence, local interconnects with few layers of SWCNT can reduce the capacitance by 50 thereby significantly decreasing the electrostatic coupling between adjacent interconnects. MWCNTs are suitable for semiglobal and global interconnects because of their very large mean free paths, which increases with increase in the diameter of the nanotube [53]. Recently MWCNT interconnects operating at gigahertz frequency range have been demonstrated. Conductivity of MWCNTs in these experiments, however, has been considerably lower than the theoretical models, mainly because of large defect density and a small ratio of outer to inner diameters [17].

CNT-based vias have been proposed as an effective solution to the current problem of VLSI circuits that originate from stress and electromigration of Cu interconnects, in particular vias and their surrounding [55]. Figure 5.15(a) shows a schematic of CNT vias in future CMOS-CNT hybrid technology. Vertically aligned CNTs in via holes with the aim of forming vias of CNT-bundles directly on Cu wiring and MOSFETs has been demonstrated. Figure 5.15(b) shows the SEM images of an array of 1 μm-diameter via holes filled with bundles of MWCNTs.

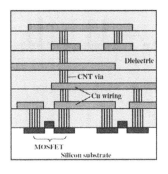
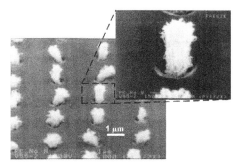

(a) Schematic cross section   (b) SEM image of a 1 μm via

FIGURE 5.15
A CMOS-CNT hybrid interconnect technology with bundles of MWCNTs grown in via holes: (a) Schematic cross section. (b) SEM image of a 1μm via [55].

5.5.3 Synthesis of CNT: A Technology Outlook

For the CNT-based-interconnects to reach fruition, methods for controlled deposition of CNT networks, with controlled density and orientation, must be developed. Chemical vapor phase deposition (CVD) is a common technique used to grow CNT networks. However, this high-temperature CNT growth process occurs at temperatures ($> 500°$C) that can cause serious damage to the integrated circuits. Lowering the CNT growth temperatures to less than $500°$C, limits the length of the CNTs to less than 1 μm [4]. An alternative method to overcome the thermal budget is by first synthesizing CNTs on a separate substrate at high temperature, and then transferred to a CMOS wafer by polymer [15]. For CNT based vias, Kreupl et al. [41] demonstrated a method for growing MWCNT bundles by CVD from the bottom of vias and contact holes decorated by an iron-based catalyst. The high-temperature step during the CNT growth has been lowered to $540°$C in [55] by employing a hot-filament (HF-CVD) technique.

In an alternative bottom-up process, MWCNTs are first grown using HF-CVD at prespecified locations, then gap-filled with oxide and finally planarized. Figure 5.16 shows a schematic of the sequence for this process that may alleviate the traditional problems associated with etching high-aspect ratio vias. A Si (100) wafer covered with 500 nm thermal oxide and 200 nm Cr (or Ta) lines is used to deposit 20 nm thick Ni as a catalyst. Ion beam sputtering is used to deposit Ni on patterned spots for local wiring or contact hole applications. For global wiring, Ni can be deposited as a 20 nm thick micronscale film. Plasma enhanced chemical vapor deposition (PECVD) is then used to grow a low-density MWCNT array by an inductively coupled plasma

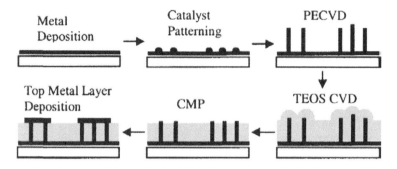

FIGURE 5.16
Schematic of process sequence for bottom-up fabrication of CNT bundle vias [45].

process or dc plasma assisted hot-filament CVD. Next, the free space between the individual CNTs is filled with SiO_2 by CVD using tetraethylorthosilicate (TEOS). This is followed by CMP to produce a CNT array embedded in SiO_2 with only the ends exposed over the planarized solid surface. Bundles of CNTs offer many advantages for on-chip interconnects, but a number of hurdles must be overcome before CNTs can enter mainstream VLSI processing. The major issue is the maturity of the CNT synthesis techniques, which still cannot guarantee a controlled growth process to achieve prescribed chirality, conductivity, diameter, spacing, and number of walls. The second major challenge that has yet to be solved is the hybrid fabrication of CNTs and CMOS components with optimal thermal budgets without degradation, and in realizing high-quality contacts. An overview of CNT interconnect research has been presented in this section. This provides an early look at the unique research opportunities that CNT interconnects provide, before they see widespread adoption.

5.6   3D Integration Technology

Considering cost, complexity, and interconnect, 3D wafer-bonding technology is perhaps the best way to mitigate the fundamental physical problem. It is developing quite fast in recent years, and is now believed to be the future interconnection technology. Wafers or chips can be bonded face-to-face (F2F) or face-to-back (F2B). As it can be seen in Figure 5.17, actually, F2F does not need TSVs, but it allows only two layers to be stacked. For more-than-two-layer stacking, F2B has to be the choice.

The main bonding technology can be categorized to three main approaches, adhesive bonding, microbumps bonding, and direct covalent bonding. Bonding can be used on wafer level (wafer-to-wafer) or chip level (chip-to-wafer or chip-to-chip). Wafer-level bonding has much higher accuracy and allows for larger density of interconnects comparing to chip-level bonding nowadays. In mid-2006, Tezzaron's wafer-level process consistently achieves alignment accuracy of less than a micrometer. At the same time, chip-level placement accuracy is about 10 μm [59]. Today, the most used method for chip-level bonding is flip chip using solder bumps. It requires one by one manipulation that means higher cost. Some novel ideas were proposed to solve the low through-put limitation of flip chip technology and enhance the alignment accuracy. Tohoku University's fluidic self-alignment method claims to have achieved a high-chip alignment accuracy of 1 μm [22].

Another crucial technology in 3D integration is thinning. Lower-aspect ratio TSVs are always more preferred because of less fabrication difficulties. Silicon wafers can be thinned to less than 50 μms. The process usually starts with mechanical grinding, followed with chemical mechanical polishing, then finishes with unselective etching to achieve smooth and planar surface. Alternatively, some groups such as MIT Lincoln Lab choose Silicon-on-Insulator (SOI) wafers. The thinning process is automatically done by etching away the oxide layer in the SOI wafers to strip the backside thick silicon part [11]. 3D integration technology with TSVs can be categorized into in-processing and postprocessing approaches. In-process makes the TSVs ready before the fabrication of metal wires on chips, meaning that generally this approach can tolerate high temperature of above 1000°C. Another advantage of this approach is that the interconnection length can be minimized with 3D place and routing design. In-process approach is suitable for Integrated Circuit (IC) fabs. Anther approach is postprocessing. The fabricated chips are formed with TSVs before/after dicing, and then stacked. The temperature budget is limited under 350°C to avoid any degeneration in preprocessed circuits. Wafer-level and chip-level bonding can be used in both approaches.

5.6.1 Metal/Poly-Silicon TSVs

Various 3D vertical interconnection technologies have been explored, e.g., contact-less coupling (capacitive and inductive), optical and metallic. Metallic TSV has become the most popular solution because of its great potential to get the highest density. Circular or square holes are etched on the silicon wafers; the holes' sidewalls are deposited with an isolation layer, such as silicon dioxide and parylene; seed layers for electroplating are deposited and then continued with electroplating of metal, usually copper or tungsten.

5.6.2 Carbon Nanotube TSVs

Through-wafer interconnects by aligned carbon nanotube for three dimensionally stacked integrated chip packaging applications have been reported [75]. Two silicon wafers are bonded together by tetra-ethyl-ortho-silicate. The top wafer (100 μm thick) with patterned through-holes allows carbon nanotubes

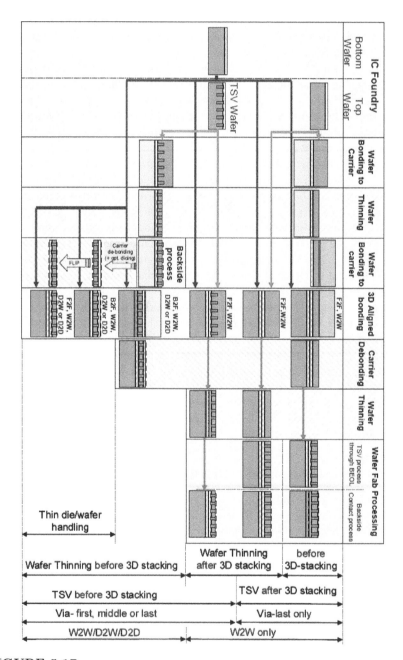

FIGURE 5.17
3D stacking technologies [74].

to grow vertically from the catalyst layer Fe on the bottom wafer. By using thermal chemical vapor deposition technique, the authors have demonstrated the capability of growing aligned carbon nanotube bundles with an average length of 140 μm and a diameter of 30 μm from the through holes.

5.6.3 Optical TSVs

Optical TSV could empower an enormous bandwidth increases, increased immunity to electromagnetic noise, a decrease in power consumption, synchronous operation within the circuit and with other circuits, and reduced immunity to temperature variations. The constraints for developing optical TSVs are the fact that all the fabrication steps have to be compatible with future IC technology and that the additional cost incurred remains affordable. Obtaining high optical-electrical conversion efficiency, decreasing optical transmission losses while allowing for a sufficient density of photonic waveguides on the circuit and reduction of the latency when the chips are operating above 10 GHz are the difficulties. Various technological solutions may be proposed. Materials chosen should be able to ensure efficient light detection, efficient signal transport and technological compatibility with standard CMOS processes. Vertical Cavity Surface Emitting Lasers (VCSELs) are without doubt the most mature emitters for on-chip or chip-to-chip interconnections. Light is emitted vertically at the surface, by stimulated emission via a current above a few microamperes. The active layer is formed by multiple quantum wells surrounded by III–V compound materials, and the whole forms the optical cavity of the desired wavelength. VCSELs have a very low threshold current, low divergence and arrays of VCSELs are easy to fabricate. However, internal cavity temperature can become quite high and this is important because both wavelength and optical gain are dependent on the temperature. Thin-film metal-semiconductor-metal (MSM) photodetectors are under research, due to their improved area per unit capacitance, or in other words, allowance for higher circuit speed.

How to fabricate the waveguides and how to integrate the active optical devices are also important decisions. Similarly to the metallic TSVs fabrication process, the integration of the silicon waveguides can be done either at the front end of the CMOS process or after the whole CMOS processes are finished. The former is possible but other considerations have to be taken into account. When the optical link is used for 2D interconnection, at the transistor level, the routing of the waveguide is extremely difficult. But if it turns to be a vertical 3D optical link, the problem for 2D does not exist. The crucial problem is the active optical device integration. The sources and detectors have to be bonded by flip chip, but high number of individual bonding operations is not realistic. And silicon-based devices can only work at low wavelengths (850 nm), which translates to higher attenuation in the waveguides. This solution requires an extraordinary mutation in the CMOS process, and as such is highly unattractive from an economic point of view. Hybrid integration of the optical layer on top of a complete CMOS IC is much more practical and promising. The source and detector devices are no longer bound to be realized in the host material. One possible solution can be: the waveguides are fabricated penetrating the CMOS metal connection layers and the silicon substrate; microsources and photodetectors are fabricated in a thinned GaAs wafer; this GaAs wafer is then wafer-bonded to the bottom CMOS IC. Since active GaAs layer is stacked on top of a thinned CMOS IC substrate, bonding precision and quality are more critical than metal TSVs.

The bit error rate (BER) is commonly not considered in IC design circles, because metallic interconnects typically achieve BER figures better than 10^{-45}. For optical interconnect network, a BER of 10^{-15} is acceptable [56]. As operating frequencies are likely to rise up to over 10 GHz in high-speed digital system, the combination of necessarily faster rise and fall times, lower supply voltages, and higher crosstalk increase the probability of wrongly interpreting the signal that was sent. This may increase design complexity. Metallic TSVs can already support sufficient throughput for current VLSI circuits. Concerning fabrication, metallic TSV is the easiest and most straightforward method for 3D interconnection. As the system frequency requirement goes higher than the traditional VLSI circuit can support, optical TSVs are promising for future circuits that exploit the novel functionalities coming from photonic and/or plasmonic devices. So far, optical on-chip or chip-to-chip interconnections have been proposed, but optical TSVs are rarely discussed.

5.7 Summary and Conclusions

Novel functionalities driven by the device innovation will require new interconnect concepts. Among several possibilities, optical signaling is the most promising in terms of bandwidth/latency for global interconnects. Future developments of plasmonic devices would bring photonic signal at the device level, thus solving the common issues of electromigration and crosstalk of standard electrical conductors. From a technological point of view, the optical interconnects are still not mature for On-Chip signaling, mainly for the difficulties of integrating the optical components at low cost. Cheaper solutions would probably come from new materials with electrical links, such as nanowires or CNTs. Silicon nanowires are interesting due to their superior performance as field effect transistors. The 1D nature is also compatible with ultradense interconnects, and several bottom-up and top-down approaches are nowadays available for the fabrication of Si nanowire arrays. High dense crossbars and arrays of nanowires built with spacer technique or DRIE are capable of overcoming the obstacle of the lithographic pitch. This inherent CMOS-compatibility with the scalability of Si nanowires set a promising radical solution for future technology nodes. Finally, the ballistic properties of

CNTs can be exploited for disruptive interconnect concepts, achieving lower resistivity and higher robustness toward electromigration issues than copper. Still, CNTs can only be grown by means of bottom-up approaches. This aspect implies major difficulties for CNTs to be utilized in the Back End of Line. Very good control on the growth conditions must be achieved at specific positions at low temperature. All the approaches presented in the chapter have the potential as copper wires replacement for future technology nodes. Further research would have to solve the disadvantages associated with each particular technology in order to enable fabrication at the industrial scale.

5.8 Glossary

BER: Bit Error Rate

BHF: Buffered Hydrofluoridric acid

CMOS: Complementary Metal Oxide Semiconductor

CMP: Chemical Mechanical Polishing

CVD: Chemical Vapor Deposition

DRIE: Deep Reactive Ion Etching

EUV-IL: Extrema Ultraviolet Interference Lithography

FET: Field Effect Transistor

FIB: Focused Ion Beam

F2B: Face to Back

F2F: Face to Face

HF-CVD: Hot Filament Chemical Vapor Deposition

IC: Integrated Circuits

IST: Iterative Spacer Technique

LPCVD: Low Power Chemical Vapor Deposition

MOSFET: Metal Oxide Semiconductor FET

MSM: Metal Semiconductor Metal

MSPT: Multiple Spacer Patterning Technique

MWCNT: Multiple-Walled CNT

NIL: NanoImprint Lithography

NW: Nanowire

PDMS: Polydimethylsiloxane

PECVD: Plasma Enhanced Chemical Vapor Deposition

PEDAL: Planar Edge Defined Alternate Layer

RF: Radio Frequency

SBFET: Schottky Barrier FET

SEM: Scanning Electron Microscope

SNAP: Superlattice Nanowire Pattern Transfer Technique

SOI: Silicon on Insulator

SWCNT: Single-Walled CNT

TEM: Transmission Electron Microscope

TEOS: Tetraethylorthosylicate

TSV: Through Silicon Via

VCSEL: Vertical Cavity Surface Emitting Lasers

VLS: Vapor Liquid Solid

VLSI: Very Large Scale of Integration

WDM: Wavelength Division Multiplexing

5.9 Bibliography

[1] I. Ahmed, C. E. Png, E.-P. Li, and R. Vahldieck. Electromagnetic wave propagation in a Ag nanoparticle-based plasmonic power divider. *Optics Express*, 17:337+, January 2009.

[2] A. T. Voutsas and M. K. Hatalis. Surface treatment effect on the grain size and surface roughness of as-deposited LPCVD polysilicon films. *Journal of the Electrochemical Society*, 140(1):282–288, 1993.

[3] V. Auzelyte, H. H. Solak, Y. Ekinci, R. MacKenzie, J. Voros, S. Olliges, and R. Spolenak. Large area arrays of metal nanowires. *Microelectronic Engineering*, 85(5-6):1131–1134, 2008.

[4] E. J. Bae, Y. S Min, D. Kang, J. H. Ko, and W. Park. Low-temperature growth of single-walled carbon nanotubes by plasma enhanced chemical vapor deposition. *Chemistry of Materials*, 17:5141–5145, 2005.

[5] K. Banerjee, H. Li, and N. Srivastava. Current status and future perspectives of carbon nanotube interconnects. *Proceedings of the 8th Int. International Conference on Nanotechnology*, 432–436, 2008.

[6] R. Beckman, E. Johnston-Halperin, Y. Luo, J. E. Green, and J. R. Heath. Bridging dimensions: demultiplexing ultrahigh density nanowire circuits. *Science*, 310(5747):465–468, 2005.

[7] M. H. Ben Jamaa, D. Atienza, K. E. Moselund, D. Bouvet, A. M. Ionescu, Y. Leblebici, and G. De Micheli. Fault-tolerant multi-level logic decoder for nanoscale crossbar memory arrays. *ICCAD 2007. IEEE/ACM International Conference on Computer-Aided Design*, 765–772, 4-8 Nov. 2007.

[8] M. H. Ben Jamaa. *Fabrication and Design of Nanoscale Regular Circuits*. PhD thesis, Lausanne, 2009.

[9] F. Bergamini, M. Bianconi, S. Cristiani, L. Gallerani, A. Nubile, S. Petrini, and S. Sugliani. Ion track formation in low temperature silicon dioxide. *Nuclear Instruments and Methods in Physics Research Section B: Beam Interactions with Materials and Atoms*, 266(10):2475–2478, 2008.

[10] J.-P. Bourgeon. *Nanostructures—Fabrication and Analysis*. Springer, 2007.

[11] J. A. Burns, B. F. Aull, C. K. Chen, C.-L. Chen, C. L. Keast, J. M. Knecht, V. Suntharalingam, K. Warner, P. W. Wyatt, and D.-R. W. Yost. A Wafer-Scale 3-D Circuit Integration Technology. *IEEE Transactions on Electron Devices*, 53:2507–2516, Oct. 2006.

[12] K. Byon, D. Tham, J. E. Fischer, and A. T. Johnson. Systematic study of contact annealing: ambipolar silicon nanowire transistor with improved performance. *Applied Physics Letters*, 90(14):143513, 2007.

[13] K. C. Cadien, M. R. Reshotko, B. A. Block, A. M. Bowen, D. L. Kencke, and P. Davids. Challenges for on-chip optical interconnects. In J. A. Kubby & G. E. Jabbour, editor, *Society of Photo-Optical Instrumentation Engineers (SPIE) Conference Series*, volume 5730 of *Society of Photo-Optical Instrumentation Engineers (SPIE) Conference Series*, 133–143, March 2005.

[14] G. Cerofolini. Realistic limits to computation. II. The technological side. *Applied Physics A*, 86(1):31–42, 2007.

[15] C.-C. Chiu, T.-Y. Tsai, and N.-H. Tai. Field emission properties of carbon nanotube arrays through the pattern transfer process. *Nanotechnology*, 17:2840–2844, June 2006.

[16] Y.-K. Choi, J. S. Lee, J. Zhu, G. A. Somorjai, L. P. Lee, and J. Bokor. Sublithographic nanofabrication technology for nanocatalysts and DNA chips. *Journal of Vacuum Science Technology B: Microelectronics and Nanometer Structures*, 21:2951–2955, 2003.

[17] G. F. Close, S. Yasuda, B. Paul, S. Fujita, and H.-S. P. Wong. A 1 GHz integrated circuit with carbon nanotube interconnects and silicon transistors. *Nano Letters*, 8:706–709, February 2008.

[18] Y. Cui, X. Duan, J. Hu, and C. M. Lieber. Doping and electrical transport in silicon nanowires. *The Journal of Physical Chemistry B*, 4(22):5213–5216, 2000.

[19] M. S. Dresselhaus, G. Dresselhaus, and P. Avouris. *Carbon Nanotubes: Synthesis, Structure, Properties, and Applications*. Springer, 2001.

[20] X. Duan, Y. Huang, R. Agarwal, and C. M. Lieber. Single-nanowire electrically driven lasers. *Nature*, 421:241–245, January 2003.

[21] J. Fujikata, K. Nose, J. Ushida, K. Nishi, M. Kinoshita, T. Shimizu, T. Ueno, D. Okamoto, A. Gomyo, M. Mizuno, T. Tsuchizawa, T. Watanabe, K. Yamada, S. Itabashi, and K. Ohashi. Waveguide-integrated Si nano-photodiode with surface-plasmon antenna and its application to on-chip optical clock distribution. *Applied Physics Express*, 1(2):022001+, February 2008.

[22] T. Fukushima, Y. Yamada, H. Kikuchi, and M. Koyanagi. New three-dimensional integration technology using chip-to-wafer bonding to achieve ultimate super-chip integration. *Japanese Journal of Applied Physics*, 45:3030+, April 2006.

[23] C. Girard, E. Dujardin, R. Marty, A. Arbouet, and G. C. Des Francs. Manipulating and squeezing the photon local density of states with plasmonic nanoparticle networks. *Physical Review B*, 81(15):153412+, April 2010.

[24] J. E. Green, J. W. Choi, A. Boukai, Y. Bunimovich, E. Johnston-Halperin, E. Deionno, Y. Luo, B. A. Sheriff, K. Xu, Y. Shik Shin, H.-R. Tseng, J. F. Stoddart, and J. R. Heath. A 160-kilobit molecular electronic memory patterned at 10^{11} bits per square centimetre. *Nature*, 445:414–417, 2007.

[25] M. S. Gudiksen, L. J. Lauhon, J. Wang, D. C. Smith, and C. M. Lieber. Growth of nanowire superlattice structures for nanoscale photonics and electronics. *Nature*, 415:617–620, 2002.

[26] H. A. Atwater, S. Maier, A. Polman, J. A. Dionne, and L. Sweatlock. The new "p-n junction": Plasmonics enables photonic access to the nanoworld. *MRS Bulletin*, 30:385–389, 2005.

[27] O. Hayden, M. T. Björk, H. Schmid, H. Riel, U. Drechsler, S. F. Karg, Emanuel Lörtscher, and W. Riess. Fully depleted nanowire field-effect transistor in inversion mode. *Small*, 3(2):230–234, 2007.

[28] R. He and P. Yang. Giant piezoresistance effect in silicon nanowires. *Nature Nanotechnology*, 1(1):42–46, 2006.

[29] A. I. Hochbaum, R. Fan, R. He, and P. Yang. Controlled growth of Si nanowire arrays for device integration. *Nano Letters*, 5(3):457–460, 2005.

[30] T. Hogg, Y. Chen, and P. J. Kuekes. Assembling nanoscale circuits with randomized connections. *IEEE Transactions on Nanotechnology*, 5(2):110–122, 2006.

[31] J. D. Holmes, K. P. Johnston, R. C. Doty, and B. A. Korgel. Control of thickness and orientation of solution-grown silicon nanowires. *Science*, 287(5457):1471–1473, 2000.

[32] J. Hone, M. Whitney, C. Piskoti, and A. Zettl. Thermal conductivity of single-walled carbon nanotubes. *Phys. Rev. B*, 59(4):R2514–R2516, Jan 1999.

[33] J. Hållstedt, P.-E. Hellström, Z. Zhang, B. G. Malm, J. Edholm, J. Lu, S.-L. Zhang, H.H. Radamson, and M. Östling. A robust spacer gate process for deca-nanometer high-frequency MOSFETs. *Microelectronic Engineering*, 83(3):434–439, 2006.

[34] J.-Fu Hsu, B.-R. Huang, and C.-S. Huang. The growth of silicon nanowires using a parallel plate structure. Volume 2:605–608, July 2005.

[35] Y. Huang, X. Duan, Q. Wei, and C. M. Lieber. Directed assembly of one-dimensional nanostructures into functional networks. *Science*, 291(5504):630–633, 2001.

[36] S. Iijima. Helical microtubules of graphitic carbon. *Nature*, 354:56–58, November 1991.

[37] G.-Y. Jung, E. Johnston-Halperin, W. Wu, Z. Yu, S.-Y. Wang, W. M. Tong, Z. Li, J. E. Green, B. A. Sheriff, A. Boukai, Y. Bunimovich, J. R. Heath, and R. S. Williams. Circuit fabrication at 17 nm half-pitch by nanoimprint lithography. *Nano Letters*, 6(3):351–354, 2006.

[38] J. Kedzierski and J. Bokor. Fabrication of planar silicon nanowires on silicon-on-insulator using stress limited oxidation. *Journal of Vacuum Science and Technology B*, 15(6):2825–2828, 1997.

[39] A. Kohno, T. Sameshima, N. Sano, M. Sekiya, and M. Hara. High performance poly-Si TFTs fabricated using pulsed laser annealing and remote plasma CVD with low temperature processing. *IEEE Transactions on Electron Devices*, 42:251–257, February 1995.

[40] S.-M. Koo, A. Fujiwara, J.-P. Han, E. M. Vogel, C. A. Richter, and J. E. Bonevich. High inversion current in silicon nanowire field effect transistors. *Nano Letters*, 4(11):2197–2201, 2004.

[41] F. Kreupl, A. P. Graham, M. Liebau, G. S. Duesberg, R. Seidel, and E. Unger. Carbon Nanotubes for Interconnect Applications. *Microelectronic Engineering*, 64(1-4):399–408, 2002.

[42] L. J. Lauhon, M. S. Gudiksen, D. Wang, and C. M. Lieber. Epitaxial core-shell and core-multishell nanowire heterostructures. *Nature*, 420:57–61, 2002.

[43] K.-N. Lee, S.-W. Jung, W.-H. Kim, M.-H. Lee, K.-S. Shin, and W.-K. Seong. Well controlled assembly of silicon nanowires by nanowire transfer method. *Nanotechnology*, 18(44):445302 (7 pp.), 2007.

[44] H. J. Li, W. G. Lu, J. J. Li, X. D. Bai, and C. Z. Gu. Multichannel ballistic transport in multiwall carbon nanotubes. *Physical Review Letters*, 95(8):086601–+, August 2005.

[45] J. Li, Q. Ye, A. Cassell, H. T. Ng, R. Stevens, J. Han, and M. Meyyappan. Bottom-up approach for carbon nanotube interconnects. *Applied Physics Letters*, 82:2491+, April 2003.

[46] K. Liu, P. Avouris, R. Martel, and W. K. Hsu. Electrical transport in doped multiwalled carbon nanotubes. *Physics Review B*, 63(16):161404, April 2001.

[47] M. M. Nihei, D. Kondo, A. Kawabata, S. Sato, H. Shioya, M. Sakaue, T. Iwai, M. Ohfuti, and Y. Awano. Low-resistance multi-walled carbon nanotube vias with parallel channel conduction of inner shells. *Proceedings IEEE International Interconnect Technology Conference*, 234–236, 2005.

[48] S. A. Maier and H. A. Atwater. Plasmonics: Localization and guiding of electromagnetic energy in metal/dielectric structures. *Journal of Applied Physics*, 98(1):011101+, July 2005.

[49] P. L. McEuen, M. S. Fuhrer, and H. Park. Single-walled carbon nanotube electronics. *IEEE Transactions on Nanotechnology*, 1:78–85, March 2002.

[50] D. A. B. Miller. Rationale and challenges for optical interconnects to electronic chips. *Proceedings of the IEEE*, 88(6):728 –749, June 2000.

[51] K. E. Moselund, D. Bouvet, L. Tschuor, V. Pot, P. Dainesi, C. Eggimann, N. Le Thomas, R. Houdré, and A. M. Ionescu. Cointegration of gate-all-around MOSFETs and local silicon-on-insulator optical waveguides on bulk silicon. *IEEE Transactions on Nanotechnology*, 6(1):118–125, 2007.

[52] K. E. Moselund. *Three-Dimensional Electronic Devices Fabricated on a Top-Down Silicon Nanowire Platform*. PhD thesis, 2008.

[53] A. Naeemi and J. D. Meindl. Carbon nanotube interconnects. *Annual Review of Materials Science*, 39:255–275, August 2009.

[54] K. Nakazawa. Recrystallization of amorphous silicon films deposited by low-pressure chemical vapor deposition from Si_2H_6 gas. *Journal of Applied Physics*, 69(3):1703–1706, 1991.

[55] M. Nihei, H. Masahiro, A. Kawabata, and A. Yuji. Carbon nanotube vias for future lsi-interconnects. *Proceedings IEEE International Interconnect Technology Conference*, 251–253, 2004.

[56] I. O'Connor and F. Gaffiot. Advanced research in on-chip optical interconnects. In *Lone Power Electronics and Design*, editor. C. Pignet. Boca Raton: CRC Press, 2004.

[57] R. F. Oulton, V. J. Sorger, D. A. Genov, D. F. P. Pile, and X. Zhang. A hybrid plasmonic waveguide for subwavelength confinement and long-range propagation. *Nature Photonics*, 2:496–500, August 2008.

[58] R. F. Oulton, V. J. Sorger, T. Zentgraf, R.-M. Ma, C. Gladden, L. Dai, G. Bartal, and X. Zhang. Plasmon lasers at deep subwavelength scale. *Nature*, 461:629–632, October 2009.

[59] R. S. Patti. Three-dimensional integrated circuits and the future of system-on-chip design. *Proceedings of IEEE*, 94:1214–1224, 2006.

[60] V. Pott. *Gate-all-Around Silicon Nanowires for Hybrid Single Electron Transistor/CMOS Applications.* PhD thesis, Lausanne, 2008.

[61] P. Ramm, D. Bollmann, R. Braun, R. Buchner, U. Cao-Minh, M. Engelhardt, G. Errmann, T. Graul, K. Hieber, H. Hubner, G. Kawala, M. Kleiner, A. Klumpp, S. Kuhn, C. Landesberger, H. Lezec, W. Muth, W. Pamler, R. Popp, E. Renner, G. Ruhl, A. Sanger, U. Scheler, A. Schertel, C. Schmidt, S. Schwarzl, J. Weber, and W. Weber. Three dimensional metallization for vertically integrated circuits: Invited lecture. *Microelectronic Engineering*, 37-38:39–47, 1997.

[62] D. Sacchetto et al. Fabrication and Characterization of Vertically Stacked Gate-All-Around Si Nanowire FET Arrays. In *ESSDERC*, 2009.

[63] D. Sacchetto, M. H. Ben-Jamaa, G. De M., and Y. Leblebici. Fabrication and characterization of vertically stacked gate-all-around Si nanowire FET arrays, 62, 2009.

[64] V. Schmidt, H. Riel, S. Senz, S. Karg, W. Riess, and U. Gösele. Realization of a silicon nanowire vertical surround-gate field-effect transistor. *Small*, 2(1):85–88, 2006.

[65] S. R. Sonkusale, C. J. Amsinck, D. P. Nackashi, N. H. Di Spigna, D. Barlage, M. Johnson, and P. D. Franzon. Fabrication of wafer scale, aligned sub-25 nm nanowire and nanowire templates using planar edge defined alternate layer process. *Physica E: Low-Dimensional Systems and Nanostructures*, 28(2):107–114, 2005.

[66] W. Steinhögl, G. Schindler, G. Steinlesberger, and M. Engelhardt. Size-dependent resistivity of metallic wires in the mesoscopic range. *Physics Review B*, 66(7):075414+, August 2002.

[67] S. D. Suk, S.-Y. Lee, S.-M. Kim, E.-J. Yoon, M.-S. Kim, M. Li, C. W. Oh, K. H. Yeo, S. H. Kim, D.-S. Shin, K.-H. Lee, H. S. Park, J. N. Han, C.J. Park, J.-B. Park, D.-W. Kim, D. Park, and B.-I. Ryu. High performance 5nm radius twin silicon nanowire MOSFET (TSNWFET): Fabrication on bulk Si wafer, characteristics, and reliability. 717–720, December 2005.

[68] O. Vazquez-Mena, G. Villanueva, V. Savu, K. Sidler, M. A. F. van den Boogaart, and J. Brugger. Metallic nanowires by full wafer stencil lithography. *Nano Letters*, 8(11):3675–3682, 2008.

[69] A. T. Voutsas and M. K. Hatalis. Structure of as-deposited LPCVD silicon films at low deposition temperatures and pressures. *Journal of the Electrochemical Society*, 139(9):2659–2665, 1992.

[70] A. T. Voutsas and M. K. Hatalis. Deposition and crystallization of a-Si low pressure chemically vapor deposited films obtained by low-temperature pyrolysis of disilane. *Journal of the Electrochemical Society*, 140(3):871–877, 1993.

[71] R. S. Wagner and W. C. Ellis. Vapor-liquid-solid mechanism for single crystal growth. *Applied Physics Letters*, 4(5):89–90, 1964.

[72] B. Q. Wei, R. Vajtai, and P. M. Ajayan. Reliability and current carrying capacity of carbon nanotubes. *Applied Physics Letters*, 79:1172+, August 2001.

[73] W. Wu, G.-Y. Jung, D. L. Olynick, J. Straznicky, Z. Li, X. Li, D. A. A. Ohlberg, Y. Chen, S.-Y. Wang, J. A. Liddle, W. M. Tong, and R. S. Williams. One-kilobit cross-bar molecular memory circuits at 30-nm half-pitch fabricated by nanoimprint lithography. *Applied Physics A: Materials Science and Processing*, 80(6):1173–1178, 2005.

[74] www.itrs.net.

[75] T. Xu, Z. Wang, J. Miao, X. Chen, and C. M. Tan. Aligned carbon nanotubes for through-wafer interconnects. *Applied Physics Letters*, 91(4):042108+, July 2007.

[76] C. Yang, Z. Zhong, and C. M. Lieber. Encoding electronic properties by synthesis of axial modulation-doped silicon nanowires. *Science*, 310(5752):1304–1307, 2005.

[77] R. P. Zingg, B. Hoefflinger, J. A. Friedrich, and G. W. Neudeck. Three-dimensional stacked MOS transistors by localized silicon epitaxial overgrowth. *IEEE Transactions on Electron Devices*, 37:1452–1461, June 1990.

6

HeTERO: Hybrid Topology Exploration for
RF-Based On-Chip Networks

Soumya Eachempati, Reetuparna Das, Vijaykrishnan Narayanan

Pennsylvania State University

Yuan Xie

Pennsylvania State University

Suman Datta

Pennsylvania State University

Chita R Das

Pennsylvania State University

CONTENTS

6.1 Introduction

Future microprocessors are predicted to consist of 10s to 100s of cores running several concurrent tasks. A scalable communication fabric is required to connect these components and thus, giving birth to networks on silicon, also known as Network-on-Chip (NoC). NoCs are being used as the de facto solution for integrating the multicore architectures, as opposed to point-to-point global wiring, shared buses, or monolithic crossbars, because of their scalability and predictable electrical properties.

The network topology is a vital aspect of on-chip network design as it determines several power-performance metrics. The key challenge to design a NoC topology is to provide *both* high *throughput* and low *latency* while operating under *constrained power* budgets. The 2D-mesh topologies are popular for tiled Chip Multi-Processores (CMPs) [36, 29, 32] due to their simplicity and 2D-layout properties. 2D-meshes provide the *best network throughput* [11] albeit their scalability limitations in terms of *latency and power*. The scalability setbacks in MESHes is due to the large network diameter that grows linearly. To address the scalability of 2D-meshes, researchers have proposed concentration [3], richly connected topologies [15, 13] and hierarchical topologies [11]. Concentration achieves smaller network diameter by sharing a router among multiple injecting nodes and thus resulting in fewer routers. However, concentrated topologies trade-off achievable network throughput and bandwidth for lower latency. In addition, concentrated networks also consume higher power as they need larger switches and fatter physical channels. Richly connected topologies achieve lower latency by trading-off throughput at moderate power consumption. Hierarchical topologies take advantage of communication locality in applications to achieve low latency and low power. However, these topologies provide suboptimal throughput because the global network becomes the bottleneck in the proposed clustered communication architecture [11].

The proposal of express paths (virtual and physical) has been shown to enhance latency and throughput [17]. For the throughput constrained hierarchical and concentrated topologies, adding express paths could provide substantial throughput benefit. As the global interconnect delay exasperates in future technologies, express paths will be very challenging to implement with traditional RC interconnect technologies. Alternate interconnect technologies such as optical networks, radio-frequency (RF) based signal transmission and low-dimensional materials (LDM) such as nanowires, nanotubes etc., are being explored [1]. Low-dimensional materials are considered far-term solutions while optical and RF-based technologies are predicted as near-term solutions due to their Complementary Metal-Oxide-Semiconductor (CMOS)-compatibility. These emerging technologies have one thing in common, which is low-latency for long distance communication. For on-chip networks, this property of the emerging technology translates to cheaper express paths.

In this chapter, we will explore network topology designs that facilitate high throughput *and* low latency while operating under tight power constraints by using radio-frequency interconnect technology. RF-based interconnect (RF-I) incur lower energy and higher data rate density than their electrical counterparts [7]. Radio-frequency (RF) mm-wave propagation modulates data onto a carrier electromagnetic wave that is guided along a wire. Such a propagation has the least latency that is physically possible as the electromagnetic wave travels at the speed of light. As a result, high data rates limited by the speed of the modulator can be achieved in RF-interconnect. This high RF bandwidth can be multiplexed among multiple carriers using techniques such as frequency division multiple access (FDMA), leading to higher throughput as well. Thus, for the distances on-chip, RF-I can provide high bandwidth low latency super-express paths (from one-end of the chip to the other end). In addition, RF-interconnect components, namely the transmitter consisting of modulators, mixers, and receivers can benefit from CMOS technology scaling. An RF integration of mesh topology has been explored in [9, 6]. Even though RF technology requires significant design efforts to mature before becoming mainstream, assessing the benefits it offers architecturally will be an important factor in determining their usage model in future.

In this chapter, we use RF-interconnect in various state-of-art topologies. Das et al., grouped the network nodes into logical clusters and showed that a hierarchical network made up of bus network for intracluster communication and a global mesh network for intercluster communication achieved the best performance and power trade-offs when compared to state-of-art topologies [11]. This hierarchical design, however, had lower throughput and the global network was the throughput bottleneck. We will adopt this hierarchical philosophy for our study as the high bandwidth of RF could address the low throughput problem of the global network. On replacing the global mesh network with RF-enhanced mesh network, a energy delay product reduction of upto 30% is obtained while providing upto 40% higher throughput than the base hierarchical design. The main insights of our chapter are

- Hierarchical networks provide superior delay and power trade-offs but suffer in throughput. RF-I when applied to hierarchical network enhances the throughput and also result in lower latency at approximately the same power.

- The throughput improvement obtained by using RF-I increases with increase in concentration.

- For medium-sized networks RF-I enhanced concentrated network is attractive.

The rest of this chapter is organized into 8 sections. Section 6.2 gives a brief background of RF-interconnect. Section 6.3 describes the state-of-art topologies and Section 6.4 describes the RF-enhanced topologies. Section 6.5 describes the simulation setup and the results are presented in Section 6.6. Section 6.7 outlines some of the prior work in this area and Section 6.8 concludes the chapter.

6.2 RF-Interconnect

Current and future technology generations are faced with interconnect delay problems due to the high latency of charging and discharging the repeated RC wires [1]. Differential signaling techniques can help in decreasing the time as well as the power of RC wires. Yet, these techniques are not sufficient for mitigating the high global interconnect delay as the differential wires are typically not buffered (they require differential buffers). Technological alternatives to address this problem are being thoroughly explored. A promising near-term solution is the use of through-silicon-vias (TSVs) as an enabling technology for three-dimensional (3D) stacking of dies. 3D-integration leads to smaller die area and thus, leads to reduced wire length. Long-term solutions that are being examined include fundamental changes such as using different material for interconnection wires, and using novel signaling techniques. Single-walled carbon nanotubes bundled together are being investigated as a possible replacement for copper in the future. The ballistic transport of electrons in Single-Walled Carbon Nanotube (SWCNT) makes them highly resistant to electromigration and have lower resistivity than copper at the same technology. While there are several factors that will influence the deployment of these new materials into mainstream integrated circuits, CMOS compatibility is viewed as an overriding factor. Radio-frequency based mm-wave signal propagation (RF-I) is an attractive option due to their CMOS compatibility. Photonics based on-chip signaling is also being considered as a viable option. Optical networks use components such as waveguides, ring resonators, detectors, and laser source. With significant research efforts, many of these components have been (re)designed to be placed on-die. The laser source is still kept off-die. The leakage power and temperature sensitivity of the optical components need to be optimized for making photonic communication viable on chip [33].

As the CMOS device dimensions continue to scale down, the cut-off frequency (f_t) and the maximum frequency (f_{max}) of CMOS devices will exceed few hundreds of GHz. At such high frequencies, conventional line has very high-signal attenuation. The loss increases with increase in length of the interconnect [8]. Traditional RC signaling can be run at a few GHz, resulting in a wastage of bandwidth. The RF-concept is that data to be transmitted is modulated as an electromagnetic wave (EM), which is guided along the transmission line (waveguide). Microwave transmission in guided mediums such as microstrip transmission line, or coplanar waveguides is known to have very low loss of about -1.6 db at 100 GHz [8]. As the EM wave travels at the speed of light, radio-frequency based communication is an attractive option providing high bandwidth and low latency independent of the length of the transmission line (for the distances on-chip).

A RF circuit from a high level is made up of a transmitter, frequency generator, receiver, and a transmission line (in case of wired) (see Figure 6.4). These components are made up of amplifiers, low-noise amplifiers (LNA), attenuators, filters, oscillators, and mixers. Oscillators are used to generate signals. The data is modulated onto a high-frequency carrier signal using a mixer in the transmitter and demodulated in the receiver. Modulating the base data band onto high frequencies result in lower variation in signal attenuation. The reader is encouraged to study more detailed radio frequency material for better understanding. Many tough practical problems are being tackled by researchers for realizing RF-I on-chip. Impedance matching is one such tough issue. Laskin et al. demonstrated CMOS-integrable 95 GHz receiver design at 65 nm technology node [19] using a fundamental frequency quadrature voltage controlled oscillator (VCO), Intermediate Frequency (IF) amplifier, Low-Noise Amplifier (LNA), and a mixer. The 3 dB bandwidth of the receiver is 19 GHz from 76 GHz to 95 GHz at room temperature. Chang et al., demonstrated a highly-CMOS integrable 324 GHz frequency generator with 4 GHz tuning range using linear superposition technique [14]. Multiple phase shifted ($\frac{2*\pi}{N}$) and rectified fundamental signals (w_0) are used to produce the superposed output signal at the intended frequency of Nw_0. Thus, there is increasing effort to build CMOS integrable, area and power efficient RF components for on-die communication.

We'll now present a brief overview of the implementation of RF-technology on chip. The high RF bandwidth can be multiplexed among multiple communicating nodes by using multiple carrier frequencies using FDMA. These multiple access algorithms can be used to effectively alleviate the cross-channel interference in the shared medium. Such multiband communication requires multiple carrier frequencies to be distributed to various transmitters and receivers on chip. A simultaneous subharmonic injection locked frequency generator for supplying multiple carrier frequencies has been demonstrated in 90 nm CMOS [7, 10]. This is especially interesting for multiband communication on-chip primarily because of two reasons. First, it incurs significantly lower area and power overhead as it eliminates the requirement of multiple Phase Loops Locked (PLLs) for multiple frequencies. Second, the harmonic injection locking technique allows the VCOs to be physically distributed on the chip and only the low-reference frequency to be distributed. Such demonstrations of CMOS integrable RF components with reasonable area and power are encouraging for realizing RF-I on-chip. Even though there are several factors that influence the commercial realization of RF-I on-chip, understanding the usage model of such a novel technology and the performance and power impacts from an architecture standpoint is important. In this work, we apply RF-I to the state-of-art on-chip interconnect topologies and evaluate the network latency and power for both synthetic and application workloads.

Next, we lay out the technology assumptions for RF-I. These assumptions were made through comprehensive literature survey of current RF-interconnect models [14, 19, 9, 7]. The area overhead, energy per bit, and the aggregate data rate of RF-I are dependent on the technology node. Table 6.1 show these parameters for various technology nodes obtained from [9]. The total data rate per transmission line is determined as the product of the number of carriers and the data rate per carrier. The number of carriers and the maximum carrier frequency in a given technology are based on the cut-off frequency (f_t), maximum oscillation frequency (f_{max}), and the intercarrier separation. With technology scaling, the carrier frequency increases due to the rise in f_t for the CMOS transistor. Such an increase in the carrier frequency implies the increase in the data rate per carrier and increase in the number of carriers. In addition, the area overhead of passive components and inductor area was also shown to decrease at higher carrier frequencies. Similar to any other power management schemes if the RF link is idle, i.e., there is no data transfer, the RF-components such as frequency generator, modulator, and demodulator can be shut off. The granularity at which this shut-off occurs and the turning on/off cost determine the power savings. In this chapter, we assume that whenever there is no data sent of the RF link it can be turned off immediately, in order to keep the discussion simple. Consequently, the energy per bit shown in Table 6.1 is dynamic energy consumed.

The multiband RF-I can be abstracted to multiple logical point-to-point connections between communicating nodes. If there are N frequencies then, there can be N bidirectional point-to-point logical concurrent connections. Based on the total RF allocated bandwidth BW, the logical link bandwidth is $\frac{BW}{N}$. Note that the actual number of physical wires required to realize these N logical links is determined by the data rate per wire as given in Table 6.1. For example, if we assume the per band data rate to be 8 Gbps, 8 data carriers, the communication network is operating at 2 GHz. Let us say that the total RF bandwidth we allocate is 256 B. The bandwidth allocation per band is 256 B/8, which is 32 B.

TABLE 6.1

RF parameters for a wire for various technology nodes

Parameter	70 nm	50 nm	35 nm
No. of carriers	8	10	12
Data rate per carrier (Gb/s)	6	7	8
Total data rate (Gb/s)	48	70	96
Energy per bit (pJ)	1	0.85	0.75
Area(all Tx+Rx) mm^2	0.112	0.115	0.0119

6.3   State-of-Art Topologies

Topology defines the connectivity, the bisection bandwidth (for a equal bi-section comparison), and the path diversity of the network. All of these in turn are responsible for the observed latency, power consumption, and the throughput. In this section, we briefly describe the state-of-art topologies and their characteristics.

6.3.1   Mesh

2D-Meshes have been a popular topology for on-chip networks because of low complexity and compact 2D-layout. They also provide the best throughput because of plenty of network resources and higher path diversity. Meshes, however, have poor latency and power scalability because of rapidly increasing diameter, yielding it unsuitable for larger network sizes.

6.3.2   Concentrated Mesh

A concentrated mesh(*CMESH*), as its name suggests, shares a router among a small number of nodes as shown in the Figure 6.1. Thus, it has a many-to-one mapping of processing elements (PEs) to routers (meshes have one to one mapping of PEs to routers). This results in reduced hop count that

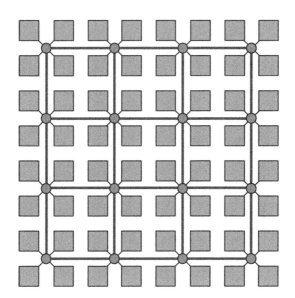

FIGURE 6.1
Concentrated mesh.

translates to lower latency. There is a dedicated port for each PE and four ports for the cardinal directions. Though *CMESH* provides low latency, it is energy inefficient because of the high-switch power of high-radix routers and fatter channels. The crossbar power is increases with the number of ports and the port width. Thus, concentration trades off throughput and power for lower latency. Concentration results in a smaller network size. Consequently, *CMESH* has reduced path diversity and lesser network resources (buffers, etc.).

6.3.3   Flattened Butterfly

A flattened butterfly *fbfly* has longer links to nonadjacent neighbors as shown in the Figure 6.2. (*fbfly*) reduces the hop count by employing both concentration as well as rich connectivity. The higher connectivity requires larger number of ports in the router. Further, the bisection bandwidth for *fbfly* is higher and thus, each physical channel is narrower compared to the *CMESH* and mesh topologies. Consequently, even though *fbfly* uses high radix switches, the thinner channels causes the crossbar power to be under control. The thinner channels however, result in higher serialization latency. Thus, the rich connectivity trades off serialization latency for reducing the hop count. In

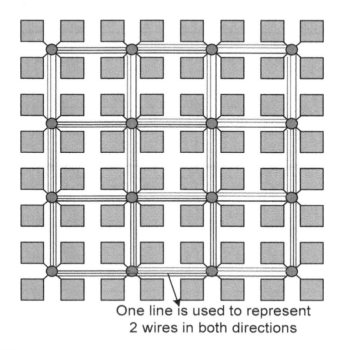

One line is used to represent
2 wires in both directions

FIGURE 6.2
Flattened butterfly.

the presence of high communication locality (nearest neighbor traffic) the reduced hop count may not compensate for the higher serialization latency and this can adversely affect the overall packet latency. *Fbfly* also suffers from poor throughput like other concentrated networks. In summary, *fbfly* topology gives lower latency at low power and reduced throughput.

6.3.4 Hierarchical Topology

In [11], the authors propose a hierarchical and heterogeneous topology (referred to as *hier* from now on) to replace prevalent 2D-meshes in NoCs. The proposal was based on three key insights: (a) communication locality, (b) concentration, and (c) power. To minimize the communication cost, future systems are likely to be designed such that data accessed by threads will be allocated to cache banks or memories closer to the processing core servicing the thread. Hence, the volume of local traffic compared to global traffic is likely to be higher in NoCs. 2D-Meshes are agnostic to communication locality and treat local and global traffic equally. Further, concentration is good for lower latency but result in lower throughput. Concentration could also result in higher power based on design choices.

A wide bus for intracluster communication became the natural choice as opposed to the crossbar for concentration in *CMESH*. This is a scalable and energy efficient option for increasing the concentration. The global network in their design was the mesh. Figure 6.3 shows the a 64-node network with *hier* topology. The authors point out that after investigating other global network options such as fat-tree, ring etc., a mesh-based global network performed the best because of its high path diversity. The packet traversal through the network can be either entirely contained in the local bus or will incur global transactions. A local transaction takes at least 3 cycles (arbitration could take more than one cycle) to get the bus and the bus transfer could take multiple cycles. In this chapter, the bus transfer is one cycle. If the destination lies on a different bus, then the packet traversal consists of three separate transactions, i.e., local bus traversal, global mesh traversal that could span multiple hops followed by another remote local transaction. Thus, a remote bus destination is quite expensive. The routing for the hierarchical network is shown in Figure 6.5.

Table 6.4 shows the network parameters for these topologies. These topologies are implemented in our in-house simulator. More details of the experimentation setup are presented in Section 6.6.2. Figure 6.6 shows the average message latency for uniform random traffic at 64-node network. *CMESH* has the lowest latency. This topology provides 2X latency reduction and 3X reduction in power-consumption over 2D-meshes [11]. As with all concentrated networks, the *hier* topology had limited throughput and less than half that of the simple mesh. The saturation throughput of the *hier* topology is the lowest without any optimizations. Figure 6.7 plots the network power with load rate. The graph shows that the hierarchical topology has the lowest power

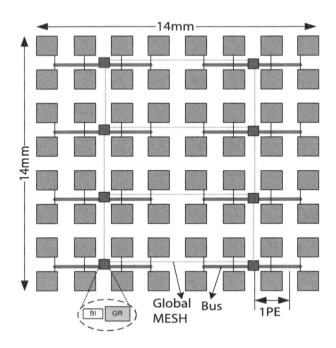

FIGURE 6.3
Hierarchical network.

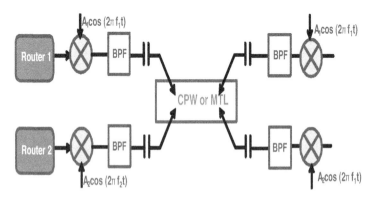

FIGURE 6.4
High-level RF circuit.

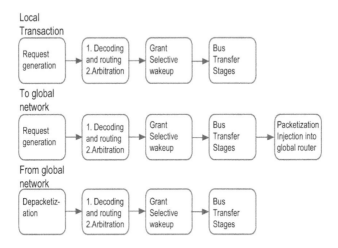

FIGURE 6.5
Hierarchical network packet traversal.

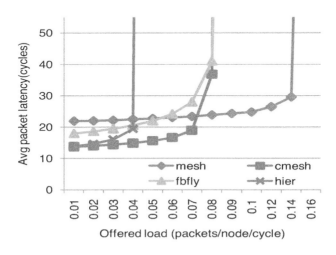

FIGURE 6.6
Average packet latency.

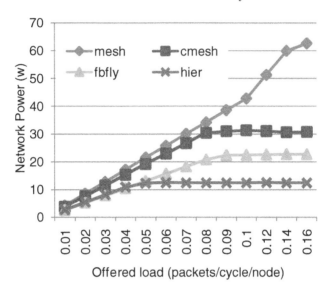

FIGURE 6.7
Network power consumption.

followed by *fbfly*. Thus, even though *CMESH* provides the best latency, it is power-hungry. The *hier* topology is desirable with respect to latency and power.

6.3.5   Impact of Locality

One of the main inspirations behind the hierarchical design is communication locality. Let us understand the performance and power of the various topologies under a mix of traffic patterns with high-communication locality. We define locality as the percentage of traffic whose destination belongs to one of the nearest neighbors of the source. The nearest neighbor is a single hop away from the node. Thus, a mesh can have up to 4 nearest neighbors. In the case of the concentrated and hierarchical networks, all the nodes sharing a router or bus are the nearest neighbors.

Figure 6.8 shows the performance of the various topologies when 75% of the traffic is local and the remaining 25% is uniform random traffic. As depicted in the figure, the hierarchical topology benefits maximum from the LC traffic by exploiting the communication locality. *Cmesh* is a close competitor for LC traffic. As the network size increases, *bfly* performs worse than simple mesh in throughput and latency. The reason being that the one hop latency of mesh is smaller than *fbfly*. The hierarchical element in *fbfly* offers no benefit because of the low-channel bandwidth leading to higher one-hop latency. The power graph is shown in Figure 6.9.

The saturation point for all topologies with LC traffic shifts towards the

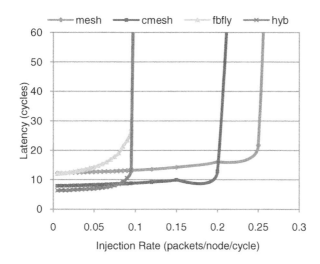

FIGURE 6.8
Average message latency for local traffic pattern at 64 node network.

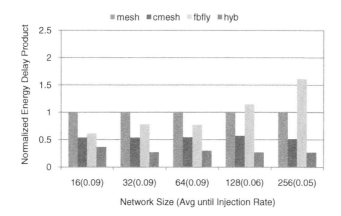

FIGURE 6.9
Energy delay product for local traffic pattern at 64-node network.

higher end compared to UR traffic by as much as 50% for large network sizes. This reveals two interesting points. First, this shows that with high local traffic, like cmesh and hierarchical topologies can offer higher throughput. Second, for uniform traffic, the the saturation bottleneck is the global network, not the bus/concentration since the local network could support an injection load rate of up to 10% with LC traffic. Theoretically, the bus can take load up to 12.5% (1 packet every cycle for 8 injecting nodes).

In summary, hierarchical topology provides low-power, fast, high-bandwidth communication that can be efficiently used in the presence of high locality. Yet, it saturates faster than the state-of-art topologies.

6.4   RF-Topology Modeling

Let us understand the implementation of RF-enhanced topologies. The routers that contain the transceivers are henceforth referred to as RF-enabled routers. Each RF-enabled router has an extra physical port dedicated to connect to the transmission line through the transceivers. This results in a larger crossbar, increase in buffers and arbiters. All of these extra resources along with the transceivers are counted as the overhead in providing RF-interconnect on chip. We base our RF modeling on the model in [6, 9]. For the sake of this chapter, the frequencies are statically allocated to the various transceivers in multi-band communication. Other proposals of dynamic frequency allocation and thus, dynamic link configuration are being investigated. These dynamic links are configured once for an application.

The total data rate is equally divided among all the frequencies. A pair of transceivers that listen and communicate at a frequency can be viewed as a point-to-point link between the end nodes. Thus, a multiband RF-interconnect logically forms a set of point-to-point links between various communicating nodes. The next natural question that arises is which routers should contain the transmitters and receivers. We adopt a simple heuristic from [6] for choosing the routers. This heuristic has a computation complexity of $O(BV^3)$ where B is the RF-link budget and V is the network size. The goal of the heuristic is to minimize the maximum network diameter. Let B be the budget of RF-links that can be inserted. Let $A(C_x, C_y)$ and $B(D_x, D_y)$ be the end points of the diameter. A RF-link is inserted between A and B, the budget of links is decremented, and the network diameter is recalculated. This process is done iteratively until no more RF budget is available, or no more links can be inserted in the network [6]. Only one RF-physical port can be added per router that accommodates one incoming and one outgoing RF-link. Such a restriction limits the total hardware overhead as well as keeps the exploration space tangible. To summarize, each RF-frequency carrier can be logically abstracted to a point-to-point link between the end points. Thus, a RF-I is a set of point-to-point links or express paths.

By enhancing a network with RF-I, we are building a network with a few express paths. This leads to an irregular network. Consequently, the routing function needs to be more generic than the simple deterministic logic. The routing function needs to be smart in using the express paths. The ideal scenario is to use a dynamic adaptive routing scheme that can detect the network conditions and traffic requirements to load-balance based on whether the resources are being underutilized, overutilized/congestion. The routing function could also target lowering the power consumption. We will adopt a simple table-based routing here for ease of understanding. Table-based routing is a generic routin. However, combined with a deterministic router and in the absense of deadlock prevention schemes, the table-based routing leads to deadlocks. Consequently, we use a deadlock detection and recovery mechanism. The deadlock is detected when there is a circular dependency in the flits waiting for each other. When a flit is blocked due to lack of buffer in the receiving router, then waitlist of the next buffer is added to the current router's waitlist. In this way, the waitlist is propagated and if any router is waiting on itself then a circular dependency is detected and a deadlock is flagged. More details can be found in [9]. If a deadlock is detected then recovery is done by using dedicated escape virtual channels. All the packets in the network are marked for recovery and follow deterministic route using the escape Virtual Channels (VCs). Only the new packets entering the network can use the RF-links and the table-based routing. As deadlocks are uncommon, the performance penalty of this mechanism is less. In addition, we also experiment with different RF-usage models. All the packets injected in the network use table-based routing (RF-100%), only 50% use table-based routing and remaining use deterministic routing, 25% packets use table-based routing, and the remaining 75% use only deterministic routing. The deterministic routing in our chapter does not use any of the RF-express links and the routing is purely through the mesh links.

For the same RF-link budget, a smaller network could possibly have lesser number of links inserted. For a equi-bandwidth analysis, if the link budget is underutilized then the data rate per RF-link/band could be increased. In such a case, the spacing between carrier frequencies should be increased to avoid interference. In the simplest case, we use the RF-link bandwidth to be the same as the flit width. When the data rate is higher, multiple flits could be sent at once. This involves extra logic to combine and split packets. The hardware overhead of such a logic is small as we need few extra buffer space and recomputation of the header flit and the state information. In this chapter for the sake of simplicity, we used a constant data rate per band at a technology node for any network size. Thus, for a smaller network size, the total RF-bandwidth is lesser.

6.5 RF-Based Topologies

In this chapter, we will analyze three topologies 1) mesh, 2) concentrated mesh, and 3) hierarchical topology for possible RF-I integration. We also quantitatively compare the proposed topologies with flattened butterfly [15]. The routing complexity in flattened butterfly topology would be exasperated if RF-interconnect is also used. Thus, we do not address applicability of RF-interconnect to butterfly topology in this chapter. Figure 6.10 shows the base case that was evaluated in [9] and used as a point of comparison. Three key metrics are used for comparison as well as performance evaluation: average packet latency, aggregate throughput, and average power consumption. We evaluate the benefits of using RF-I with *cmesh* as shown in Figure 6.11. The routers are all RF-enabled in this particular example of *RF-cmesh* and color-coded router pairs communicate via RF-link. In this example, 8 RF-links are enabled. In hierarchical design, the RF-interconnect is applied only in the global network because using one RF-I for each local network would incur significant power overhead. In addition, the global network being the throughput bottleneck can leverage maximum benefits of RF-interconnect bandwidth. Further, longer distances derive more latency benefit from RF-interconnect. Thus, we do not use RF-I for the local networks. Figure 6.12 shows RF-I overlaid on the mesh global network. The table-based routing is applied to the

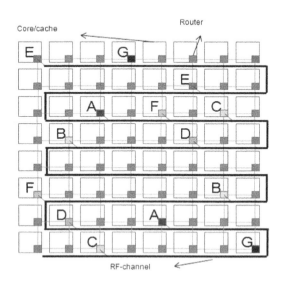

FIGURE 6.10
Base case with RF overlaid over MESH. Alphabet pairs have an RF link between them.

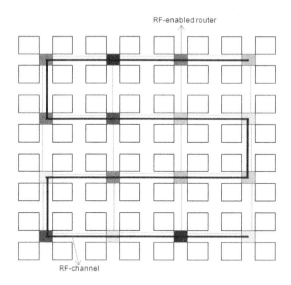

FIGURE 6.11
RF overlaid on CMESH(RF-CMESH). RF-links between routers of same color.

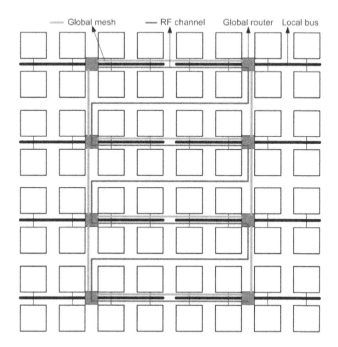

FIGURE 6.12
RF overlaid on hierarchical network.

routing in the global network. The routing on the bus is identical to the base hierarchical network. We expect the throughput to increase in all RF-enhanced networks.

6.6   Experimental Setup

6.6.1   Technology Assumptions

The design space is explored over a period of the 10 years starting from 70 nm (2007) technology (corresponds to 16-node NoC) down to 18 nm (2018) technology node (corresponds to 256-node NoC). A 14 mm x 14 mm die size is assumed, which remains constant across technologies and network node size of 3.5 mm (CPU or cache) at 70 nm that scales down by a factor of 0.7 for each technology generation. The wire parameters for global layers are obtained from Predictive Technology Model (PTM) [38] and we assume 1 metal layer to be dedicated for the network. (Wire width at 70 nm is 450 nm). Assuming cores occupy half the chip area [4] and the cores use all the available metal layers, we use a conservative estimate of 4096 horizontal tracks per die at 70 nm. The number of wires increases by 30% due to wire-width scaling with each technology generation. For a given topology, channel width is calculated from the bisection bandwidth and total number of wires that is fixed across all four topologies. These are presented in Figure 6.4 for different technologies/networks.

The delay and power model of CACTI 6.0 [23] is used for estimating the values for link and the bus. Table 6.5 gives our assumptions of lengths and the values of dynamic energy and leakage power for bus and the links. The delay and power model of bus include the higher loading due to senders and receivers. The additional overhead of dedicated point-to-point links required for request and grant lines for the bus have also been modeled. Both the network and cores run at a frequency of 2 GHz. We used the Orion power model for estimating the router power [34].

The power of the longer RF-links is used from the Table 6.1. A per band RF-link width is equal to the flit width is for all topologies and all technology nodes. For example, a per RF band of 512 bits is used for MESH, CMESH, and HYBRID is used. This results in only $2 * 128$ physical wires for the entire RF-I differential waveguide at 35 nm technology node. The fact that RF-I actually occupies a small number of wires is important for routing it on the chip as it has to be routed all through the network. Thus, the total wire resources is constant across all topologies.

6.6.2   Simulation Setup

The detailed configuration of our baseline simulation set-up is given in Table 6.2. Each terminal node in the network consists of either a core or a L2 cache bank. We used an in-house cycle-accurate hybrid NoC/cache simulator. The memory hierarchy implemented is governed by a two-level directory cache coherence protocol. Each core has a private write-back L1 cache. The L2 cache is shared among all cores and split into banks. Our coherence model includes a MESI-based protocol with distributed directories, with each L2 bank maintaining its own local directory. The network connects the cores to L2 cache banks and to the on-chip memory controllers (MC) (Figure 6.13). Application traces are collected using Simics [21].

For the interconnects, we implemented a state-of-art low-latency packet-based NoC router architecture. The NoC router adopts the deterministic X-Y routing algorithm, finite input buffering, and wormhole switching and virtual channel flow control. We also model the detail artifacts of specific topologies that can lead to increased pipeline stages in link traversal. The parameter we use across the topologies are given in Table 6.4.

For the performance analysis, we use synthetic and a diverse set of ap-

TABLE 6.2

Baseline processor, cache, memory, and router configuration

Processor Pipeline	SPARC 2 GHz processor, two-way out of order, 64-entry instruction window
L1 Caches	64 KB per-core(private), 4-way set associative, 128 B block size, 2-cycle latency, split I/D caches
L2 Caches	1 MB banks,shared, 16-way set associative, 128 B block size, 6-cycles latency, 32 MSHRs
Main Memory	4 GB DRAM,up to 16 outstanding requests for each processor, 400 cycle access
Network Router	2-stage wormhole switched, virtual channel flow control, 1024 maximum packet size

TABLE 6.3

Application workloads

SPLASH 2: Is a suite of parallel scientific workloads. Each benchmark executed one threads per processor.
SPEComp: We use SPEComp2001 as another representative workload. The results of applu, apsi, art, and swim are presented.
Commercial Applications. (1) TPC-C, a database benchmark for online transaction processing (OLTP), (2) SAP, a sales and distribution benchmark, and (3) SJBB and (4) SJAS, two Java-based server benchmarks. The traces were collected from multiprocessor server configurations at Intel Corporation.

plication workloads comprising of scientific and commercial applications. We run each application for at least one billion instructions. The commercial applications are run for at least 10,000 transactions. The workload details are summarized in Table 6.3.

TABLE 6.4
Network parameters

Topology	No. of nodes	Channel Width	Conc. Degree	Radix	VCs	Buffer Depth	No. of Routers	Total Wires	RF BW
Mesh	16	512	1	5	4	4	16	4096	448
	64	512	1	4	4	4	64	8192	768
	256	512	1	4	4	4	256	16384	1024
CMesh	64	512	4	8	4	2	16	8192	448
	256	512	4	8	4	2	64	16384	1024
Fbfly	16	512	4	7	2	8	4	4096	
	64	256	4	10	2	8	16	8192	
	256	128	4	13	2	16	64	16384	
Hyb	64	512	8	5	4	4	8	8192	256
	256	512	8	5	4	4	32	16384	1024

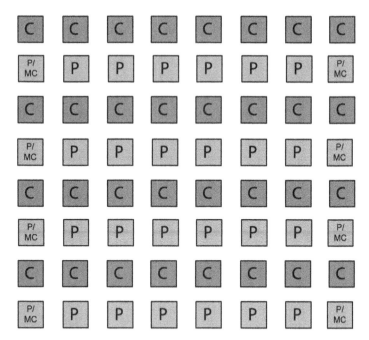

FIGURE 6.13
Physical layout.

TABLE 6.5

Energy and delay of bus and interrouter links

Parameters	Bus				
	70 nm	50 nm	35 nm	25 nm	18 nm
Length (mm)	7	4.9	3.43	2.4	1.68
Delay (ps)	498.9	442.9	353.9	247.7	173.4
Energy (pJ)	1.4	0.67	0.28	0.20	0.14
Leakage (nW)	23.5	13.3	3.5	2.4	1.7
	Link				
	70 nm	50 nm	35 nm	25 nm	18 nm
Length (mm)	3.5	2.45	1.7	1.2	0.84
Delay (ps)	233	208.8	167.5	117.3	82.1
Energy (pJ)	0.6	0.29	0.12	0.08	0.06
Leakage (nW)	10.2	5.49	1.4	0.98	0.69

6.7 Results

We will first study the mesh topology enhanced with RF alone for various network parameters. Followed by this, we will understand the effect of overlaying RF-I on CMESH and hierarchical topologies.

6.7.1 Simple Mesh

We experiment with three traffic patterns across various injection rates and four network sizes ranging from 16 to 256 nodes. Technology node is also scaled along with the network size. 16 nodes is assumed to be at 70 nm scaled down to 256 nodes are at 16 nm. The RF-parameters for the various technology nodes are detailed in Table 6.1. For each traffic pattern, we also evaluate the power and performance trade-offs for the various RF-usage models explained in Section 6.2. Energy Delay Product (EDPO) is computed as the product of total energy consumed per message and the average message latency. The throughput is computed as the fraction of total number of messages ejected and the total time taken. For these set of experiments, the per band RF-bandwidth is assumed to be equal to the flit width. Note that the total budget of RF links is not used in some cases (The 16-node network uses only 14 unidirectional RF-links).

6.7.1.1 16 Nodes

Figures 6.14, and 6.16 depict the latency for varying injection rates for uniform random (UR), and local pattern (LC) traffic patterns respectively. For this small network size, the RF-100% where all the packets are routed using table-based routing provides the lowest latency and the highest throughput

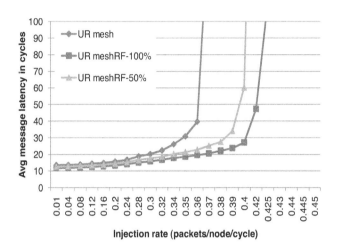

FIGURE 6.14
Average message latency for UR traffic of *Mesh+RF* at 16 nodes.

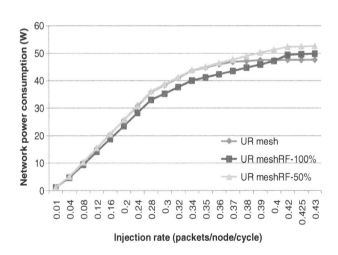

FIGURE 6.15
Average message latency of *Mesh+RF* for UR traffic at 16 nodes.

advantages for the two traffic patterns. For uniform random traffic pattern, most of the latency and throughput benefits are obtained with only 50% of the traffic using the table-based (TB) routing (implies access to RF-express paths).

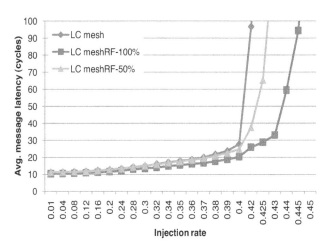

FIGURE 6.16
Average message latency of *Mesh+RF* for LC traffic at 16 nodes.

The local traffic pattern is generated as a mix of nearest neighbor and uniform random patterns. Half the number of packets have their destination as the nearest neighbors of the source and the other half have their destination chosen randomly. Note that while the base Mesh has at most 4 neighbors, the RF-enhanced topology has atmost five nearest neighbors. Thus, the RF links could be used for both the nearest-neighbor traffic component as well as for the UR traffic component. Consequently, when only 50% of the packets are allowed to route using TB routing, the latency benefits are minimal. When all the packets are allowed to route using TB, significant latency and throughput benefits. The energy delay product (EDP) for each of the traffic patterns is averaged over all injection rates just before saturation of base Mesh (as Mesh saturates first). The power plots for the two traffic patterns are shown in Figures 6.15 and 6.17 . The power plot shows that for the UR traffic, both the 100% and 50% usage models can compensate the power overhead by improved performance. With LC traffic pattern, however the 50% usage does not provide sufficient power savings due to lower hop count to compensate for the power overhead of supporting RF-I. The percentage improvement in EDP of RF-enhanced mesh over base case and the throughput improvement are summarized in Table 6.6. The latency and EDP benefits are measured just before the base MESH saturates. The throughput benefits are computed at their respective saturation points.

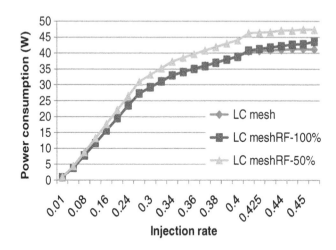

FIGURE 6.17
Total network power consumption of *Mesh+RF* for LC traffic at 16 nodes.

TABLE 6.6
Energy delay product and
throughput benefits of overlaying
RF over Mesh at 16 nodes

Traffic pattern	EDP	Throughput ratio
UR-100%	34	1.16x
UR-50%	19.7	1.10x
LC-100%	14	1.08x
LC-50%	-10	1.05x

6.7.1.2   36 Nodes

Figures 6.18, 6.19, 6.20, and 6.21 show the load latency plots for 36 nodes at 50nm technology. In addition, we also considered the conservative case when RF does not take advantage of the technology scaling. This is shown in Figure 6.18 as the EquiRFBW lines. These experiments show that for the reduced RF-bandwidth case the 50% usage lead to a higher throughput. With the equal-bandwidth scenario, the demand on RF is higher and thus, leads to RF congestion when there is no restriction and thus, in the 100% usage the RF-enhanced topology saturates faster. For this node size and UR traffic the 50% usage lead to sufficient power savings to compensate the power overhead of RF. In fact, even the 25% usage case for the EquiRFBW also has similar power as the base case. However, for LC traffic we still see that 50% usage leads to power overhead.

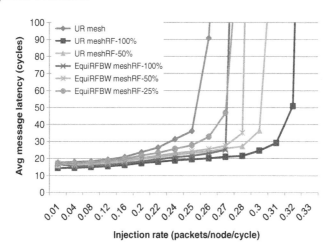

FIGURE 6.18
Average message latency of *Mesh+RF* for UR traffic at 36 nodes. EquiRFBW: RF BW is same as that used at 16 nodes.

6.7.1.3   64 Nodes

For this network size (Figures 6.22, 6.23, 6.24, and 6.25), we can see from Figure 6.22 that with all traffic patterns, the RF-I gets congested first and thus, 100% usage saturates first. The restricted usage models yield higher throughput. The power trend continues to be similar to the previous network sizes. Thus, at 64 nodes there is a trade-off between latency and throughput. If latency is the primary design goal, then the RF should be used opportunistically. Thus, as the network size increases, congestion management of RF-I is essential for extracting maximum throughput advantages. We show this using

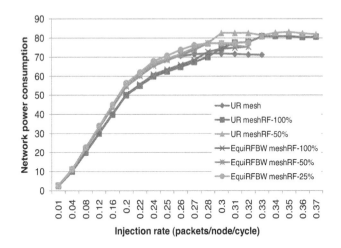

FIGURE 6.19
Network power consumption of *Mesh+RF* for UR traffic at 36 nodes.

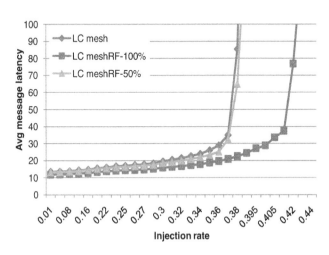

FIGURE 6.20
Average message latency of *Mesh+RF* for LC traffic at 36 nodes.

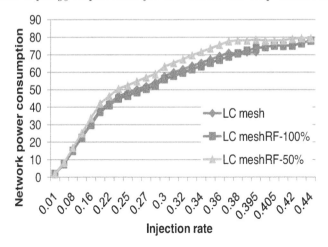

FIGURE 6.21
Network power consumption of *Mesh+RF* for LC traffic at 36 nodes.

our static congestion management and a dynamic scheme is expected to do better.

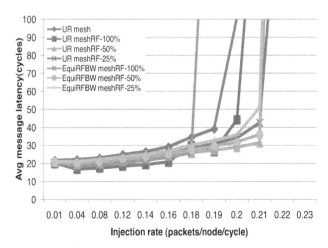

FIGURE 6.22
Average message latency of *Mesh+RF* for UR traffic at 64 nodes.

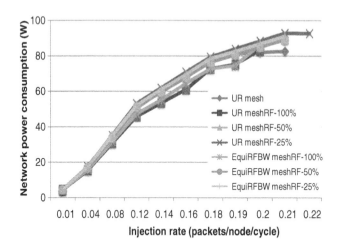

FIGURE 6.23
Network power consumption of *Mesh+RF* for UR traffic at 64 nodes.

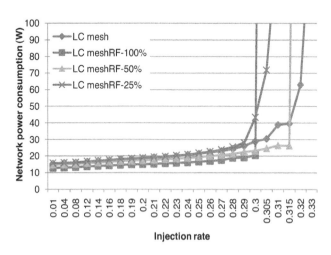

FIGURE 6.24
Average message latency of *Mesh+RF* for LC traffic at 64 nodes.

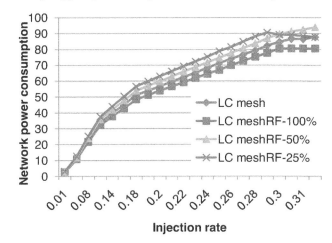

FIGURE 6.25
Network power consumption of *Mesh+RF* for LC traffic at 64 nodes.

6.7.1.4 256 nodes

Due to the large network size, the probability of deadlocks increase at high injection rates (larger dependency chains). Even the smaller-sized networks experience deadlocks at high injection rates that lie beyond the network saturation point. We did not plot those points in our figures so far. However, the larger networks in this phenomenon occurs even before the network reaches saturation.

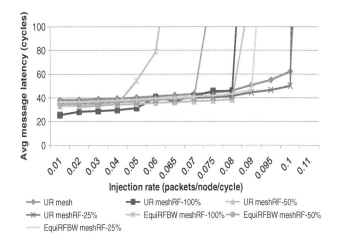

FIGURE 6.26
Average message latency of *Mesh+RF* for UR traffic at 256 nodes.

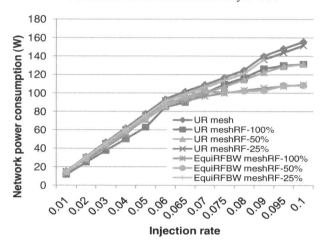

FIGURE 6.27
Network power consumption of *Mesh+RF* for UR traffic at 256 nodes.

This limits the sustainable throughput of the RF-enhanced mesh. As seen from the Figure 6.28, the spike in the latency is due to the deadlock recovery mechanism that is invoked on deadlock detection. From the Figure 6.26, it can be seen that only 25% usage model gives equivalent throughput as the base case. Thus, as the network size increases, we get diminishing returns on the throughput of the network if a simple flat mesh is used as the underlying topology. As seen from the Figures 6.27 and 6.29, even 25% usage leads to

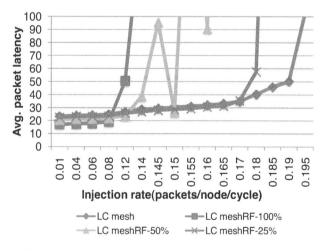

FIGURE 6.28
Average message latency of *Mesh+RF* for LC traffic at 256 nodes.

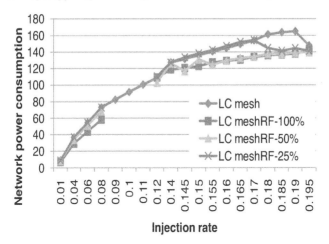

FIGURE 6.29
Network power consumption of *Mesh+RF* for LC traffic at 256 nodes.

lower power than the base case. Thus, as the network size increases, the power savings due to lower hop count can compensate for the overhead of using RF-I.

Figure 6.30 shows the energy delay product improvement of RF-enhanced mesh over the base mesh for the four sizes. This figure shows that for UR and LC traffic patterns when RF is used 100% of the time, at 36 nodes we obtain maximum EDP benefits from RF when mesh topology is considered. The reason for this trend is that initially with technology scaling, there is an increase in bandwidth and RF-I takes only one cycle for very long distances. However, as the network size increases, we observe diminishing returns in terms of throughput and EDP.

The reason for such a behavior is twofold: (a) congestion of the RF-I and (b) the occurrence of deadlock. As the network size increases and RF-I is allowed to use only half the time, then EDP advantages also increase. We expect that eventually even for 50% the EDP benefits will start to drop. Thus, at larger network sizes, there are diminishing returns in terms of throughput and EDP by using RF-I for a flat topology.

Figure 6.31 shows the power breakdown in the various router components for 0.04 injection rate. We can now clearly understand why the 50% usage does not lead to power advantages. The crossbar switch is a power-hungry component. RF-I leads to larger crossbar size, which results in higher power. In the 100% usage case, this increase in power is compensated by the decrease in the total number of hops in the network.

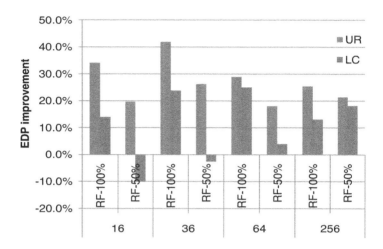

FIGURE 6.30
EDP improvement for all the network sizes over base mesh.

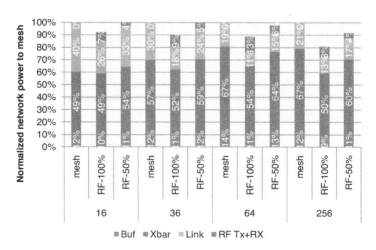

FIGURE 6.31
Power breakdown for various network sizes at 0.04 injection rate.

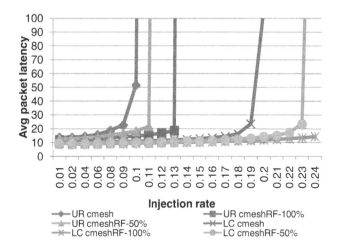

FIGURE 6.32
Average message latency of CMESH+RF for a 64-node network.

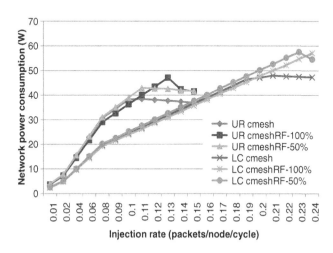

FIGURE 6.33
Network power consumption of CMESH+RF for a 64-node network.

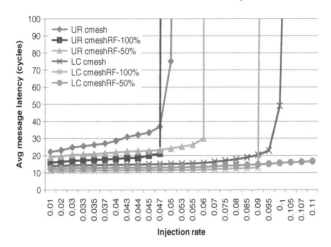

FIGURE 6.34
Average message latency of CMESH+RF for a 256-node network.

6.7.2 Cmesh and Hierarchical Topologies

In this section, we present the results of applying RF-interconnect to concentrated mesh and hierarchical topologies for 64 nodes and the 256-node network. The RF interconnect is overlaid over the CMESH and the global network of the hierarchical network. In the case of concentrated networks, the nearest neighbors for a node (A) are all the nodes that share the same router connected to node A. As a result, the nearest-neighbor component of LC traffic does not use the RF-I and only the UR component of the LC traffic pattern uses the RF-I.

6.7.2.1 CMESH

The average message latency results for 64-node CMESH and 256 nodes are shown in Figures 6.32 and 6.34, respectively. For the 64-node network, the RF 50% usage itself gives significant benefits. Unlike in the mesh, where the RF 100% usage saturated at 64 nodes for UR traffic, in concentrated mesh the 100% usage gives throughput benefits as well. This is true for LC traffic as well. At 256-network size, however, the 100% usage saturates earlier than the base case and 50% usage gives throughput benefits. We don't see the deadlock forming before saturation in the concentrated mesh for the large 256-network size. The power plots for UR and LC are shown in Figures 6.33 and 6.35. *Cmesh+RF* has the lowest latency for uniform random traffic pattern.

6.7.2.2 Hierarchical Topology

The global network size in the hierarchical topology is smallest due to the local bus. For 64 nodes the global network is 8 routers and for 256 nodes, there are

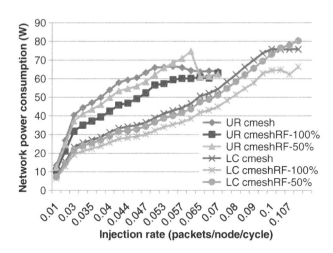

FIGURE 6.35
Network power consumption of CMESH+RF for a 256-node network.

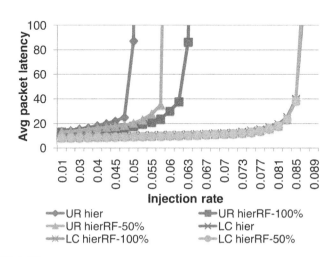

FIGURE 6.36
Average message latency of Hier+RF for a 64-node network.

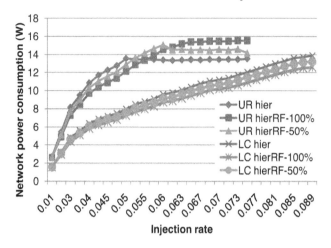

FIGURE 6.37
Network power consumption of Hier+RF for a 64-node network.

32 routers in the global network. Due to the small size, there is minimal performance impact of RF for the LC traffic case as seen in Figure 6.36. For the
UR case, we find that 100% usage gives throughput and latency benefits. For
256 nodes, the RF-enhanced topology at 100% usage does as well as the base
case under LC traffic pattern (see Figure 6.38). This means that RF decreased
the delay variability in the global network. It should be noted that at the same
technology, the RF bandwidth allocated to $MESH > CMESH > HIER$.
It is interesting to note that even with the small RF-bandwidth allocation
hierarchical network has very high throughput advantage. Thus, concentrated
and hierarchical networks take better advantage of the RF-I technology. Thus,
the hierarchical network and the concentrated mesh will have lower area overhead (even in the scenario where all the topologies use the same bandwidth).
We expect that with higher bandwidth allocation, there will be throughput
benefits and power benefits (see Figure 6.37 and 6.39) but the latency benefits
will flatten out as it saves only the serialization latency in the global network.

Figure 6.40 shows the energy delay product averaged for all load rates until
the *hier* network saturates (0.04 packet/node/cycle). This value is normalized
to the base mesh. The average packet latency is calculated in a similar fashion
and is also shown in Figure 6.40. In this plot, lower is better. The plot shows
that $CMESH+RF$ has the lowest latency and the hierarchical network provides the best performance and power trade-offs for these injection rates. By
overlaying the concentrated mesh with RF, as in $Cmesh+RF$, we obtain better
EDP than the flattened butterfly network. Providing express paths using RF
in cmesh could mimic a richly connected network. Figure 6.41 shows that the
throughput at 64 nodes for each topology normalized to mesh.

In order to understand which topology is able to take maximum advantage of RF-I, the saturation throughput improvement, average latency, and

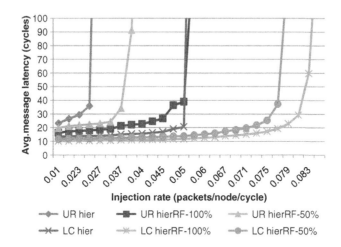

FIGURE 6.38
Average message latency of Hierarchical+RF for a 256-node network.

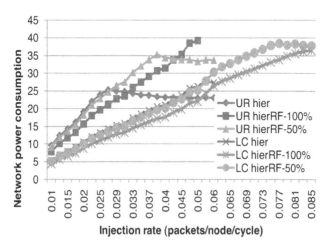

FIGURE 6.39
Network power consumption of Hierarchical+RF for a 256-node network.

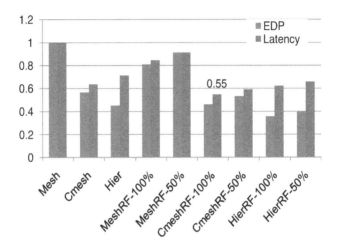

FIGURE 6.40
Energy delay product averaged up to 0.04 load rate and normalized to mesh.

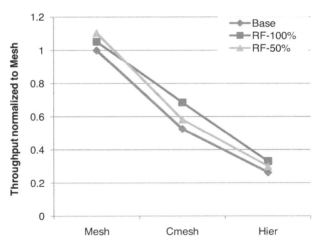

FIGURE 6.41
Throughput at 0.04 load rate of all topologies normalized to mesh.

average EDP improvements of RF-based topologies compared to their respective non-RF topologies is plotted (i.e., the RF-mesh improvement over mesh topology). Figure 6.42 shows the percentage savings in latency, EDP, and throughput by using RF-interconnect in a simple mesh, a concentrated mesh, and the hierarchical topology. The simple mesh seems to gain the most in latency. However, this gain in latency is quite small (15% latency savings) to bridge the gap between mesh and advanced topologies as seen from 6.40 (up to 45% improvement with respect to latency). The CMESH has the highest throughput benefit and the hierarchical has the highest EDP advantage. Thus, if EDP is the primary design goal then hierarchical is a good design choice.

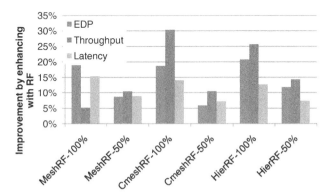

FIGURE 6.42
Improvements by applying RF interconnect.

6.8 Applications

A representative set of benchmarks from commercial, scientific, and SPLASH suites were chosen for study. A 32-way CMP with 32 cache nodes leading to the 64-node network is considered for study. The layout for the CMP is shown in Figure 6.13. All the experimental assumptions and network parameters are explained in Section 6.6. Figure 6.43 shows the instruction per cycle (IPC) metric normalized to the simple mesh topology. We do not show the mesh+rf results here for clarity. On an average, the *cmesh+rf* topology provides 37% ipc improvement. As can be seen from Figure 6.44, the hierarchical topology enhanced with RF is comparable to *cmesh+rf*. Note that the percentage edp improvement is plotted and higher is better.

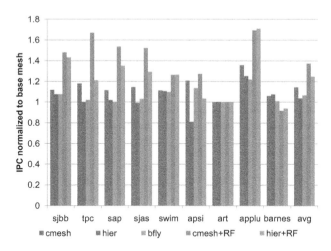

FIGURE 6.43
IPC normalized to mesh topology.

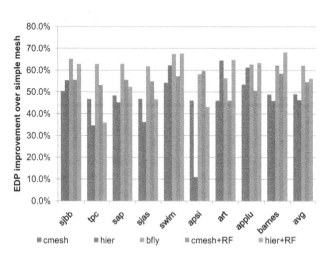

FIGURE 6.44
Energy delay product improvement for various apps over base mesh.

6.9 Related Work

We partition our related work based on topology and the network-on-chip architectures that avail these emerging technologies.

Network-on-Chip Architectures In the past decade, there have been several emerging technology proposals both for interconnect as well as for devices. As the wire width scales down, the resistivity of copper wire is leading to higher RC delay. The rise in RC delay does not significantly impact local and intermediate layers as these wires are used for small lengths. The rise in RC delay has an adverse effect on the performance of global wires due to their long lengths. The traditional RC delay of an optimally repeated 2 cm wire has a latency of 800 ps at 90 nm and this worsens to 1500 ps in 22 nm technology [9]. As a result, there has been significant research effort dedicated to the design of emerging technologies [1]. The most promising among the proposals for on die interconnect are 3D-integration, RF-signaling, and optical interconnects.

Three-dimensional-integration results in smaller form-factor of the chip resulting reduced interconnect lengths. [27] explored different 3D-network topologies for network on chip. [12] explored 3D-mesh based and tree-based topologies and showed significant performance and power advantages. [37] describes 3D-stacked memory and builds a 3D-CMP with a 3D-network using the TDMA bus. [26] proposes a true 3D-router architecture for 3D-multilayered CMP. Carloni et al. provide a summary of current work in using emerging technology in network-on-chip architectures and enumerate the advantages and challenges [5].

Kumar et al. used express virtual paths and show performance advantages [17]. Ogras and Marculescu use application-specific long range link insertion and show significant advantage in the achievable network throughput [24]. Modarressi et al. use adaptive virtual point to point links and show significant reduction in power and latency [22]. Ogg et al. propose using serialized asynchronous links with the same performance as the synchronous links but with fewer wires. Current RF proposals use RF-I as physical express paths that leads to lower power and latency. Chang et al. propose using multiband RF interconnect with FDMA techniques in on-chip networks [9]. They evaluated mesh overlaid with RF for application traffic (low injection rate) and show significant performance benefits. A simultaneous triband on-chip RF interconnect was demonstrated in [10]. In [6], they propose reconfigurable dynamic link insertion and show that for a small RF-bandwidth allocation, significant power reduction can be achieved when adaptive shortcuts and RF-multicast is used. Their proposal shows a 65% power reduction for application traffic. In this work, we overlay RF on mesh, cmesh, and hierarchical networks and present the latency, power, and throughput results.

Kirman et al. proposed a hierarchical opto-electrical bus that uses electrical network at the local level and a optical ring in the global level. Their proposal was evaluated for bus-based 64-way CMP and showed up to 50% latency improvement for some applications and up to 30% power reduction over a baseline electrical bus [16]. Their key insight was that an electrical network is a must for achieving high performance from optical networks. Shacham et al. proposed a hybrid optical circuit-switched and a packet-switched electrical network for decreasing the power consumption of optical networks. A folded torus topology was shown to provide the least latency and the highest bandwidth. They obtain 576 Gbps and a total power consumption of 6 W at 22 nm technology [31]. Vantrease et al. proposed a nanophotonic interconnect for throughput optimization for a many core CMP [33]. Their evaluations were targeted at 16 nm and use WDM and an all-optical arbitration scheme. They also modeled optical interconnect off-die. They show that for an optical crossbar network and 3D-integration, a performance improvement of up to 6 times for memory-intensive applications was observed when compared to an all-electrical network. Petracca et al. explore topologies of photonic network design for a single application [28]. Pan et al. propose a hierarchical network consisting of electrical local communication and an all optical crossbar for global communication and compare it to concentrated mesh [25]. Zheng et al. provide a low latency multicast/broadcast subnetwork and throughput optimized circuit switched optical network [20].

Topology Kumar et al. [18] presented a comprehensive analysis of interconnection mechanisms for small scale CMPs. They evaluate a shared bus fabric, a cross bar interconnection, and point to point links. Pinkston and Ainsworth [2] examined the cell broadband engine's interconnection network, which utilizes two rings and one bus to connect 12 core elements. The mesh network-on-chip topology has been prototyped in Polaris [32], Tile [36] and TRIPS [30] for medium sized(50+ nodes) on-chip networks. Wang et al. [35] did a technology-oriented, and energy-aware topology exploration of mesh and torus interconnects with different degrees of connectivity. Recently to address the power inefficiency and scalability limitations of mesh, concentrated mesh topologies [3], high radix topologies [15], and topologies with express channels [13] have been proposed. This chapter focuses on applying RF-I technology to current topologies. In [11] the authors propose a hierarchical network with a global mesh connecting local buses (serving small sets of cores). While such a hierarchical network solves the power-inefficiency and latency-scalability limitations of mesh topologies, it provides lower throughput than mesh.

6.10 Conclusions

The number of cores on-die is predicted to grow for a few technology generations. The exacerbated global interconnect delay in future technologies call for novel emerging technology alternatives. Radio-frequency transmission is one such technology alternative that holds promise due to its CMOS compatibility. As a result, RF circuitry can also take advantage of CMOS scaling. Using multiplexing techniques and some circuit tricks, multiband RF transmission has been demonstrated. While several factors determine the feasibility of such a solution on-die, it is imperative to revisit the architecture and understand the benefits and implications of the new technology. In this chapter, we evaluated RF-I when applied to various on-chip network topologies.

Topology forms the backbone of the on-chip network and is one of the main factors in determining delay, energy consumption, throughput, and scalability of the network. We consider three state-of-art topologies namely mesh, concentrated mesh and hierarchical network and enhance them using RF-I. The base hierarchical and concentrated networks reduce latency and power as compared to mesh at the expense of network throughput. Our results show that by overlaying RF on the hierarchical topology 25% energy delay product savings, 12% latency savings, and up to 21% improvement in throughput can be obtained. For concentrated mesh the throughput improvement is higher (30%). Our main contributions are:

- Concentrated and hierarchical topologies have lower throughput and thus, high bandwidth in RF can be leveraged to improve throughput.

- The *hier+RF* topology that has RF-I overlaid onto the global mesh gives the least energy delay product.

- Among the three topologies, the concentrated mesh topology benefited most by the integration of RF both for latency and throughput at 64-node network size. As the concentration increases, the network diameter decreases and thus, the smaller RF bandwidth allocation is shared among fewer routers leading to lower throughput advantages in hierarchical network.

In summary, we understand that concentrated and hierarchical networks are scalable in terms of latency and power but can sustain lesser throughput. Adding express paths by using RF-I technology can lead to throughput benefits and thus making these scalable topologies a good design option to be deployed in future systems. Further, a smarter RF-I usage models and/or deadlock prevention schemes can be explored at higher injection rates. It would be interesting to study RF-I as the only global network by eliminating the mesh in the global network of hierarchical topology. Considering that power is a very important constraint in future systems, this design option would be extremely power-efficient as long as the performance is reasonable. This chapter explained how RF-I could tackle the low throughput of the hierarchical topology and thus, making hierarchical topology an energy efficient, moderate throughput, and low latency topology.

Acknowledgments

This work is supported in part by the National Science Foundation award CCF-0903432 and CCF-0702617. We would like to thank Aditya Yanamandra for insightful comments and discussions.

6.11   Glossary

Communication locality: Percentage of traffic whose destination lies within one hop of the source.

Deadlock: A circular dependency where two or more competing actions are waiting on each other.

Path diversity: Having multiple routes to reach a destination.

Saturation point: The load rate at which the network latency grows exponentially.

Topology: Graphical mapping of configuration of (physical/logical) connections between network components (nodes).

6.12   Bibliography

[1] International Technology Roadmap for Semiconductors (ITRS), 2008 edition, http://www.itrs.net/.

[2] T. W. Ainsworth and T. M. Pinkston. Characterizing the cell EIB on-chip network. *IEEE Micro*, 27(5):6–14, 2007.

[3] J. Balfour and W. J. Dally. Design tradeoffs for tiled cmp on-chip networks. In *ICS '06: Proceedings of the 20th Annual International Conference on Supercomputing*, 187–198, New York, NY, USA, 2006. ACM.

[4] S. Borkar. Networks for multi-core chips: A contrarian view. In *Special Session at ISLPED 2007*.

[5] L. P. Carloni, P. Pande, and Y. Xie. Networks-on-chip in emerging interconnect paradigms: Advantages and challenges. In *NOCS '09: Proceedings of the 2009 3rd ACM/IEEE International Symposium on Networks-on-Chip*, 93–102, Washington, DC, USA, 2009. IEEE Computer Society.

[6] M.-C. F. Chang, J. Cong, A. Kaplan, C. Liu, M. Naik, J. Premkumar, G. Reinman, E. Socher, and S.-W. Tam. Power reduction of cmp communication networks via rf-interconnects. In *MICRO '08: Proceedings of the 2008 41st IEEE/ACM International Symposium on Microarchitecture*, 376–387, Washington, DC, USA, 2008. IEEE Computer Society.

[7] M.-C. F. Chang, E. Socher, S.-W. Tam, J. Cong, and G. Reinman. Rf interconnects for communications on-chip. In *ISPD '08: Proceedings of the 2008 International Symposium on Physical Design*, 78–83, New York, NY, USA, 2008. ACM.

[8] M.-C. F. Chang, V. P. Roychowdhury, L. Zhang, H. Shin, and Y. Qian. Rf/wireless interconnect for inter-and intra-chip communications. In *Proceedings of the IEEE*, 89:456–466, 2001.

[9] M. F. Chang, J. Cong, A. Kaplan, M. Naik, G. Reinman, E. Socher, and S.-W. Tam. Cmp network-on-chip overlaid with multi-band rf-interconnect. In *High Performance Computer Architecture, 2008. HPCA 2008. IEEE 14th International Symposium on High Performance Computer Architecture*, 191–202, Feb. 2008.

[10] J. Cong, M.-C. F. Chang, G. Reinman, and S.-W. Tam. Multiband rf-interconnect for reconfigurable network-on-chip communications. In *SLIP '09: Proceedings of the 11th International Workshop on System level Interconnect Prediction*, 107–108, New York, NY, USA, 2009. ACM.

[11] R. Das, S. Eachempati, A. K. Mishra, V. Narayanan, and C. R. Das. Design and evaluation of a hierarchical on-chip interconnect for next-generation CMPS. In *High Performance Computer Architecture, 2009. HPCA 2009. IEEE 15th International Symposium on High Performance Computer Architecture*, 175–186, Feb. 2009.

[12] B. S. Feero and P. P. Pande. Networks-on-chip in a three-dimensional environment: A performance evaluation. *IEEE Trans. Comput.*, 58(1):32–45, 2009.

[13] B. Grot, J. Hestness, S. W. Keckler, and O. Mutlu. Express cube topologies for on-chip interconnects. In *High Performance Computer Architecture, 2009. HPCA 2009. IEEE 15th International Symposium on High Performance Computer Architecture*, 163–174, Feb. 2009.

[14] D. Huang, T. R. LaRocca, L. Samoska, A. Fung, and M.-C.F. Chang. 324GHz CMOS frequency generator using linear superposition technique. In *Proceedings of Solid-State Circuits Conference, Digest of Technical Papers.* 476–629, 2008.

[15] J. Kim, J. Balfour, and W. J. Dally. Flattened butterfly topology for on-chip networks. *Microarchitecture, 2007. MICRO 2007. 40th Annual IEEE/ACM International Symposium on Microanchitecture*, 172–182, Dec. 2007.

[16] N. Kirman, M. Kirman, R. K. Dokania, J. F. Martinez, A. B. Apsel, M. A. Watkins, and D. H. Albonesi. Leveraging optical technology in future bus-based chip multiprocessors. In *MICRO 39: Proceedings of the 39th Annual IEEE/ACM International Symposium on Microarchitecture*, 492–503, Washington, DC, USA, 2006. IEEE Computer Society.

[17] A. Kumar, L.-S. Peh, P. Kundu, and N. K. Jha. Toward ideal on-chip communication using express virtual channels. *IEEE Micro*, 28(1):80–90, 2008.

[18] R. Kumar, V. Zyuban, and D. M. Tullsen. Interconnections in multi-core architectures: Understanding mechanisms, overheads and scaling. *SIGARCH Comput. Archit. News*, 33(2):408–419, 2005.

[19] E. Laskin, M. Khanpour, R. Aroca, K. W. Tang, P. Garcia, and S. P. Voinigescu. 95GHz receiver with fundamental-frequency VCO and static frequency divider in 65nm digital CMOS. In *Proceedings of Solid-State Circuits Conference*, 180–181, 2008.

[20] Z. Li, J. Wu, L. Shang, A. R. Mickelson, M. Vachharajani, D. Filipovic, W. Park, and Y. Sun. A high-performance low-power nanophotonic on-chip network. In *ISLPED '09: Proceedings of the 14th ACM/IEEE international Symposium on Low Power Electronics and Design*, 291–294, New York, NY, USA, 2009. ACM.

[21] P. S. Magnusson, M. Christensson, J. Eskilson, D. Forsgren, G. Hallberg, J. Hogberg, F. Larsson, A. Moestedt, and B. Werner. Simics: A full system simulation platform. *Computer*, 35(2):50–58, 2002.

[22] M. Modarressi, H. Sarbazi-Azad, and A. Tavakkol. Performance and power efficient on-chip communication using adaptive virtual point-to-point connections. In *NOCS '09: Proceedings of the 2009 3rd ACM/IEEE International Symposium on Networks-on-Chip*, 203–212, Washington, DC, USA, 2009. IEEE Computer Society.

[23] N. Muralimanohar, R. Balasubramonian, and N. Jouppi. Optimizing NUCA organizations and wiring alternatives for large caches with cacti 6.0. *Microarchitecture, 2007. MICRO 2007. 40th Annual IEEE/ACM International Symposium on Microarchitecture*, 3–14, 1–5 Dec. 2007.

[24] U. Y. Ogras and R. Marculescu. Application-specific network-on-chip architecture customization via long-range link insertion. In *ICCAD '05: Proceedings of the 2005 IEEE/ACM International Conference on Computer-Aided Design*, 246–253, Washington, DC, USA, 2005. IEEE Computer Society.

[25] Y. Pan, P. Kumar, J. Kim, G. Memik, Y. Zhang, and A. Choudhary. Firefly: illuminating future network-on-chip with nanophotonics. *SIGARCH Comput. Archit. News*, 37(3):429–440, 2009.

[26] D. Park, S. Eachempati, R. Das, A. K. Mishra, Y. Xie, N. Vijaykrishnan, and C. R. Das. Mira: A multi-layer on chip interconnect router architecture. In *Proceedings of the 35th International Symposium on Computer Architecture, ISCA-2008, 251–261*.

[27] V. F. Pavlidis and E. G. Friedman. 3-d topologies for networks-on-chip. *IEEE Trans. Very Large Scale Integr. Syst.*, 15(10):1081–1090, 2007.

[28] M. Petracca, B. G. Lee, K. Bergman, and L. P. Carloni. Photonic NoCs: System-level design exploration. *IEEE Micro*, 29(4):74–85, 2009.

[29] K. Sankaralingam, R. Nagarajan, H. Liu, C. Kim, J. Huh, D. Burger, S. W. Keckler, and C. R. Moore. Exploiting ILP, TLP, and DLP with The Polymorphous TRIPS Architecture. In *Proceedings of the 30th Annual International Symposium on Computer Architecture*, 422–433, 2003.

[30] K. Sankaralingam, R. Nagarajan, R.McDonald, R. Desikan, S. Drolia, M. S. Govindan, P. Gratz, D. Gulati, H. Hanson, C. Kim, H. Liu, N. Ranganathan, S. Sethumadhavan, S. Sharif, P. Shivakumar, S. W. Keckler, and D. Burger. Distributed microarchitectural protocols in the trips prototype processor. In *MICRO 39: Proceedings of the 39th Annual IEEE/ACM International Symposium on Microarchitecture*, 480–491, Washington, DC, USA, 2006. IEEE Computer Society.

[31] A. Shacham, K. Bergman, and L. P. Carloni. Photonic networks-on-chip for future generations of chip multiprocessors. *IEEE Transactions on Computers*, 57(9):1246–1260, 2008.

[32] S. Vangal, J. Howard, G. Ruhl, S. Dighe, H. Wilson, J. Tschanz, D. Finan, P. Iyer, A. Singh, T. Jacob, S. Jain, S. Venkataraman, Y. Hoskote, and N. Borkar. An 80-tile 1.28tflops network-on-chip in 65 nm CMOS. In *Solid-State Circuits Conference, 2007. ISSCC 2007. Digest of Technical Papers. IEEE International*, 98–589, 11-15 Feb. 2007.

[33] D. Vantrease, R. Schreiber, M. Monchiero, M. McLaren, N. P. Jouppi, M. Fiorentino, A. Davis, N. Binkert, R. G. Beausoleil, and J. H. Ahn. Corona: System implications of emerging nanophotonic technology. *SIGARCH Comput. Archit. News*, 36(3):153–164, 2008.

[34] H. Wang, X. Zhu, L.-S. Peh, and S. Malik. Orion: A Power-Performance Simulator for Interconnection Networks. In *ACM/IEEE MICRO*, Nov. 2002.

[35] H. Wang, L.-S. Peh, and S. Malik. A technology-aware and energy-oriented topology exploration for on-chip networks. In *DATE '05: Proceedings of the conference on Design, Automation and Test in Europe*, 1238–1243, Washington, DC, USA, 2005. IEEE Computer Society.

[36] D. Wentzlaff, P. Griffin, H. Hoffmann, L. Bao, B. Edwards, C. Ramey, M. Mattina, C.-C. Miao, J. F. Brown III, and A. Agarwal. On-chip interconnection architecture of the tile processor. *IEEE Micro*, 15–31, 2007.

[37] Y. Xie, N. Vijaykrishnan, and C. Das. *Three-Dimensional Network on Chip Architectures*. Springer, 2010.

[38] W. Zhao and Y. Cao. New generation of predictive technology model for sub-45 nm design exploration. In *ISQED '06: Proceedings of the 7th International Symposium on Quality Electronic Design*, 585–590, Washington, DC, USA, 2006. IEEE Computer Society.

7

Intra-/Inter-Chip Optical Communications: High Speed and Low Dimensions

Braulio García-Cámara

Optics Group. Department of Applied Physics. University of Cantabria

CONTENTS

The recent and very fast advances in computation have led to an important improvement in the design and manufacture of chips. However, the high-speed needs and the very high quantity of information that must be shared have been constrained by physical limitations. The impedance and size of metallic wires are not enough for future technologies, hence new devices must be developed. Some alternatives have been proposed to overcome this challenge. One of them is the use of optical components for inter- and intra-chips communications, substituting electrical signals by light. New advances in photonic devices, new engineered materials, photolithography, and nano-manufacture techniques have approached this alternative and, in the near future, it could be a real counterpart to actual chips. Then, optoelectronics can be considered as the natural evolution of the microelectronic integrated circuits.

In this chapter, the principles of optical links will be analyzed. Different designs and implementations of optical devices for intra- and inter-chip interconnects are summarized in other to show a state-of-the-art of this new technology.

7.1 Introduction

It is well known that the evolution of microelectronics and in particular the evolution of chips follows the famous Moore's law [76]. The exponential evolution of electronic devices established by this law has been able to follow the amount of data that must be processed nowadays. However, data evolution continues to increase and microelectronic technologies are approaching their physical limits. The so-called "interconnection bottleneck" can be reflected in the increasing differences between the gate delay (switching speed) of a transistor and the propagation delay along wires between transistors of a integrated circuit, as is shown in Figure 7.1.

In a few years, the actual electronics devices will not be able to transport and process the quantity of data that we will need. Until now, the physical characteristics of electrons make them suitable for processing information via discrete logic. Also, they could be sent very fast through metal wires, therefore they can be also useful for transferring information. However, the recent innovations need to transfer a high amount of information very fast and this speed is limited in metal lines by the resistance, capacitance, and reliability of them. In addition, the increase of the number of elements in the chip obliges to have a high number of communications and not much space for wires. This high density of interconnections, high speed of transmission, and power limitations cannot be obtained using the existent metallic wires. For this reason, one of the best options is the use of electronics for processing the information while photons are used as carriers for bits. As an example, Figure 7.2 shows the evolution of the telecommunications technologies and their relative capacity. In this figure, it can be observed that the capacity rates presented by systems that use optical fibers as transmission channels cannot be obtained using metallic channels.

The International Technology Roadmap for Semiconductors (ITRS) of 2007 [1] stated that optical interconnections (OI) present some advantages with respect to electric interconnections (EI) as, for example,

- High speed propagation of light.

- Large bandwidth of optical waveguides.

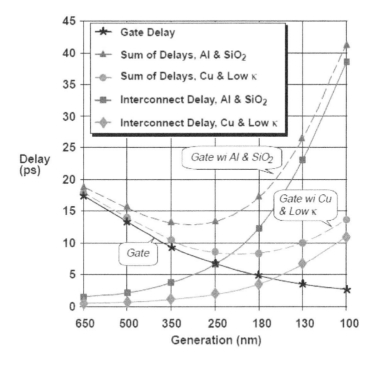

FIGURE 7.1
Comparison of technology trends in transistor gate delay versus Al/SiO_2 interconnect propagation delay [101].

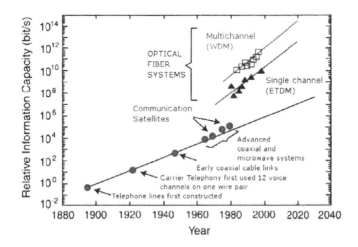

FIGURE 7.2
Increase in relative information capacity of communication channels versus time [52].

- Possibility to minimize the crosstalk between signal transmission paths.

- Design flexibility due to the facility to be transmitted using waveguides or free-space configurations.

- Capability to transmit several wavelengths through a single optical path providing bandwidth densities even higher than those with electrical interconnections.

These improvements in communications will be explained in more detail later. However, optical interconnections exhibit important disadvantages or difficult challenges than must be overcome. The integration of optical systems on chips involves an important design complexity. The size of the actual optical components is often larger than nanometer scales of modern electronic devices producing important penalizations in area, noise, and power consumption. In addition, special attention must be taken of the fact that new devices should be low-power. One on-chip global interconnect consumes typically 1 pJ. To be competitive, an optical system must consume a power 10 times lower [110]. The cost of this technology is also an important issue and, nowadays, it still causes that chips with optical interconnections are not competitive with respect to electrical communications.

Finally, the election of materials for manufacturing these optical devices has to ensure that they are Complementary Metal-Oxide-Semiconductor (CMOS) compatible. As it is well-known, silicon is the base material of the recent microelectronic technologies [24]. This is the second most abundant material on Earth after oxygen, because of that it is relatively inexpensive. Also its properties as semiconductor are quite stable and well-understood. These reasons made silicon a good candidate to develop photonic components for optical interconnections [85]. In the last years, as it will be explained later, several works have been devoted to study the photonic properties of silicon devices [53, 96]. However, the characteristics of bulk silicon, and in particular its indirect gap, have made the study to be extended also for silicon-base engineered materials. Additionally, other materials have been analyzed. The plasmon scattering properties presented by metals produce an important confinement of light and also a high enhancement of the scattered intensity. Both facts have been used for designing photonic devices. Also, in the last years, new nanostructured materials, called metamaterials, have arisen. The main advantage of these new materials is the possibility to modify their global optical properties. Although they are still in a design phase, they would be the main component of future integrated circuits.

Upon these considerations, in this chapter, two major items will be discussed: i) the last aggressive developments of photonic devices required for intra-/inter-chip optical communications based on very different materials and ii) the integration and fabrication of these new devices on a chip ensuring their compatibility with CMOS technologies. Also, the advantages and disadvantages that already have or must still be overcome, will be analyzed in detail.

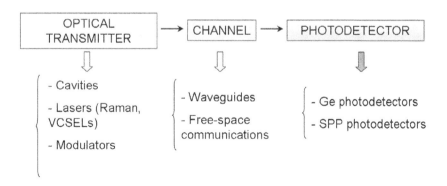

FIGURE 7.3
Scheme of the main components of an OI and the different devices that can perform those tasks.

In order to deal with each of these items, the present chapter is organized as follows. Section 7.2 is devoted to discuss the main optical components of a typical optical interconnect (OI). Although, as will be explained, an OI can be formed of several elements, this section is only focused on the three main ones (transmitter, channel, and receiver). Also, there are many devices that can perform these tasks, as can be seen in Figure 7.3. In this section, the newest devices and their main characteristicts will be analyzed. In Section 7.3, a comparison between the electric and the optical interconnects is carried out. The main advantages and disadvantages of both are analyzed for several important parameters: the power consumption, the propagation delay, the bandwidth density, the crosstalk noise, and the fan-out. In this section, it has been explained why optical interconnects are adequate for inter- and/or intra-chip communications. Photonic networks are also an interesting case, whose number of applications in system on-chip increases day per day. For this reason, Section 7.4 is devoted to the study of them. In the last years, researchers are trying to improve the characteristics of the actual photonic components and to widen their applications. Metamaterials allow doing that by tuning the optical properties of materials. Thus, in Section 7.5, the possible applications of metamaterials to perform new optical nanocircuits are discussed. Finally, an analysis of the current development of this new type of on-chip communications is included in the final section.

7.2 Photonic Components for On-Chip Optical Interconnects

7.2.1 Introduction

There are several models and designs for an optical interconnection. Each of these alternatives has its advantages and disadvantages. However, the common structure for an optical link is composed of an **optical transmitter**, a **channel** that transfers photons from one point to other or from one chip to other chip and finally an **optical receiver**. These three elements can be included or not into the digital logic chip. The first case is known as Integrated Architecture [1] and it implies that the optical devices are in the chip in such a way that any communication between the chip and the rest of the world are through optical signals. The total integration is quite interesting because the space requirements and power consumption are low. However, the design constraints produce that we are still far from these integrated architecture. The second case, that is more realistic nowadays, is the Discrete Architecture. In this implementation, the logic and the communication task are clearly separated. While the logic operations are made using electric signals, light is used for communications. Because of that, auxiliary circuitry is necessary to transform the electrical signal, which results from the logic part of the chip, into light, and then this light must be again transformed into an electric signal that can be treated in other parts of the chip. This architecture needs, obviously, more space and the power consumption increases with respect to the integrated architecture. The actual technology and the advances in optical devices are not sufficient to create total integrated optical chips that can be the counterpart to the electric devices. For this reason, discrete optical chips will be considered in the rest of the chapter. A basic and general scheme of an optical connection for intra- or inter-chip with discrete architecture is described in Figure 7.4.

The design and manufacture of the optical components of the link should be compatible with microelectronics. Because of that the materials and processors must be CMOS compatible and allow a monolithic integration. Silicon is possibly the best candidate for that. Then, in the last years, several studies have been devoted to the study and design of silicon optical components [41, 53, 122]. Also, other metallic nanostructures can be suitable to perform these activities. This can be explained by the high light-confinement capability that they present due to the excitation of surface plasmon resonances (**SPR**) or localized surface plasmon resonances (**LSPR**), as it will be described later.

In this section, a general analysis of each part of the optical interconnection (emitter, channel, and receiver) and their implementations will be performed. Several devices that are designed as emitters, channels, or receivers for an OI are described. Also, their main characteristics and their fabrication techniques are discussed in detail.

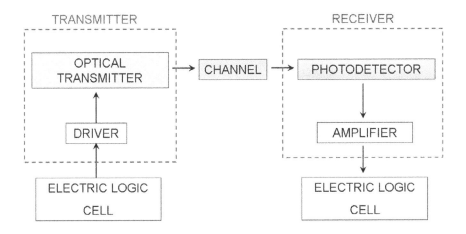

FIGURE 7.4
Scheme of a discrete optical interconnection.

7.2.2 Optical Sources

Discrete optical links can also be classified into two kinds of architectures depending on the light source. Integrated light source and external light source architectures can be distinguished. It is easy to see that the main difference between the two types of architectures is if the light emitter is in the chip or if it is an external source. In order to warrant sufficient intensity of the signal signal, the light transmitter must have an adequate power, what implies high power consumption and also high heat dissipation. Because of that, an external light source could be a good candidate. However, it also presents important disadvantages. Special attention must be taken into account to the coupling between the incident light and the chip. Introducing light into a chip is not an easy task and it could produce important losses. In general, power restrictions are more important in the chip design than light losses. For this reason external designs are more common. In this kind of architecture, the incident light is focused into a modulator, integrated in the chip, that controls the optical flux. However, in the last years, new advanced lasers have been developed decreasing their size and their power consumption.

7.2.2.1 In-Chip Light Sources: Small Emitting Devices

Following the idea of integration that dominates in microelectronics, some researches have proposed light emitters that can be included on the chip, instead of an external source. The first and possibly the most important characteristic of these devices is their size. They must be as small as possible. Also, as it will be noted, they must be CMOS compatible. Some devices have been developed

FIGURE 7.5
Whispering gallery mode in a droplet with a diameter of 130μm. From the web of the Ultrafast and Optical Amplifiers Group of the ITT Madras (India).

with these characteristics. In this section, a few of them are analyzed in order to give a brief view of this kind of light emitters.

Emitting Cavities

Very small optical structures can present high quality factor (Q-factor) resonances. In general, these resonant modes are produced by geometrical factors and are called *morphology-dependent* resonances (MDR) or *whispering gallery modes* (WGMs) [28] (see Figure 7.5). The physical explanation of these MDRs is based on the propagation of the light rays inside the resonant structure. The rays approach the inner surface with an angle beyond the critical angle and are totally internally reflected (TIR) [4]. Then, light is confined inside the structure. After propagating around the cavity, rays return, in phase, to their respective point which they entered and follow again the same path without attenuation due to destructive interference [10]. The frequency or wavelength at which these resonances are excited depends on the size, shape and refractive index of the structure. MDRs have been observed in spheres [74], cylinders, [67] and other complex geometries [3].

Some applications have been proposed in the last years using these MDRs [69]: filters [97], tunable oscillators [61], optical switching [124], sensing [89], or even high-resolution spectroscopy [95]. However, the main application in which we are interested is the possibility to generate lasers based on them [113].

The simplest geometry that presents these resonances is a sphere. Also, silicon is a well-known material with adequate characteristics for the integration on systems-on-chip. Also, the size of that resonators is perfect for inter- or intra-chip optics interconnections. For this reason, these silicon spheres could be ideal for microphotonics. Some works have been focused on their application in optical communications. As an example, in reference [96], the authors

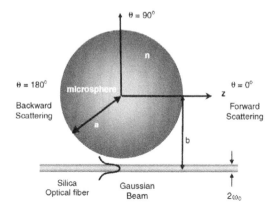

FIGURE 7.6
Scheme of the coupling between a silicon microsphere and a silicon optical fiber to excite the corresponding MDRs. The coupling distance is b [96].

made a complete study about microsphere resonators and their spectrum in the telecom wavelengths.

The excitation of MDRs in a silicon microsphere must be produced by an external signal. This fact could be a disadvantage of this kind of source, because an optical fiber must be coupled with the spherical particle in order to introduce light in the geometry. However, this coupling is through the evanescent part of the field. The confinement of light in the optical fiber is in such a way that it is not needed a direct contact between the fiber and the microsphere. A scheme of that is showed in Figure 7.6.

The resonant condition in microspheres, as it was mentioned, depends on the incident light and on the sphere characteristics, in particular on the refractive index of the material ($n = 3.5$ in the case of the silicon) and on the size of the sphere through the size parameter, x. This parameter is defined as $x = 2\pi a/\lambda$, a being the radius of the particle and λ the incident wavelength in the vacuum [14]. Each resonant mode is labeled with two numbers: the radial mode order, l, and an angular mode number, m.

Resonant modes in these geometries are well observed at two certain directions, the forward direction ($\theta = 0°$) and the 90° direction, as it was described in Figure 7.7 [96]. The confinement of light inside the particle for certain wavelengths produces abrupt minimums in the forward direction while enhanced maximums are observed for the same values of λ at 90°. These important maximums are those that can be used as input for an optical interconnection due to their spectral characteristics and also because light from these modes can be coupled easily to optical fibers or waveguides.

The spectral width of the resonance ($\delta\lambda$) is also a quite important factor

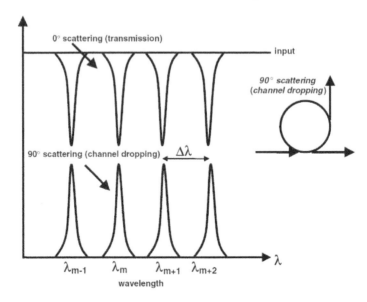

FIGURE 7.7
Spectra of light scattering by a silicon microsphere in the forward direction
($\theta = 0°$) and at 90° when MDRs are excited [96].

because it is directly related with the Q-factor through the relation

$$Q = \frac{\lambda}{\delta\lambda} \tag{7.1}$$

where λ is the resonant incident wavelength. For silicon particles of size around
$15\mu m$, authors of [96] reported Q-factors on the order of 10^7 and modes-
separation around 7 nm, which made this device good for the goal of being
optical source or even for Wavelength Division Multiplexing (WDM).

Sometimes, and in particular in the design of system-on-chips, the space
requirements are more important than input power requirements and the size
of sources must be decreased. The metallic counterpart of dielectric micro-
cavities are the called surface-plasmon cavities [75] that are optimized for
subwavelength-scale minituarization. Light confinement in these cavities is
produced by other kind of resonances, that are called *localized surface plasmons resonances (LSPR)*. These particular modes are excited due to a strong
interaction between the incident electromagnetic field and the conduction electrons that are enclosed inside the structure. When free electrons start to oscil-
late coherently at the same frequency that the one of the incident beam, the
electron cloud is displaced from the nuclei giving rise to a surface distribution
[88]. This effect produces high enhancement in light scattering and/or absorption. Based on this feature, some plasmonic-resonator devices have been
developed for several applications, emerging the idea of a "plasmonic laser"

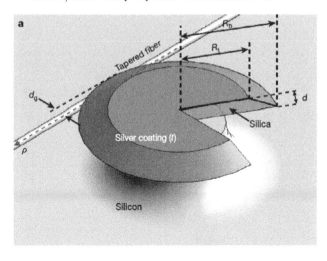

FIGURE 7.8

Schematic representation of the surface-plasmon-polariton (SPP) microcavity proposed in [72]. Light is pumped into the cavity through a tapered optical fiber passing under its edge. A transverse cross-section of the cavity is shown for clarity.

as was described by B. Min and coauthors in [72]. The system proposed in this work is quite similar to the dielectric microcavity, described above, but with a more complex geometry and a metallic coating, as can be seen in the scheme (Figure 7.8). This particular cavity has an outer radius of 10.96 μm, the inner one is 7.89 μm and the height of silica is 2 μm with a silver coating of a width equal to $100nm$ on it.

The structure presents LSPRs and also dielectric modes due to the presence of a dielectric wavelength channel. Then, a laser based on this geometry could have several lasing modes. Metals present high losses due to light scattering. However, the system is optimized to minimize them, reaching quality factors (Q-factors) around 2000. These values are much smaller than Q-factors obtained in dielectric materials, but they are still valid for laser cavities, demonstrating the possibility to develop surface-plasmon devices. The advantage of metallic devices in comparison with the dielectric ones falls in the fact that the first ones can confine light in volumes much more smaller than the incident wavelength. In [44], authors show a nanolaser with dimensions around of few hundreds of nanometers based on the excitation of surface plasmons resonances inside the device. As it is described in [44], the basic structure is that of a semiconductor laser formed by a InP/InGaAs/InP pillar with a conventional double heterostructure surrounded by a thin layer of silicon nitrade and finally encapsulated in gold for the metallic behavior (see Figure 7.9).

The dimensions of the height, h, and the diameter of the pillar are around 300 nm and 200 nm respectively, and the excitation of the modes is made

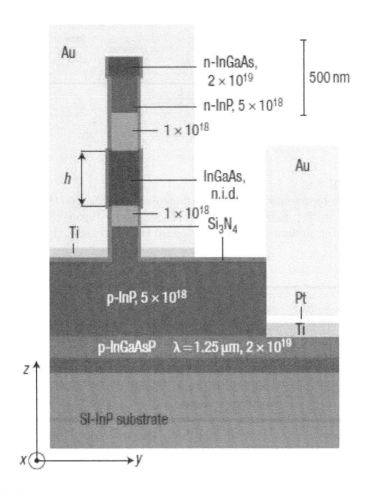

FIGURE 7.9
Description of the metallic nanocavity implemented experimentally in [44]. The values correspond to the doped levels in cm^{-3} and SI means semi-insulating.

through the injection of an electric signal (electrons at the top of the pillar and holes through the p-InGaAsP layer). This small laser active medium volume gives rise, unfortunately, to small quality factors and temperature-dependencies of the laser operation. The value of Q-factor obtained for these devices is only around 48 at room temperature. However, it could be improved to values around 180 replacing the gold capsule for a silver one and extending the InP regions. Also, operating at lower temperatures increases the quality factor with values like $Q = 268$ at $T = 10K$ or $Q = 144$ at $T = 77K$. These small parameters produce the kind of lasers that cannot be yet implemented for optical interconnects. Nevertheless, it could be considered as a first step for the development of nanolasers, because dimensions could be even decreased considering other geometries for the cross-section of the pillar, as authors remarked in their work [44].

Raman Lasers

Another physical feature that has applications in the generation of ultra-small optical sources is the so-called Stimulated Raman Scattering (SRS). This effect was described by Profs. Raman and Krishnan several years ago [90]. It is an inelastic scattering process that consists in the absorption of one photon for the diffuser. If this photon has enough energy, it could produce the excitation or desexcitation of vibrational, rotational or electronic levels, producing scattered photons with an energy higher (Anti-Stokes) or lower (Stokes) that the incident photons [5]. Recently, it was demonstrated that this kind of process can produce a lasing operation in ring cavities formed by a silicon waveguide incorporated to a 8-*m-long* optical fiber [15]. In this work, the authors show that the silicon waveguide acts like a gain medium producing laser pulses at the Stokes wavelength of 1675 *nm* (the pump laser emits at 1540 *nm*.) with a gain around 3.9 *dB* in the threshold limit (9 *W*). For pump powers larger than the threshold, the gain increases, almost linearly, with a slope efficiency $\sim 8.5\%$.

Following this idea, H. Rong and coworkers showed in [92] an experimental Raman laser integrated in a single silicon chip, which is one of the first steps to the integration of a laser for optical communications. The proposed device is similar to that in [15] following the same physical principles but optimized to its integration in a system-on-chip. The proposed device is formed by a single-mode rib waveguide with a reverse biased p-i-n diode structure in order to minimize the losses. Its geometrical cross section is schematically plotted in Figure 7.10 with the following experimental dimensions: W (waveguide width) = 1.5 μm, H (rib height) = 1.55 μm and d (etch depth) = 0.7 μm, which gives rise to an effective area around 1.6 μm^2 for the core of the waveguide. However, the global device has larger dimensions. The laser cavity is formed by an S-shape curve of this waveguide, as can be seen in Figure 7.11. One of the sides of the waveguide is coated with a high-reflectivity multilayer ($\sim 90\%$) for the pump (1536 *nm*) and the Raman (1670 *nm*) wavelengths while the other one leaves uncoated presenting a reflectivity $\sim 30\%$ for both wavelengths. The

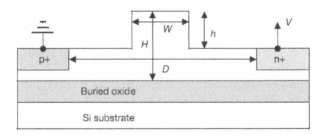

FIGURE 7.10
Diagram of the cross section of a silicon-on-insulator (SOI) rib waveguide with
a reverse biased p-i-n diode structure as it is presented in [92].

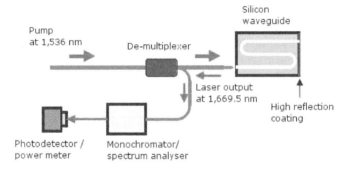

FIGURE 7.11
Experimental set-up of a silicon Raman laser integrated on a silicon chip [92].

spectra of this device presents a very narrow peak, which means high quality
factors (Figure 7.12).

The Raman scattering process presents an important source of losses, the
two-photon absorption (TPA), which produces a considerable amount of free
carries. To reduce this effect, researchers have included the p-i-n diode structure. Under an applied electric field, the pairs electron-hole generated by TPA
are guided to the p- or n- doped regions, reducing the losses generated by them
in the silicon waveguide and increasing strongly the output power of the system. This scheme presents several advantages with respect to other similar
devices like that exposed before [15]. The first one and more important is
that it is designed for its integration on a chip, then the materials and the
experimental techniques are CMOS compatible. Also, the laser characteristics
are quite interesting with a gain ~ 5.2 dB for a threshold ~ 0.4 mW and
a slope efficiency of 9.4%. These results correspond to the output power of

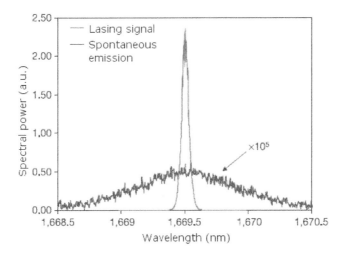

FIGURE 7.12
Silicon Raman laser spectra for a pump power of 0.7 mW compared with the spontaneous emission from a similar waveguide without cavity [92].

the uncoated facet of the waveguide. If both facets are considered, the slope efficiency can be increased to 10%.

Semiconductor Sources

The main disadvantages of the emitting cavities is that they have to be pumped with other optical signal. Therefore, an optical fiber must be coupled to the system. Some of the Raman lasers, explained before, are electrically pumped and this solves the problem of having an optical source. However, the optimized optoelectronic systems transforming an electric signal to an optical one are the *semiconductor sources of photons*. These devices are based in a p-n junction with a direct-gap semiconductor in the middle (see Figure 7.13) emitting photons due to electron-hole recombination. The generated photons can be reabsorbed producing a new electron-hole pair (nonradiative recombination) or be radiated (radiative recombination) (see Figure 7.14). Usually, an external potential is applied to the p-n junction to produce a population inversion in the semiconductor with a high amount of pairs electron-hole. This is the physical explanation of a semiconductor source [87], that includes light emitting diodes (LEDs), semiconductor optical amplifiers, and semiconductor injection lasers. All of them are very efficient transformers of electric energy into optical ones. Also, their small sizes, realizability, and compatibility with other electronics devices produce that they are ideal for several electronic applications.

LEDs are quite simple and cheap systems. Their common characteristics with other semiconductor devices make them good candidates for output

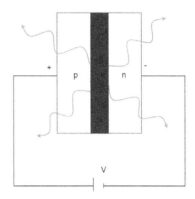

FIGURE 7.13
Basic scheme of pn junction.

sources in optical interconnects. However, they present, at least, two substantial problems that must be weighed carefully for each application. The first one is related to the fact that the output signal of a LED is incoherent and not-directional light, in contrast to a laser. Then, problems arise due to the collection of light. In addition, emitting diodes are usually limited by the carrier recombination times, which also limit the operation speeds of the source. For this reason, LED can be used for optical interconnections with harsh demands.

Despite the utility of each of these devices for a intra- or inter-chip communications, Vertical Cavity Surface Emitting Lasers (VCSELs) are probably the most adequate. The interconnection efficiency (the relation between the current at the receiver and the electric power dissipation of the transmitter), of these lasers is around 0.5 while that for LEDs or other kinds of semiconductor sources is poorer, around 0.05 and 0.1 [82].

VCSELs are classified as one kind of semiconductor injection laser. It consists of a quantum well, typically made of InGaAs, that acts as active region. As the dimensions of this region are very small (~ 70 Å), it is necessary that light goes through it several times. To achieve that, a stack of high-reflectivity mirrors are positioned at the top and bottom parts of the active region, as it can be observed in Figure 7.15. The mirrors have a double function: first, reflecting the photons in order to enlarge their path through the active layer and second, they contribute to the distribution of the electron-holes pairs because the top stack is p-doped and the bottom one is n-doped. In general, they are made of slides of width $w = \lambda/4$ and combining materials with high-contrast refractive indexes. This kind of arrangement is called Distributed Bragg Reflector (DBR) [114]. The advantages that these lasers present are quite interesting for electronic devices:

1. Operation with ultralow power consumption

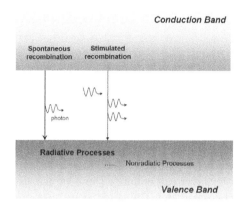

FIGURE 7.14
Diagram of the recombination processes in a semiconductor material.

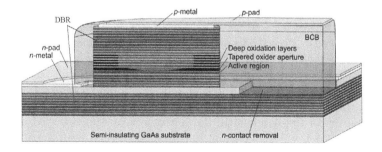

FIGURE 7.15
Schematic cross section of a VCSEL obtained from [19].

2. Circular and symmetric beam

3. Low threshold current (<6 mA)

4. Low temperature dependence

5. Narrow beam divergence

6. High-density $2D$ arrays

The implementation of VCSELs as an optical source for optical communications is very common due to their characteristics and that they are cost-effective [117, 93, 59]. This growing interest in VCSELs has produced an important improvement in their operation parameters. In the last years, vertical-cavity-surface-emitting lasers that operate at data rates of 30 Gb/s or even up to 40 Gb/s have been reported. The careful design of the DBRs and the aperture of the active layer can enhance these values, as well to improve other parameters as the bandwidth or the power consumption. As an example, Y.C. Chang and L.A. Coldren described recently [19] a high-efficiency and high-speed VCSEL operating at 850 nm and whose structure is compatible with actual manufacturing processes. This device is bottom-emitting and allows a direct integration on silicon electronics without the use of bond wires that reduce losses. The system is made up on a semi-insulating GaAs layer by molecular beam epitaxy followed by a 14-period undoped AlAs/GaAs DBR. The n-contact is composed of a five-quarter wavelength thick n-GaAs layer with another 4-period n-type $Al_{0.9}Ga_{0.1}As/GaAs$ DBR above it. The active layer has three InGaAs/GaAs quantum wells (QWs) embedded in $Al_{0.3}Ga_{0.7}As$ separate confinement heterostructure (SCH) layer. On top of it, there is the oxide aperture, that in this case is optimized to achieve low-power dissipation and a high bandwidth. It has an increased thickness from $\lambda/4$ to $\lambda/2$ with a 4 μm length. Finally, to complete the device, a 30-period AlGaAs/GaAs DBR and a p-contact are located on it, as can be seen in Figure7.15.

The previously described source is only 3 μm in diameter and it lases at a 850 nm wavelength. It could operate at 35 Gb/s with only a 10 mW power dissipation, which means a data-rate/power-dissipation ratio of 3.5 $Gbps/mW$. The system shows also a low threshold current of only 0.144 mA and a high efficiency of 0.67 W/A, which corresponds to a quantum efficiency of 54%. With respect to the bandwidth, the authors have demonstrated a 15 GHz bandwidth at a 1 mA bias current but it could be improved up to 20 GHz.

7.2.2.2 Out-of-Chip Light Sources: Modulators

For OI interconnects where power requirements are critical, it is possible to reduce the total power consumption by removing the light source from the on-chip system. This can be seen as an important advantage, but the exclusion of the optical source makes necessary extra optics to couple the light into the chip. The data encoding on the light is made, in these systems, by a modulator

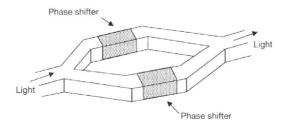

FIGURE 7.16
Schematic diagram of an asymmetric Mach-Zhender interferometer including a phaser shifter in the two arms [65].

that is also an optical device and is integrated on-chip. The main parameter considered in the modulation tasks is the variation of the effective refractive index (Δn_{eff}) in the active area (or the phase shifter) of the modulator. At this point, two kinds of modulators can be differentiated: *Mach-Zhender interferometer-based* (MZI) modulators and *micro-ring resonator-based* modulators.

Mach-Zhender Interferometer Modulators

This kind of device uses phase shifters located in both arms of a Mach-Zhender interferometer geometry [18], as it is shown in Figure 7.16, where the input beam is split and then joined using multimode interference (MMI) couplers to transform a phase modulation into an intensity modulation. The phase-shift ($\Delta\varphi$) produced in the active region is consequence of a change in the effective refraction index (Δn_{eff}) of the area in such a way that $\Delta\varphi$ is directly proportional to Δn_{eff} and the length of the phase-shifter (L) and inversely proportional to the incident wavelength [65]

$$\Delta\varphi = \frac{2\pi\Delta n_{eff}L}{\lambda} \tag{7.2}$$

There are many commercial high-speed optical modulators based on electro-optical materials or in III–V semiconductors [79, 112]. These devices have reported operation speeds of 40 Gb/s, much higher that those achieved by a silicon modulator. Although the weak electro-optical effects on silicon [107] limit the modulation capability of it, this section is focused on silicon modulators due to its potential in optical communication and its compatibility with the CMOS process.

Modulation in silicon is possible through the free carrier plasma dispersion effect [65]. Any change in the free carrier density produces a change in the refractive index of silicon and then a phase-shift can be induced. There are three possible configurations for these devices depending on the free-carrier manipulation: a) forward biased p-i-n diode (carrier injection), b) metal-oxide-semiconductor (MOS) capacitor, and c) reverse biased pn junction (carrier

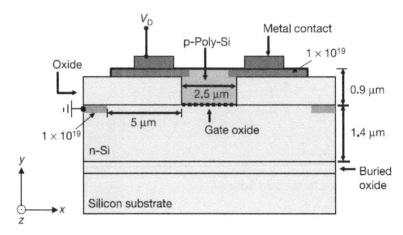

FIGURE 7.17
Scheme of the cross section of the MOS capacitor proposed by A. Liu and coworkers in [65]. The gate oxide thickness is 120 Å. The polysilicon rib and gate oxide widths are both ~ 2.5 μm. The doping concentrations of the n-typed and p-type polysilicon are ~ 1.7×10^{16} cm^{-3} and ~ 3×10^{16} cm^{-3}, respectively.

depletion). The first configuration is strongly speed limited. Although p-i-n diodes provide large index changes, then high modulation efficiency, their speed is limited by the carrier lifetime with values of few gigahertz unless complicated circuits are employed. For this reason, only the two other cases are discussed.

In Figure 7.17, schematic cross section of a MOS capacitor phaser shifter is shown as it was proposed first by A. Liu and coworkers [65]. It consists of an n-type doped crystalline silicon slab and a p-type doped polysilicon rib with a gate oxide between them. The structure (extra p-poly-Si layers, oxide regions) and their dimensions were briefly optimized in order to reduce the loss sources and it is completely manufactured using existing CMOS techniques. The proposed structure contains a single mode in the near-infrared part of the spectrum ($\lambda \sim 1.5$ μm) and its response strongly depends on the polarization with seven times larger values for a TE than for a TM polarization.

While the n-type Si layer is grounded, a driven voltage V_D can be applied to the p-type poly-Si region producing the accumulation of charge at both sides of the oxide gate. This charge-density change, either for electrons (ΔN_e) and for holes (ΔN_h), is directly proportional to V_D and inversely proportional to the thickness of the oxide gate (t_{ox}) through the equation [65]

$$\Delta N_e = \Delta N_h = \frac{\epsilon_0 \epsilon_r}{e t_{ox} t}[V_D - V_{FB}] \tag{7.3}$$

ϵ_0 and ϵ_r being the vacuum and low-frequency relative permittivity, respec-

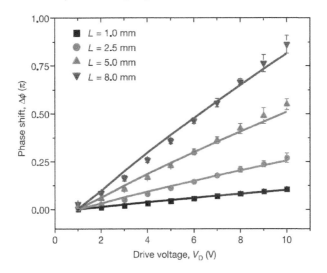

FIGURE 7.18
Phase shift produced by a MOS capacitor as a function of the driven voltage (V_D) and for several lengths at $\lambda = 1550\ nm$ (symbols represent experimental results while lines correspond to simulated values). Figure obtained from [65].

tively, e the electron charge, t the effective charge layer thickness, and V_{FB} the flat-band voltage. This change in the charge density with the presence of free carriers produces changes in the refractive index, that will be higher as higher driven voltage. And consequently, changes in the effective refractive index of the material induce a phase shift through Equation (7.2). In Figure 7.18, the phase shift is plotted as a function of the driven voltage for different capacitor lengths showing the previous dependencies.

The device proposed initially by A. Liu et al. [65] presented a 3-dB bandwidth exceeding 1 GHz for a only 2.5-*mm-long* capacitor and a data rate that is up to 1 Gb/s. Another interesting figure of merit of these devices is the product $V_\pi L = (V_D - V_{FB})L$ whose value is ~ 8 *V-cm* for this model. Actually, these values can be easily improved with a refined optimization of the capacitor parameters, as the authors recognized. In particular, the doped and undoped poly-Si area present high losses that are the main source of losses in the chip, with values $\sim 6.7\ dB$. An improved version of the capacitor was developed in [64] using crystalline Si for both the n-type and p-type regions, which is significantly less lossy, using for its growth a technique called epitaxial lateral overgrowth (ELO). Also, the doping concentrations are influent in the response of the device; because of that, this enhanced version also presents higher concentrations of dopers. The overall device is 15 mm long but with a phase shifter of only $1.6 \times 1.6\ \mu m^2$. This reduction of the shifter length gets better the characteristics with a $V_\pi L$ parameter of only 3.3 *V-cm*, a 3-dB bandwidth around 10 GHz and a data rate of 10 Gb/s, which is an im-

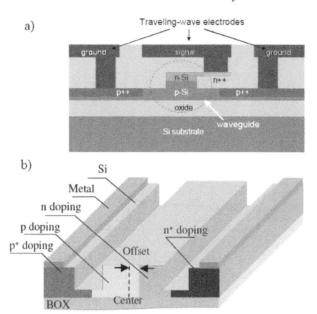

FIGURE 7.19
Schematic cross section of the phase shifter based on carrier depletion proposed by (a) A. Liu et al. [66] and (b) N.N. Feng et al. [33].

portant improvement with respect to previous MZI modulators using a MOS capacitor.

Despite these results, the analyzed device can be upgraded with a deeper research in their parameters. These can be optimized to improve the bandwidth, losses, and mainly the operation speed, bringing these modulators to the characteristics showed by other no-silicon modulators.

Other possible configuration for a MZI-based optical modulator is that in which the effective refractive index change is obtained through a free carrier depletion. This modification of the free carrier density is achieved using a reverse biased pn junction located at both arms of the MZI structure (see Figure 7.16). Then, the main structure is quite similar to the previous configuration with the unique difference in the phase shifter. Two different conformations for the depletion-based shifters are shown in Figure 7.19.

For a reverse biased pn diode, the carriers density is decreased in an area with a width (W_D) depending on both the doping concentrations and the driven voltage (V_D). A change in W_D means a change in the free carrier density and then a phase shift, as before. However, while in the MOS capacitor the charge density depends on the applied voltage linearly (see Equation (7.3)), in this case the dependence follows the relation

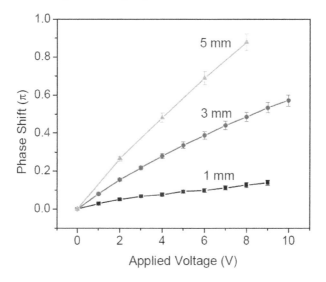

FIGURE 7.20
Phase shift as a function of the driven voltage for an individual shifter similar to that explained in [66] for different device lengths and an incident wavelength of $\lambda = 1550 \ nm$.

$$W_D = \left(\frac{2\epsilon_0 \epsilon r (V_{Bi} + V_D)}{eN_A} \right)^{1/2} \tag{7.4}$$

ϵ_0 and ϵ_r being the vacuum and low-frequency relative permittivity, respectively, e the electron charge, N_A the acceptor concentration, and V_{Bi} the built-in voltage. As can be seen, as the depletion width and then the free charge density change with the square root of the applied voltage or driven voltage (to be consistent with the previous configuration), the phase shift changes in a nonlinear way with W_D, unlike what happened in a MOS modulator. In Figure 7.20 the phase shift versus the applied voltage is plotted to show the nonlinear dependency that follows the relation (7.4). Although this figure corresponds to the device proposed in [66], similar curves are observed for other devices as those presented in [33]. Usually, W_D is much smaller than the waveguide height in which the shifter is integrated. For this reason, the pn junction position is an important parameter that can optimize the phase modulation.

The phase shift produced in the device depends also on other parameters. The following expression shows the main dependencies of the change in the phase induced in a carrier-depletion shifter [33]

$$\Delta\varphi = \frac{2\pi\Delta\Gamma\Delta nL}{\lambda} \tag{7.5}$$

where $\Delta\Gamma$ is the optical confinement change, Δn is the effective refractive index change, both induced for the free charge density change, L is the phase shifter length, and λ is the incident wavelength, as usual. It is important to remark that any change of the optical confinement is proportional to a depletion width change. From this expression, it can be concluded that the phase modulation can be optimized in two main ways: increasing $\Delta\Gamma$ (that can be achieved either with a smaller waveguide or optimizing the pn junction) or increasing the change of the refractive index, Δn (that is possible increasing the doping concentrations, however it also implies an increment of the losses).

A typical geometrical configuration of depletion-based MZI-based modulator is like those shown in Figure 7.19 composed by a p-typed doped silicon rib waveguide with a n-type doped silicon cap layer forming the pn junction. The dimension of these and their doping concentrations can be chosen carefully in order to enhance the modulation efficiency without damage for the global size and losses. However, it is usual that the n-doping concentration is higher than the p-doping one because, as it was demonstrated, a change in the hole density produces a larger effective refractive index change according to Kramers-Kroning analysis [107]. To ensure a proper ohmic contact, a heavily doped region is located under the metal contacts, as can be seen in Figure 7.19. Another important parameter is the distance between the buried oxide layer and the waveguide being important in the overlap between the depletion area and the optical mode.

The main characteristics of the two MZI-based modulators using a reverse biased pn junction discussed in this section are summarized in the next table (Table 7.1). As can be seen, the main parameters, as size or losses, of the most recent devices are prioritized against others, as the data rate.

TABLE 7.1
Main characteristics of the MZI-based modulators presented in references [66] and [33]

	A. Liu et al. [66]	N.-N. Feng et al. [33]
Waveguide dimensions (height × width)	$0.5 \times 0.6 \ \mu m$	$0.25 \times 0.5 \ \mu m$
Typical Phase Shifter length (L)	$5 \ mm$	$1 \ mm$
p-doped concentration	$1.5 \cdot 10^{17} \ cm^{-3}$	$5 \cdot 10^{17} \ cm^{-3}$
n-doped concentration	$3 \cdot 10^{18} \ cm^{-3}$ at the top to $1.5 \cdot 10^{17} \ cm^{-3}$ near the junction	$1 \cdot 10^{18} \ cm^{-3}$
Buried oxide thickness	$4 \ \mu m$	$3 \ \mu m$
$V_\pi L$	$4V - cm$	$1.4V - cm$
3-dB bandwidth	$\sim 20 \ GHz$	$\sim 12 \ GHz \ (L = 1mm)$ and $\sim 30 \ GHz \ (L = 0.25 \ mm)$
Data rate	$\sim 30 \ Gb/s$	$\sim 12.5 \ Gb/s$
On-chip Losses	$\sim 7 \ dB$	$\sim 2.5 \ dB$

In the recent years, optical modulators based on a reverse biased pn junction that also use Raman scattering effects have been designed [48]. However,

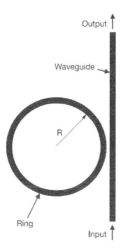

FIGURE 7.21
Scheme of the structure of a micro-ring resonator-based modulator similar to that analyzed in [2].

their characteristics are not competitive yet except for the losses that are quite small ($\sim 0.5 - 2\ dB$).

Micro-ring Resonator Modulators

As can be seen in Table 7.1, the MZI modulator dimensions are still high for several on-chip optical communications. This is one of the parameters that have to be improved. One of the ways to improve it is using high-confinement fields in the area where the refractive index is modified. The confinement of the light produces that the sensitivity of light to small changes of the refractive index (Δn) enhances significantly without requiring large devices.

Micro-ring resonator (MRR) based modulators use the resonant condition of a ring waveguide to modulate the signal traveling along an adjacent waveguide. A scheme of this structure is shown in Figure 7.21. A ring waveguide possesses, due to their geometry, a morphology-dependent resonance (MDR) that follows the condition

$$m\lambda = 2\pi R \times n_{eff} \tag{7.6}$$

where m is the resonant mode, R is the radius of the ring, n_{eff} its refractive index, and λ is the wavelength of the incoming optical beam. When the optical signal traveling through the adjacent waveguide fulfills the resonant condition, its light is coupled to the ring resonator and a minimum in the transmission of the adjacent waveguide is reached. If, however, the incoming light does not satisfy Equation (7.6), the beam passes through the waveguide

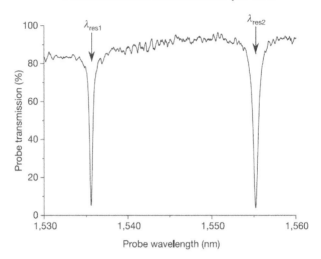

FIGURE 7.22
Quasi-TM transmission spectrum of a single-couple ring resonator [2].

without coupling into the ring. In Figure 7.22, the transmission of the adjacent waveguide to a micro-ring resonator reported in [2] is shown. As can be seen, when the incident wavelength fulfills Equation (7.6), light transmission through the adjacent waveguide drops strongly.

Modifying the resonant condition of the micro-ring, the signal at the end of the adjacent waveguide can be intensity-modulated. Seeing Equation (7.6), it is easy to conclude that by tuning the effective refractive index of the ring waveguide (n_{eff}), the resonant wavelength is also tuned and the transmitted light is modulated. The resonant behavior is quite sensible to changes in n_{eff}, then a small change in n ($\sim 10^{-3}$) can produce a modulation depth ($MD = (I_{max} - I_{min})/I_{max}$) of $\sim 80\%$.

The adjustment of n_{eff} can be made by injecting free carriers. This task can be accomplished optically using a pump beam as in [2] or electrically using a p-i-n junction as in [120, 119]. The pump beam, in the first case, could be very low-power ($\sim 25\ pJ$) because low energies are necessary to induce a free-carrier concentration of $\Delta N = \Delta P = 1.6 \times 10^{17} cm^{-3}$, which in turn produces a $\Delta n_{eff} = -4.8 \times 10^{-4}$ that shifts the resonant peak. However, the inclusion of a pump beam involves more complex optics. On the contrary, an electro-optical modulator that uses a p-i-n junction embedded in the ring resonator could be less complex than the previous one. This device (see Figure 7.23) has the same geometry of that shown in Figure 7.21 with the difference that neighbor $n+$ and $p+$ regions are defined with photolithography and implanted with boron and phosphorus, as was described by Q. Xu et al. in [120]. This configuration is quite similar to the forward-biased p-i-n junction MZI modulators discussed above. As it has been remarked, this kind of configuration produces high

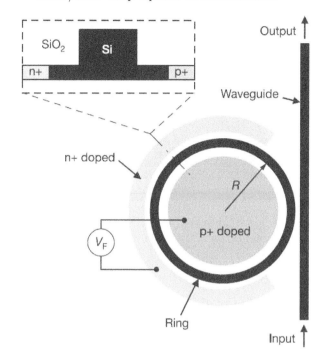

FIGURE 7.23

Scheme of the electro-optic modulator described in [120].

modulation depths but low speeds. However, the resonant behavior of the recent devices overcomes this challenge eliminating any speed restriction.

The main advantage of these devices is their size. In both discussed cases [2, 120], the ring waveguide is made with a rectangular silicon waveguide of 250-*nm-high* and 450-*nm-wide* forming a ring with a radius of 5–6 *μm* and located at 200 *nm* from the adjacent waveguide. A SEM-image of the device in [120] is shown in Figure 7.24. Then, its integration in on-chip OI is possible, in particular for WDM interconnects.

With this kind of device, high-depth modulation was reported at operation speeds of 1.5 *Gb/s*, and even 12.5 *Gb/s* has been achieved with an improved device [119]. Also, losses are quite small in these devices with $4 \pm 1 \ dB/cm$. However, they present two important disadvantages: i) they operate over very narrow bandwidths due to the resonant condition, and ii) they are quite dependent on the temperature. These items involve two challenges that should be studied in detail to be solved.

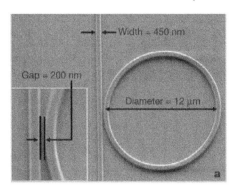

FIGURE 7.24

SEM image of the device fabricated by Q. Xu et al [120]. In the inset, a zoom of the coupling region is shown.

7.2.3 Optical Channels

The primary requirement of an optical interconnection (OI) for its consideration as a possible counterpart of electric interconnections (EI) is the minimization of the propagation delay. This fact is made possible by the higher propagation speed of light through an optical channel than that of electrons through a metallic wire. Optical fibers are probably the most common and widespread light propagation medium. They are used for several years for long-distance communications. However, for chip scales, fibers are not a good candidate and other alternatives have to be envisioned.

In a recent study, L. Yan and G.W. Hanson described, in detail, the different mechanisms of wave propagation that can be observed in on-chip communications [123]. In this work, the authors considered a dipolar antenna located on the silicon-based layer of a multilayer chip structure following previous analysis [51, 22]. In it, they stated the existence of three different wave components associated with as many propagation paths. The first one is related with the radiation modes of the multilayer including the propagated and non-propagated modes and it forms the *continuous part of the spectrum.* In this continuous part it can be also included the so-called *direct space-wave radiation* component that appears when there is a direct line-of-sight from the emitter to the receiver. However, the importance of its presence justifies its differentiation. The last component is *the discrete part of the spectrum* that is directly related with the surface modes that are guided on the layer. For the far-field, that is for large distances between the emitter and the receiver, the typical exponential attenuation produces that the surface waves can be neglected and the continuous part of the spectrum dominates as expenses of the discrete one. However, for intra- and inter-chip communications distances, it must be in the near- or intermediate-field range, in which surface modes are probably the better way for an enhanced transmission in a large range of frequencies. These modes can be suppressed or, what is more important, enhanced by the inclusion of an extra layer below or above the silicon layer, as was discussed by the authors and in other works included in it as references [123].

Upon consideration of the results related to wave components and the dominant position of surface waves in the propagation methods of waves on a chip, it is easy to see that the principal and natural counterpart of metallic wires in OI are waveguides that run on the silicon layers and are based on surface modes. Some kinds of waveguides and their implementation on chips will be discussed below. However, and as it will be seen, they present important limitations and disadvantages for certain applications. For this reason, other optical channels have been developed as the free-space configurations, also studied in here.

7.2.3.1 In-Waveguide Configurations

A rough explanation of the concept of waveguide is that these components are a structure with a given geometry that produces modes in which light is confined and where it can propagate. The simplest structure capable of presenting a guiding mode is an elementary structure that consists of two layers, a high-resistivity one above and a low-resistivity below, with a large contrast in refractive indexes between them. The modes are directly related to the geometry while the propagation properties strongly depends on the refractive index of the waveguide material and its surroundings. Typically, these optical paths for on-chip applications are made of silicon or a polymer. A summary of the general characteristics of waveguides with influence on their optical response and their typical values are included in Table 7.2 as an example. Each material presents several advantages and disadvantages. For example, while the polymer refractive index is smaller than that of silicon (that translates into higher propagation speed) its pitch is higher, reducing the bandwidth density. Also, while modulators for silicon waveguides are very developed, this is not the case of modulators for polymer waveguides. Because of that, they are feasible for systems with VCSELs as sources, where light is directly modulated by the laser. For this reason, the election of the material will be coupled with the specific characteristics of each application.

The use of waveguides leads inherent losses in the energy of the propagated wave. These losses can be divided into three different components. The first one (P_{loss}) is waveguide propagation losses that is directly related with the properties and design on the waveguide and it depends on the length of the connection. This is the only one that will be noted in what follows. The second one is the bending loss (B_{loss}) that is related with the bending angle (θ) of the waveguide and the radius of the bend (r). And the last one is the coupling loss, C_{loss}, that appears when light through the waveguide should be coupled to other optical elements as the emitter or the detector, for instance.

Mathematical expression for these dependencies were reported by D. Ding and D. Pan in reference [29].

TABLE 7.2

General characteristics for silicon and polymer waveguides as it was shown in [54]

Waveguide Materials	*Si*	*Polymer*
Refractive index	3.5	1.5
Width (μm)	0.5	5
Separation (μm)	5	20
Pitch (μm)	5.5	25
Loss (dB/cm)	1.3	1

Silicon Waveguides

Silicon waveguides are a very attractive option for optical lines in on-chip communications due to their simple fabrication by chemical or plasma etching, the low-absorption of silicon in the wavelength range 1.3–1.5 μm and/or the facility to manipulate its refractive index, and then the propagation modes, through free carrier injection/generation. As for many cases in this chapter, there are many different configurations and techniques for the fabrication of silicon waveguides [46]. Here, for space requirements, only two of them can be briefly discussed: silicon on silicon dioxide [100] and doped silica-based waveguides [91]. However, other alternatives are still valid as, for example, silicon on sapphire (SOS) waveguides.

Silicon on silicon dioxide waveguides are also known as silicon-on-insulator (SOI) waveguides and are made of a crystalline silicon on an oxide layer, ensuring a fair guiding medium. Recent silicon technologies allow to obtain high-quality SOI wafers that are ideal for the implementation of planar waveguides. The very different refractive indexes of silicon ($n = 3.45$) and silica ($n = 1.45$) ensure an important light confinement. Although this large index contrast sometimes induces nonlinear optical effects (Raman scattering, Kern effects) that can be useful for amplifier applications (amplifier, silicon laser), it is also an important source of losses.

One of the main problems associated with the design of SOI waveguides is related to the election of the thickness of the silica cladding. The core of the waveguide made of silicon must be isolated to minimize the crosstalk with other substrates. SiO_2 thicknesses larger than a few nanometers are incompatible with the CMOS layer [100]. For this reason, new techniques devoted to optimize it have to be developed. Another fabrication challenge is related with the integration of optics with electronics and the temperatures at which optical elements are subjected to during the CMOS process. Usually, temperatures higher than $1000°C$ are achieved in CMOS techniques, but this temperature can damage optical elements. To overcome these challenges, recently, N. Sherwood-Droz and coworkers have developed a new method based on usual CMOS techniques for waveguide formation and integration [100].

SOI waveguides usually have a rectangular geometry and are built through micro-lithography, or a negative-photo method, and then etching. The authors of reference [100] proposed the use of thermal oxidation to generate the necessary silica layer beneath the waveguide but protecting the original crystalline silicon with a three-sided cap of a very high oxidation temperature and compatible material, Si_3N_4. This technique is similar to other known CMOS techniques as LOCOS or SWAMI. The complete fabrication process, described in Figure 7.25, starts with a crystalline silicon wafer (Figure 7.25 (a)) that is covered with a Si_3N_4 layer using a plasma-enhance or a low-pressure chemical vapor deposition (PECVD or LPCVD) (Figure 7.25 (b)). On it, the waveguide structure is defined using a micro-lithographic technique, typically it is a rectangular strip 450 nm wide (Figure 7.25 (c)). A Induced-Coupled Plasma (ICP) is used for etching the structure vertically forming tall pillars (Figure 7.25 (d)). A new protection of Si_3N_4 is, then, deposited and etched in the same amount (Figure 7.25 (e) and (f)) in such a way that the pillar presents finally the same amount of nitrade in its three walls. At this point, the authors include a new etch step in order to reduce the silicon substrate, decreasing the oxidation time and also the necessary spacing between adjacent waveguides, which enhances the bandwidth density (Figure 7.25 (g)). The oxidation process to create the silica layer is made in a wet oxidation chamber at high temperature ($T \sim 1100°C$) (Figure 7.25 (h)). This process is extended in order to obtain the desired planar shape at the bottom part (Figure 7.25 (i)). Finally, a final clad of 3 μm PECVD SiO_2 completes the buffer (Figure 7.25 (j)).

The final result is a 740 nm in width and 245 nm in height SOI waveguide that presents a guiding mode at $\lambda = 1550$ nm with 2.9 dB/cm losses and is polarization independent. The buried oxide (SiO_3) thickness was estimated in 3.3 μm ensuring radiation losses to the substrate < 0.001 dB/cm. The unique lithographic step improves the reproducibility of the device but the process can be even improved with a better etching process, for example.

Although the process is long, the system is quite simple. However, sometimes lower losses are necessary and then more complex waveguides, what involves more integration problems, are necessary. An example of it is the case of the Er^{3+}-doped silica-based waveguides described in [91] where losses for a 1550-nm-mode do not exceed 1 dB/cm for a guide made by RF-sputtering and they have similar losses to the previous case when it is made by ion-exchange ($\sim 3 - 4$ dB/cm). Modes at lower wavelengths ($\lambda \sim 630$ nm) can also be supported by these devices with similar losses for the first technique and only 0.5 dB/cm for ion-exchange waveguides. For modes in these low wavelengths, another technique is available for doped silica waveguides that is called Sol-gel. Under this technique a Er-doped planar waveguide can be obtained with one mode at 632.8 nm with a confinement factor of 0.85 and 0.5 dB/cm losses.

The two main sources of propagation losses in silicon waveguides are the

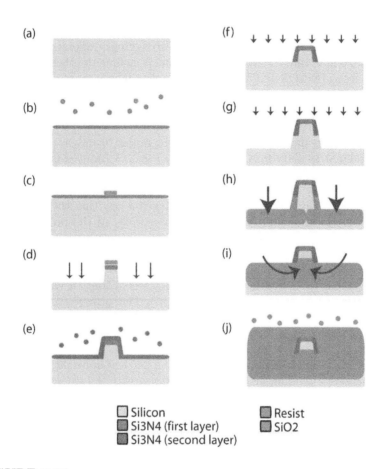

FIGURE 7.25
Fabrication process, step by step, of a SOI waveguide as it was proposed in
[100]: (a) bulk silicon wafer, (b) Si_3N_4 deposition, (c) lithographic waveguide
definition, (d) ICP etch, (e) Si_3N_4 deposition, (f) ICP cap etch, (g) extended
etch for quicker oxidation, (h) wet oxidation for buffer layer growth, (i) extended wet oxidation for waveguide underlayer flattening, and (j) upper layer
oxide deposition to complete optical buffer.

generation of free carriers due to two-photon absorption processes (TPA) and the scattering of photons due to the roughness of the waveguide surface. This second source is directly related to the fabrication process and can be reduced using thermal oxidation with large times and high temperatures [46]. The first one is more complicated to reduce because it is an inherent process associated to silicon properties. However, it is less crucial than the other sources of loss except for the design of silicon amplifiers [63].

Polymer Waveguides

Polymers are versatile materials whose use for optical devices is quite interesting. Their good properties (thermal, mechanical or environmental stability), the choice among a large number of materials and, what it is more important, their low material and production costs make them an excellent alternative for the fabrication of single-mode or multimode planar waveguides. They can be made of polymers, oligomers, monomers, thermoplastic, or thermosets. This wide range of materials along with the number of different manufacturing processes and/or manipulation of them involves an almost complete tuning of the polymer-based waveguide properties. For instance, polymer-based waveguides can present a variable refractive index along their length, those are the so-called graded-index waveguides described in [58]. Also, they are compatible with several substrates and, in particular, with chip substrates, even if they are either rigid or flexible. The main conclusion of this entire dissertation could be the fact that polymer materials permit a mass production of photonic circuits with low cost, high-quality properties and also a high-ruggedness.

In Table 7.3, there is a summary of the main commercial polymers waveguides and their optical properties. Also other polymers are used to obtain high-quality waveguides as, for example, $SU-8$ that was used to fabricate a 10 Gbps multi-mode optical guide in [21]. From the table, the low values of optical losses can be noted. In general, the combination of different materials make polymers to provide low-loss guides and almost be polarization independent (losses for TE and TM polarizations are very similar). In addition, the wide range of manufacturing techniques (RIE, photolithography, etc.) allows better matching between the characteristics of the device and the given requirements for a certain application.

The most common fabrication technique is photolithography through the use of masks in a multistep process. However, laser direct printing, not widespread, presents several advantages from respect to photolithography. First, it is a rapid process that does not need the previous steps as the design and manufacture of a mask. This process does not affect areas of the sample different from the one where the waveguide is patterned. And finally, it allows to print new structures, impossible to obtain through the use of photolithography masks.

One important disadvantage of polymers for guiding applications is the temperature dependence of their refractive index. This dependence is usually around $\Delta n \sim -2 \cdot 10^{-4}, -3 \cdot 10^{-4}$ per $°C$. This characteristic could be use-

TABLE 7.3
Key properties of optical polymers developed worldwide. Obtained from [30]

Manufacturer	Polymer Type [Trade Name]	Patterning Techniques	Optical Loss (dB/cm) [at λ (nm)]	Other properties [at λ (nm)]
AlliedSignal	Acrylate	Photoexposure/wet etch, RIE, laser ablation	0.02[840] 0.2[1300] 0.5[1550]	Birefringence: 0.0002[1550] Crosslinked, T_g : 25°C Environmentally stable
	Halogenated Acrylate	Photoexposure/wet etch, RIE, laser ablation	< 0.1[840] 0.03[1300] 0.07[1550]	Birefringence: 10^{-5}[1550] Crosslinked, T_g : −50°C Environmentally stable
Amoco	Fluorinated Polyimide [Ultradel^{TM}]	Photoexposure/wet etch	0.4[1300] 1.0[1550]	Birefringence: 0.025 Crosslinked, Thermally stable
Dow Chemical	Benzocyclobutene [Cyclotene^{TM}]	RIE	0.8[1300] 1.5[1550]	T_g :> 350°C
	Perfluorocyclobutene [XU35121]	Photoexposure/wet etch	0.25[1300] 0.25[1550]	T_g : 400°C
DuPont	Acrylate [Polyguide^{TM}]	Photolocking	0.18[800] 0.2[1300] 0.6[1550]	Laminated sheets Excimer laser machinable
General Electric	Polyetherimide [Ultem^{TM}]	RIE, laser ablation	0.24[830]	Thermally stable
Hoechst Celanese	PMMA copolymer [P2ANS]	Photobleaching	1.0[1300]	NLO polymer
JDS Uniphase Photonics	[BeamBox^{TM}]	RIE	0.6[1550]	Thermally stable
NTT	Halogenated Acrylate	RIE	0.02[830] 0.07[1310] 1.7[1550]	Birefringence:6·10^{-5}[1310] T_g : 110°C
	Deuterated Polysiloxane	RIE	0.17[1310] 0.43[1550]	Environmentally stable
	Fluorinated Polyimide	RIE	TE : 0.3 TM : 0.7[1310]	Environmentally stable

ful for active optical devices, as filters, but for a waveguide is considered a handicap because thermal variations will involve changes in the propagation modes. Then, this is an important parameter to control during the design and fabrication process of a polymer-based waveguide for communications applications.

Metallic and Plasmonic Waveguides

Recently, waveguides made of metals have been also developed. One example of it is shown by R. Bicknell et al. in their work [13]. There, they proposed a hollow metallic waveguide that is made on silicon using different patterning techniques and then, metallized with silver. However, to ensure a good adhesion and protection of the metallic layer, the authors considered a three-layer structure formed AlN (350Å)/Ag(2000Å)/Ti(500Å). This metallic waveguide not only shows low losses, reaching values of $0.05\ dB/cm$ for certain cases, but it also presents data rates approaching to THz/cm due to its air-core. These characteristics make them an excellent candidate for optical interconnect applications in a near future.

In addition, probably, the most noticeable effect on metallic wires is the excitation of surface plasmon polaritons (SPPs). In the last years, some studies have been focused on the analysis of the potential applications of this effect in the design of optical waveguides for optical interconnects [27]. Unfortunately, their characteristics are not comparable to conventional channels or other typical optical channels yet. However, further research is being carried out and the first prototypes have already appeared [50] using gold strips with a dielectric polymer cladding at 1.3 μm.

7.2.3.2 Free-Space Optical Interconnect (FSOI)

As other devices, waveguides present several disadvantages, among which one is highlighted: the density of waveguides in complex systems increases the probability of crosses. These crosses and the short distances that appear between the waveguides produce high optical losses [93], high crosstalk, and an important increase of the complexity of the fabrication process with several layers. While global communications involve low-complexity routing and hence waveguides are adequate as optical channels for them, critical nets present high-complexity routing so they need other communication methods. In order to overcome these challenges, a new configuration was proposed inspired in wireless communications in the atmosphere, called *free-space optical interconnects (FSOI)*. The main characteristics of FSOI is the use of $3D$ integration techniques in order to eliminate the routing complexity allowing the signal paths to cross each other without interference.

The basic scheme of this configuration, shown in Figure 7.26, is a composition formed by an array of micro-lenses and/or micro-mirrors for collimation and redirection of light. In general, the operation sequence of these systems is as follows: light from a source is sent to a micro-lens that collimates it and

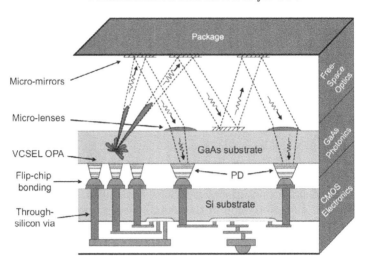

FIGURE 7.26
Scheme of a multilayer system integrating a free-space optical interconnect
[121].

guides the light to a micro-mirror. The micro-mirror changes light direction
towards another micro-lens that, finally, focuses the beam on a photodetector.
Commonly, this operation is made using arrays of lasers, lenses, mirrors, and
detectors [117]. VCSELs are the most common type of lasers used in FSOI
for their characteristics (light is directly modulated for the applied voltage,
size, facility to be integrated, cost, etc.). However, other types of sources have
been used for the design of free-space configurations as MZI-based modulators
[116].

   Each micro-lens or micro-mirror is usually made through multimask binary photolithography and reactive ion etching techniques upon silicon-based
substrates in order to ensure CMOS-compatibility. While micro-lens can be
antireflective treated with a layer of Si_3N_4, micro-mirrors have an extra metallic layer deposited by e-beam evaporation to act as a reflector [108]. To ensure
a correct alignment of the optical components two different kinds of alignment
methods are used: i) a visual adjustment of marks and ii) an imaging alignment method, both of them assisted by a mask aligner [47]. Misalignments
could be an important problem in this kind of system. For example, if the incident beam is not well centered on a microlens, the diffractive effects from the
lens borders can be nonnegligible, decreasing the signal-to-noise ratio. For this
reason, the design of a FSOI must be robust and take into account possible
misalignments due to thermal expansion of the materials or vibrations.

   As before, the free-space optical interconnects are characterized through
two important parameters: losses and crosstalk. However, other important
parameters must be controlled in these systems. For instance, the beam di-

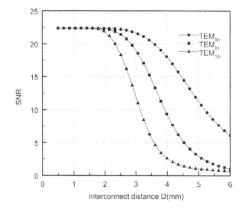

FIGURE 7.27
Evolution of the signal-to-noise ratio (SNR) as a function of the interconnect distance for a lensless FSOI. Obtained from [117].

vergence of the output light from the source is usually assured by the cluster of micro-lens. Introducing a micro-lens, the beam divergence can be corrected from 12° to only 1° [108]. However, under certain conditions the beam divergence is not a problem and the lens array only involves a more complex structure, more difficult manufacturing, higher space requirements, and higher cost. For these cases, the OI can be designed without a lens as proposed several years ago by R. Wang and coworkers [117]. Of course, this configuration is strongly limited by the interconnection distance, being valid only for $d < 10$ mm and without any complex routing. As an example, in Figure 7.27, the evolution of the signal-to-noise ratio (SNR) as a function of the distance is shown for a lensless FSOI. For the shown case, the source is a VCSEL with an aperture of $\phi_{laser} = 0.03$ mm in diameter, the corresponding photoreceptor has a diameter of $\phi_{PD} = 0.32$ mm and the channel spacing is $l = 0.40$ mm. As can be seen, the SNR depends strongly on the distance. But it also depends on the mode order of the input light in such a way that SNR is higher for short distances and low order of the mode. The order-dependence is directly related with the fact that for higher orders, the spatial-distribution of the beam is also wider.

On the contrary, if the given optical link has very restrictive conditions for SNR, losses, crosstalk, etc., the system can be complemented with other optical elements that improve its characteristics. For example, by including micro-prisms and a macro-mirror, M. McFadden and coauthors reported an increase of the channel density [70]. As was remarked above, this kind of OI was proposed inspired by the wireless communications for atmospherical or spatial applications. In these cases, adaptive optical systems should be necessary to correct the atmosphere fluctuations. Then, similar devices can be integrated on FSOI to increase the propagation quality, as was made by C. Henderson et al. in [43]. They used a ferroelectric liquid crystal on silicon spatial light modulator (LCOS-SLM) achieving communications at 1.25 Gb/s illuminating with a 850 nm light beam. However, these improvements involve high space and high power requirements and complex manufacturing techniques. For this reason, they can be observed only for very particular implementations.

Wavelength Division Multiplex (WDM) On-Chip

Multimode waveguides offer the advantange of being able to transport a complex signal formed of various wavelengths. This issue has been used for the implementation of the so-called Wavelength Division Multiplex (WDM). In WDM, multiple signals are modulated by different wavelength light beams and then transmitted through a unique multimode waveguide or via free-space propagation. WDM involves an important increase in terms of bandwidth and reduction of space requirements allowing a higher density of communications. Although there are several technological challenges that must be overcome until its real on-chip application, the last results about WDM show that it will be a feasible application for intra- and inter-chip communications in the near future [12].

7.2.4 Photodetectors

The optical receiver of an OI is composed of a light detector (photodetector, [PD]), an amplification circuit (Transimpedance Amplifier [TIA], voltage amplifier), and clock and data recovery. In this section only the photodetector will be analyzed because it is the only fully optical device.

Although there are some types of photodetectors, the most usual devices for interconnections are designed as a transducer from an optical signal to an electrical current with a p-i-n structure using a semiconductor on top of a Si substrate [9, 83]. The physical process in which these devices are based on is the generation of a pair electron-hole (eh) due to the absorption of a photon for the semiconductor material. Commonly, the semiconductor is also constrained to a potential difference that accelerates the charges, increasing the sensitivity of the detector. The semiconductor material is chosen depending on the wavelength range at which the device will work. The typical materials used in the implementation of photodiodes and the spectral ranges at which they present optimum absorption are shown in Table 7.4.

TABLE 7.4

Optimum spectral range of absorption for some semiconductors

Material	*Wavelength Range* (nm)
Silicon	*190-1100*
Germanium	*800-1700*
InGaAs	*800-2600*

For communications issues, it is usual to use germanium in the active region due its compatibility with silicon technologies [25]. Also it absorbs at the typical wavelengths ($\lambda = 1300$ and 1500 nm) where silicon is transparent. In spite of the lattice mismatch between Ge and Si in a directly grown and thin buffer structure, the defect densities ($10^7 - 10^8$ cm^{-2}) are sufficiently low for communications applications.

In order to integrate a photodetector into an optical interconnection, it must possess some important characteristics. Of course, it should be CMOS compatible, as usual, in order to allow its integration in the fabrication process of the chip. For the same reason, its size cannot be larger than a few microns or even around a few nanometers for new technologies. Also, they should be able to detect very low signals transforming them in a powerful electrical signal. And they have to make this conversion at high speed, up to 40 Gb/s. Finally, they must fulfill each of these conditions working at small voltages for not increasing the total power consumption. Some parameters are used to describe each of these items (quantum efficiency, responsivity, bandwidth, sensor area, etc.) and some others as the noise characteristics (dark current).

One example of an optical receiver, or more specifically, a normal incidence Ge-on-SOI (SiO_2) receiver is shown in [57] whose geometry is plotted in Figure 7.28. The use of the silicon oxide instead of bulk Si overcomes the problems derived from the absorption of Si at short wavelengths ($\lambda = 850$ nm). Also the presence of SiO_2 facilitates the formation of the Ge layer and decreases the diffusion of the carriers. As can be seen, the basic system consists of successive electrodes, alternating p^+ and n^+, on a grown Ge layer that is also on an ultrathin layer of SOI with widths of around a few hundreds of nanometers. The main characteristics of this type of detector are summarized in Table 7.5. One way to slightly increase the efficiency of the detector is an

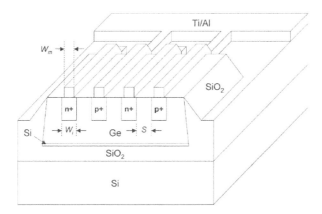

FIGURE 7.28
Scheme of a Ge-on-SOI photodectector with a normal incidence geometry as shown in [57].

antireflection-coating (ARC) layer that avoids the reflection of the incident photons and then increases the absorption of them by the semiconductor.

TABLE 7.5
Simulated characteristics for a Ge-on-SOI photodetector [57]. The labels mean: S (Finger spacing), V_B (Bias Voltage), QE (Quantum Efficiency), R (Responsivity), -3 dB-Band (-3 dB bandwidth), DC (Dark Current), and DA (Detector Area)

S (μm)	V_B (V)	**QE** λ=850 nm (%)	**R** λ=850 nm (A/W)	$3\ db - Band$ (GHz)	**DC** (nA)	**DA** (μm^2)
0.4	-0.5 -1	30	0.21	27 29	8 85	10x10
0.6	-0.5 -1	33	0.23	25 27	7 24	10x10

Although the characteristics presented by a photodetector with a normal incidence geometry are quite interesting for short wavelengths, its sensibility is not as desirable for long wavelengths, in particular for $\lambda = 1500\ nm$, which is a typical wavelength for communications, and also at a inter- or intra-chip scale. For applications in this wavelength, a Ge-on-Si detector with a waveguide geometry may be preferable. An example of this device is well explained in references [6] and [7]. This system consists of a very thin layer of Ge modeled as a waveguide. The confinement factor of the wave in the Ge layer is small. Then a combination of two layers of Ge and Si, as can be seen in Figure 7.29, are used in order to increase the responsivity of the device. As previously stated, the lattice errors produced in a direct growth of Ge on Si make necessary the interposition of another layer between them. In this case, a SiON layer. The thickness of the layers has been optimized to obtain a good guiding of the light. With a thickness of 140 nm and 100 nm and widths of 750 nm and 550 nm for the Ge and Si, respectively, the device presents two modes at 1300 nm and 1500 nm [7], for a perpendicular polarization (electric field perpendicular to the plane of the device). On the contrary, if the dimensions are chosen equal for Si and Ge and equal to 100 nm in thickness and 500 nm in width, the waveguide stack presents only a single mode at 1500 nm with a confinement factor around a 66%. A series of electrodes (W) made of tungsten and copper, are aligned on the Ge layer (diameter-150 nm and separating-300 nm). This contact geometry is optimized to minimize the contact area between the Ge layer and the electrodes reducing the device capacitance.

As can be easily deduced, the fabrication process of this kind of device into the CMOS fabrication process is not a trivial issue. S. Assefa and coworkers described it extensively in reference [6]. In summary, they have developed a rapid melt growth technique that gives as a result an ultrathin single-

FIGURE 7.29
Scheme of the germanium wavelength-integrated avalanche photodetector (APD) described in reference [7].

crystal Ge waveguide. This formation is made also before the activation of the source/drain implants in the CMOS process. This technique overcomes two important challenges. The first one is the lattice mismatches, already commented on, produced for a direct growth of Ge on S. The second one is the temperature problems derived from the fact that while the CMOS process reaches temperatures over $1000°C$ before metal contacts are created and these cannot be over $550°C$ after their formation, Ge growth requires temperatures higher than $600°C$ but lower than $1000°$. Also, it is cheaper to introduce the Ge formation process earlier in the CMOS procedure. Finally, and after the formation of the Ge waveguide, the metallic contacts W and the copper interdigitated contacts are connected.

This fabrication technique has another important advantage. It decreases the capacitance of the system, also increasing the bandwidth. The decrement in the total capacitance is also produced for the small contact area between the plugs W and the Ge layer, resulting in capacitance values around $C \cong 10 \pm 2\ fF$ [6]. This very small capacitance that the bandwidth produces is not RC limited and then it is only limited by the time spent by carries to blow from one to other contact. Then, decreasing the gap distance between contacts, the 3-dB bandwidth is strongly increased. In Figure 7.30 it can be seen that for a bias voltage of $1V$ the 3-dB bandwidth is around 5 GHz for a 400 nm contact spacing and it could reach values of 40 GHz when the spacing is only 200 nm.

The responsivity of the system is given by the relation between the photocurrent and the input power of the device. For 1 V bias voltage and an incident wavelength of 1.3 μm, this device presents a responsivity equal to

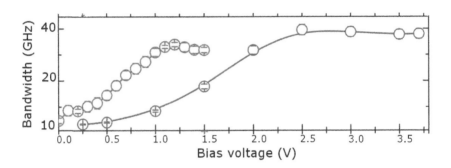

FIGURE 7.30
3-dB bandwidth as a function of the bias voltage for a APD germanium photodectector [7]. The upper curve corresponds to a contacts spacing of 200 nm
while the lower one corresponds to 400 nm.

0.42 A/W, that corresponds to a quantum efficiency $Q = 39\%$. For a larger
wavelength, $\lambda = 1.5 \ \mu m$, the responsivity drops to 0.14 A/W or $Q = 12\%$
[6]. The evolution of this parameter with the bias voltage is plotted in Figure 7.31. As can be seen, the responsivity is quite stable for a range of bias
voltage from $-0.5 \ V$ to $-1 \ V$, where there is a flat area [6]. The quantum
efficiencies measured in these references are far from the 75% that is predicted
for an optimized Ge waveguide photodetector. The important differences between the values are produced mainly for the purity of the Ge layer, that is
only around 90% in the experimental device, and its roughness that produces
scattering of the incident photons instead of their absorption. However, the
purity of the layer can be improved using other deposition techniques, such
as plasma deposition [60].

The noise in this type of device is slightly large with values around 90 μA
for a photodetector under a bias voltage of 1 V. As the bias voltage increases,
the carriers are accelerated faster. However, this higher energy of the carriers
also increases the dark current for tunneling effects.

In general, photodetectors for communication applications should be complemented by a complex circuit that manipulate the output signal, improving
it. Different configurations of it can be used depending on the given applications. In this sense, the extra circuitry can improve the sensitivity, speed,
or power consumption of the detector. Although it is not analyzed in this
chapter, an example of this circuitry is shown in Figure 7.32 to remark the
complexity of these devices.

Photodetectors Based on Surface Plasmon Polaritons (SPP)

As was commented above, the operation speed of a photodetector is determined by the transit time that electrons spend drifting from one electrode

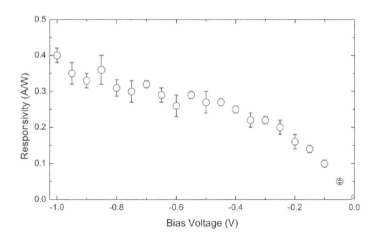

FIGURE 7.31
Responsivity of a Ge-waveguide photodetector as a function of the bias voltage for an incident wavelength $\lambda = 1.3~\mu m$ [6].

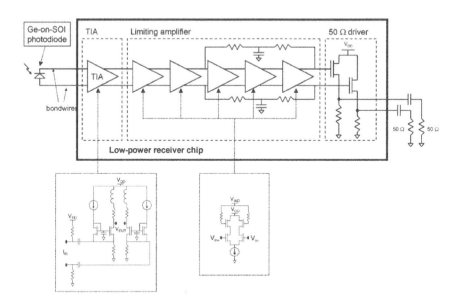

FIGURE 7.32
Low-power optimized scheme of a detector including a Ge-on-SOI photodiode and the necessary circuitry [57].

to an other. Then, the operation speed can be increased reducing the size of the active region in such a way that the electron path is reduced. However, a reduction of the active region (sizes smaller than wavelength) also means a decrement on the responsivity of the device due to the diffraction limit. This disadvantage could be overcome if high electric fields can be confined in a subwavelength area. When a surface plasmon resonance is excited in a nanoantenna [80, 34], the near field is strongly enhanced and confined in extremely small volumes. Then, as was demonstrated in reference [45], the photogeneration of carries in a semiconductor can be improved by a surface-plasmon nano-antenna. In this sense, in the last years new photodetectors that implement plasmonic nano-antennas have appeared.

The simplest example of the improvement of the characteristics of a photodetector including plasmonic features is shown in [94]. In this work the authors proposed a simple p-n junction of silicon with two aluminium contacts. In the incident window, researchers deposited gold nanoparticles of different sizes (50 nm, 80 nm and 100 nm in diameter). With this configuration, the authors reported that the response of the photodetector is globally increased for wavelengths lower than the resonant wavelengths due to the excitation of localized surface plasmon resonances in each particle. This enhancement reaches values of 50%–80% for the wavelength at which the resonance is excited, and depends on the particle size as it is well known [80]. Thus, the sensitivity enhancement and the decrease of the size of these plasmonic photodetectors make them very adequate for intra- and inter-chip links.

Other interesting configuration of plasmonic structures with direct application for photodetectors is that consisting in a subwavelength aperture in a metallic film surrounding by periodic corrugations [68, 11]. In this case, the electric field is strongly enhanced and confined in a very small area (see Figure 7.33 (a)), then the semiconductor active region can be reduced to the same size of the aperture (10–100 nm) improving drastically the operation speed and without any reduction in the responsivity of the device. A scheme of the device based on these features is shown in Figure 7.33(b).

The different ways to produce surface plasmon polaritons induce several designs for photodetectors based on them. For example, it is interesting to show the designs proposed by L. Tang et al. in their work [111] that used a single C-shaped nanoaperture in a gold film and in [109] implementing a nanoantenna formed by two gold cylinders. A scheme if these two configurations is shown in Figure 7.34. Both designs share that have been developed for larger wavelengths, that is telecom wavelengths ($\lambda \sim 1300$ nm or $\lambda \sim 1500$ nm). This is the reason why both use germanium as the semiconductor material, although other materials have been analyzed [110]. The first work stated that a C-shaped aperture as that shown in it can enhance the responsivity approximately 20%–50% with respect to a conventional photodetector and that the electric field is twice that produced by a rectangular aperture with a current of ~ 1 nA and an illumination of 1.13 μW at 1310 nm. This device is quite fast with a transit time ~ 1.5 ps and a capacitance of 0.005 fF, which means

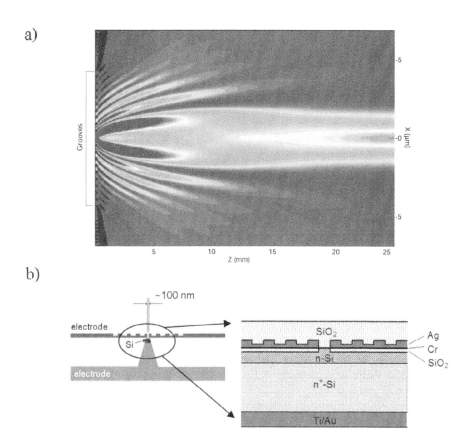

FIGURE 7.33
(a) Distribution of light emerging from a single nano-aperture surrounded by a groove array with a periodicity of 500 nm obtained from [11]. Each groove is 40 nm width and 100 nm in depth. Red color means high intensity and blue means low intensity. (b) Scheme of a nano-photodetector based on a plasmonic nanoaperture as was proposed by K. Ohashi and coworkers [81].

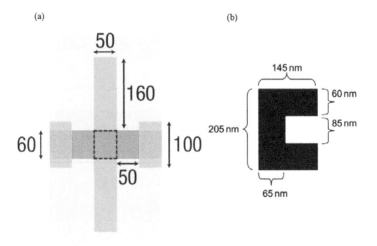

FIGURE 7.34
Scheme of the particular configurations for a photodetector based on surface plasmon antennas presented in works (a) [109] and (b) [111].

a wide bandwidth. On the other hand, the main characteristic of the C-shaped configuration [109] is its size. It has a volume of $\sim 150\ nm \times 60\ nm \times 80\ nm$ with an active volume that is $\sim 10^{-4}\lambda^3$, being probably the smallest photodetector recently manufactured. Other important feature that both devices share is their polarization dependence. Their shapes are mainly anisotropic. Then, the polarization of the incident field is quite important producing photocurrents up to 20 times lower for one polarization that for the orthogonal one.

The surface plasmon resonance is quite sensible to the influence of a substrate underneath [77]. As higher the refractive index of the substrate is, the higher perturbation of the SPR is. For this reason, the design of these devices usually includes an oxide layer between the metallic layer or nano-structure and the substrate.

The last model for a photodetector treated in this part of the chapter is probably the SPP-based one whose real application is more developed. This device was presented by J. Fujikata and coworkers in [35] and it consists of a surface plasmon antenna made as a periodic silver nano-scale metal-semiconductor-metal (MSM) electrode structure embedded in a Si layer. Then the complete set is inserted in the interface between the core of a SiON waveguide and its SiO_2 clad. A scheme of the system and a micro-photograph of it are shown in [35] and reproduced here in Figure 7.35. The grooves are 90 nm in width and 30 nm in height, their are on a 240-nm-thick Si layer that acts as an absorber, and the waveguide was chosen with a single mode

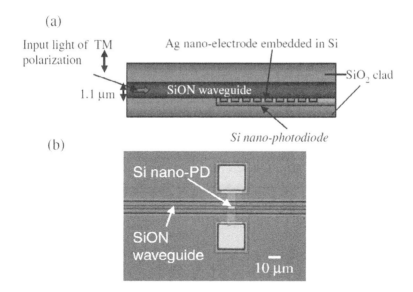

(a)

Input light of TM polarization

Ag nano-electrode embedded in Si

SiO$_2$ clad

1.1 μm

SiON waveguide

Si nano-photodiode

(b)

Si nano-PD

SiON waveguide

10 μm

FIGURE 7.35
(a) Schematic cross section of the Si nano-PD proposed by J. Fujikata et al. in [35] and (b) Micro-photograph of the fabricated device.

at 850 *nm*. Under these conditions, the authors reported an 85% coupling of the propagation light through the waveguide into the Si absorption layer. The main characteristic of the model is, of course, its integration in the optical channel reducing the coupling losses. In addition, it is also a high-speed and a very low-capacitance detector with transit times around 17 *ps* for a 1 *V* bias voltage and only 4 *pF* of capacitance. On the contrary, it does not have a prominent responsivity with a 10% quantum efficiency at an incident wavelength of 850 *nm* and a TM polarization, so a deeper analysis is needed to optimize it. The system is, as the previous ones, polarization-dependent with twice the lower quantum efficiency for a TE polarization. The authors have already implemented this device in a on-chip optical clock distribution showing a good response of the optical device to the clock operation, hence it could be a commercial device in the near future.

7.2.5 Summary

In this section several types of photonic components for an OI have been discussed. With such an amount of information, it is easy to forget the general scheme of the issue. To summarize and refresh the main types of the considered components, Table 7.6 goes over them and their principal characteristics.

TABLE 7.6
Summary of the different types of photonic elements with applications in optical communications on-chip

		Type	*Main Characteristics*
OPTICAL SOURCES	In-chip light sources	Dielectric cavities	High Q Small size
		Metallic cavities	Smaller Q-factors than dielectric cavities Ultra-small sizes ($\sim 100nm$)
		Raman lasers	Easy on-chip integration Effective area $\sim 1\ \mu m^2$
		LEDs	Quite small size Simple and cheap Incoherent and nondirectional output light
		VCSELs	Ultra-low power consumption Convenient for packing (VCSEL's arrays)
	Modulators	MZI	Based on refractive index variations An external light source is needed Transforms a phase modulation into an intensity modulation
		Micro-ring resonator	Smaller sizes than MZI modulators
OPTICAL CHANNEL	Integrated guides	Silicon waveguides	Low absorption for telecom wavelengths Easy manipulated propagation modes Several fabrication techniques
		Polymer waveguides	Versatile materials Large number of fabrication processes
		Plasmonic waveguides	Important size decrease More studies are needed
	Free-space optical interconnects		Ideal for $3D$ integration Beam divergence must be controlled
PHOTODETECTORS		Ge photodetectors	Normal-incidence and waveguide configurations Large noise Optimized dimensions and fabrication techniques for high sensitivity
		SPP-based photodetectors	Smaller than Ge PD High sensitivity and detection speed Several configuration (C-shaped, nanoparticles, etc.)

7.3 Why Optical Links? Comparison of Electrical and Optical Interconnects

New advances in microelectronics force communications to be faster and faster, and capable to transport high amounts of information, throttling the limits of actual technologies. This constraint or "communication bottleneck" could be partially overcome with the use of light and optical waveguides, as researchers have stated [40]. Integrated optical waveguides present important advantages in the propagation delay over electric wires, regardless of the waveguide material, as it is being shown in this chapter. Optical waveguides do not present any *RLC* impedances, and light propagation is intrinsically faster than electrical propagation. However, the conversion of the electric signal, that goes out from the logic cells, into an optical signal and vice versa implies certain disadvantages that should be measured. It is interesting to analyze several parameters to compare the rewards and handicaps that present both communication mechanisms: optical and electrical. In particular, in this section the bandwidth density, the power consumption, the propagation delay, and the crosstalk noise have to be considered for comparison.

In this section, each of these parameters, either for optical and electrical interconnects, are explained and discussed. Also, systems-on-chip with fan-out are considered because optical communications can improve their response.

7.3.1 Power Consumption

The energy dissipated by a communication system is probably one of the main parameters to take into account in this comparison. The power consumption of a typical electric interconnection (EI) is given in reference [32] as the sum of the effect of three different capacitances times the square of the voltage power source: C_{in} that corresponds to the input capacitance of the receiving gate, C_o, the output capacitance of the CMOS driving gate, and C_L, the capacitance of the line that depends in a linear way on the width and length of it. Then, the total energy necessary to switch a receiving inverter from one state to the other and back (switching energy) can be written as

$$E_E = V^2(C_{in} + C_o + C_L) \tag{7.7}$$

This energy is related to the power consumption (P) through the expression $E_E = 2\tau P$, being τ the rise time or the time needed for the receiving gate's input voltage to rise from a 10% to a 90% of its final value [32].

In the same way, the total power consumption of an optical link depends on the characteristics of their components. In particular, it depends on the steady current of the emitter, I_{ss}, the photodetector current, I_{ph}, the quantum efficiencies of the emitter, η_E and the detector, η_D and the efficiency of the transmission through the optical channel, η_C.

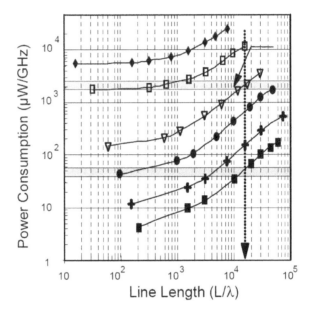

FIGURE 7.36
Power consumption of a single-piece electric interconnection across the different technologies as a function of the line length. The curves (from bottom to
up) correspond to 0.05 μm, 0.07 μm, 0.12 μm, 0.18 μm, 0.25 μm and 0.7 μm
technologies, respectively. Also the upper grey curve corresponds to the total
power consumption of an optical interconnection in a 0.18 μm technology [26].

$$P(I_{th}) = V \left(I_{ss} + \frac{I_{ph}}{2\eta_E \eta_C \eta_D} \right) \tag{7.8}$$

In both cases, V denotes the power supply voltage.

In Figure 7.36, J. Collet et al. represented the power consumption of a
single-piece electrical interconnect for several technologies. And, for comparison, they showed the value that corresponds to an OI in a 0.18 μm technology
(upper horizontal gray bar). From this figure, one important conclusion can be
deduced: the OI presents advantages, regarding the power consumption, only
for long-distance interconnects. This is, for line lengths larger than 5–10 mm.
Similar results were shown by M. Haurylau and coworkers in reference [42]
and shown here in Figure 7.37. In that case, authors showed both the power
consumption and the propagation delay (that will be analyzed later) of an EI
as a function of the length of the communication line and those values for an
OI with a length equal to 17.6 mm (edge length of the chip projected in the
International Technology Roadmap for Semiconductors (ITRS) [1]). Again,
OIs are only considered for long interconnections. The main conclusion of this
figure regards the values of power consumption. Competitive optical communi-

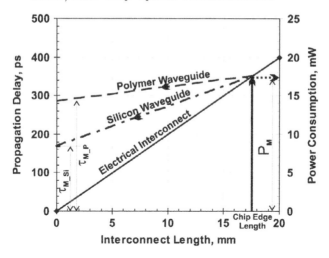

FIGURE 7.37
Power consumption (right axis) and signal propagation delay (left axis) for an EI as a function of the length. Also the data for two types of OI considering as the propagation channel a polymer or a silicon waveguide and a interconnection length equal to 17.6 mm (chip edge length in the ITRS projected chip) [42] are shown.

cations will consume power lower than 17–18 mW for technologies of 0.18 μm. Furthermore, smaller integration technologies (0.05 μm) are selected, the reduction in power should be reduced drastically to values around 10 μW. These values require the design and development of optical sources with ultra-small threshold currents and detectors with very high quantum efficiencies as those in which researchers are working now [56, 106].

7.3.2 Propagation Delay

As was commented on before, the use of an optical signal and waveguides has the intrinsic advantage of being independent of RLC impedances as those in electrical wires. This fact implies an important increment of the propagation speed and then an important decrement in the propagation delay, as can be seen in Figure 7.38. EIs commonly use repeaters to decrease the line length and then improve the propagation delay. However, this inclusion produces a fixed increment both in delay and in power consumption. The conversion steps of an OI, an electric signal into an optical one and vice versa, also include a fixed delay in the global value. Even so, for longer connections where the delay due to the propagation channel dominates, these new contributions are almost negligible in the overall delay, and OIs show better results than EIs.

The projection onto the propagation delay axis of an optical interconnect of 17.6 mm length (chip edge length) in Figure 7.37 shows τ_{M_Si} and

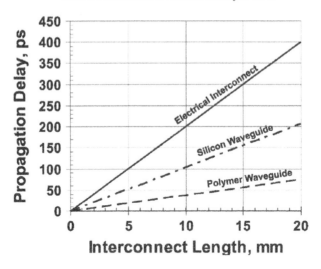

FIGURE 7.38
Propagation delay as a function of the link length for a silicon (0.34 μm wide
and refractive index 3.4) and a polymer (1.36 μm wide and refractive index
equal to 1.3) waveguide in comparison with an EI. Waveguides have square
cross section and are surrounded by a cladding with refractive index of 1.1
[42].

τ_{M_P}. These parameters are the maximum allowed delay of the conversion
process (electrical-optical-electrical conversion of the signal) for an optical
interconnect that uses a silicon or a polymer waveguide, respectively. This
plot gives us an idea about the order of values for the propagation delay due
to the conversion step, in order to obtain a competitive OI. Then, the combined transmitter-receiver delay should be lower than approximately 300 ps
for polymer waveguides and lower than 180 ps for silicon ones; which implies
total delays lower than 370 ps.

7.3.3   Bandwidth Density and Crosstalk Noise

The bandwidth density characterizes the information throughput for a unit
cross section of an interconnect [42]. This parameter depends on the aspect
ratio of the line, that is S/L = cross section/length [26]. The inclusion of
repeaters in EIs increases the bandwidth density in such a way that the design
of the electric link with repeaters (number and distances between them) can
be optimized to obtain a minimum delay or a maximum bandwidth. Besides
this, the needs in bandwidth for new chips are increasing year per year and
the density of wires in communications cannot be equally increased, due to
the limited space and to the crosstalk noise. This noise is generated by the
closeness of wires.

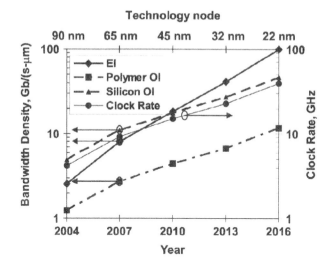

FIGURE 7.39
Comparison of the bandwidth density of electric wires and optical waveguides
made either of silicon or a polymer as a function of the year and the technology
node [42].

In optical communications using waveguides, the bandwidth density, as it
occurs with the propagation delay, depends on the geometry of the waveguide,
but also in the refractive index difference between the core and its cladding
material. In addition, geometrical parameters, and in particular the width
of the waveguide, should take also into account the crosstalk considerations.
If a waveguide is huge its bandwidth increases, but this also implies that it
is closer to adjacent waveguides, increasing the crosstalk noise. In a similar
way, if the link is very narrow, the optical signal is poorly confined allowing a
larger overlap between adjacent optical modes. This fact is well explained in
reference [42].

Although an optical signal presents less crosstalk noise and better values
for the bandwidth density than a metallic wire, OI is still worse than a delay-
optimized electric link with repeaters in terms of bandwidth density. This
can be clearly seen in the comparison shown in Figure 7.39 for an EI and
two OIs as a function of the technologies. However, optical lines present a
very important advantage in terms of bandwidth density: the possibility to
use wavelength division multiplexing (WDM) [118]. This feature consists of
the fact that an optical channel can transport more than one signal coded at
different wavelengths. This effect can improve the bandwidth density of an
OI to match the EIs or even more. However, the use of WDM to increase
the bandwidth density involves power and delay penalties that should be also
taken into account.

7.3.4   Fan-Out

It was previously shown that OIs present several advantages with respect to EIs only when the length of the connection is large enough. In other words, optical communications can be the counterpart of electric ones only for the global on-chip or inter-chip communications. However, this is true only for point-to-point communications. A system including OIs can be a good alternative to electric communications with high fan-out, even for short distances.

The fan-out of a digital system can be defined as the connection of the output signal of one logic gate to the input of various gates. Three different cases can be distinguished in fan-out connections [32]

1. when the receiving gates are close

2. when the receiving gates are far, one to the other, and need different communications lines

3. when the receiving gates are along the same line

Only for the second case, optical links can provide advantages in its characteristics (power, delay, bandwidth). On the contrary, the proximity of the receiving gates in the first and third cases makes that the increment in the propagation delay and in the power consumption due to the inclusion of the optical transmitter and the receiver cannot be compensated by the advantages in the optical channel. However, a hybrid system that combines an optical and an electrical interconnection gives better values for propagation delay and power consumption in any case. In Figure 7.40(a) a fully electrical fan-out system of the third type is plotted. As a comparison, Figure 7.40(b) represents an alternative optoelectronic system that implements optical and electrical fan-out. In this hybrid system, the first division of the signal is made optically using a transmitter (T) that sends its signal to two optical receivers ($R1$ and $R2$). Then, they transform light into an electric signal and this is divided again, but now using an electric system.

The inclusion of fan-out in a fully electrical communication implies a deterioration on the power and delay characteristics. Delay inconveniences can be solved partially using larger-sized transistors or optimally-spaced repeaters. In these cases, the propagation delay (τ) scales as the logarithmic of the number of fan-out (N). A hybrid system that includes optical fan-out can reduce these values if the electro-optical conversion is efficient in terms of energy and speed. This can be observed in Figure 7.41, where the propagation delay for a hybrid link is plotted as a function of the load capacitance and for several numbers of fan-outs (N). The delay for a fully electrical link also has been included, for comparison purposes. The load capacitance is defined as the number of inverters driven times the load capacitance of each inverter.

As was remarked above, for a point-to-point communication ($N = 1$) the optical delay is worse than the electrical one. However, as the number of fan-outs increases, the delay of the interconnection with optical fan-out tends to

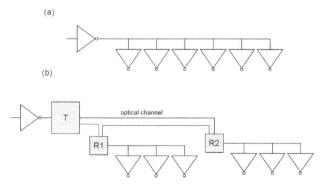

FIGURE 7.40
(a) Schematic plot of a fully electrical fan-out system in which the receiving gates (6 in this case) are along one line and (b) an equivalent optoelectronic system with an optical fan-out consisting in one transmitter and two receivers.

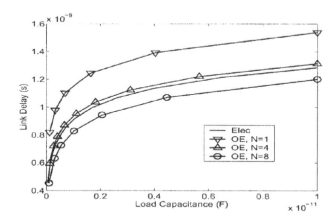

FIGURE 7.41
Propagation delay for a system with optical fan-out versus the load capacitance. Three number of fan-outs have been considered. Also the delay for an electrical system have been included as comparison.

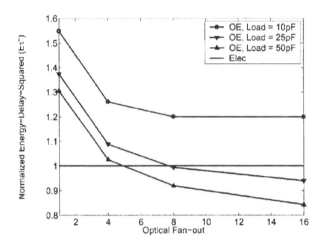

FIGURE 7.42

Normalized $E\tau^2$ as a function of the fan-out number of an optoelectronic system with respect to the electric one. Some load capacitances are considered [84].

decrease acquiring values lower than that of the electrical system. Even for small values of N ($N \sim 8$) the delay advantages of an optical fan-out can be observed. These results can be extended to very-short on-chip links around 200–300 μm.

The implementation of an OI leads to a high power consumption due to the electric-optical signal conversion. Because of that, an optoelectronic solution for a fan-out is generally worse in terms of energy than an electric one. Also, it is important to remark that an optical link with fan-out is limited in the number of driven devices due to the limited energy emitted by the source (laser, cavity, etc.). Besides this, it is necessary to take into account the global advantages and disadvantages. In order to analyze it, some authors have introduced the parameter $E\tau^2$ [32], that includes the influence of the two main parameters: the propagation delay and the power consumption of the interconnection. In Figure 7.42, the considered parameter is shown as a function of the fan-out for a hybrid optoelectronic system with respect to the electric one and for several load capacitances. As can be seen, as the load capacitance and the number of fan-outs increase, the disadvantages related to the power consumption are overcome by the advantages induced in the propagation delay. A. Pappu and coworkers [84] stated that a hybrid optoelectronic link scales 3.6 times better in delay and 1.5 times worse in energy than an electrical one. Then, the implementation of optical devices in on-chip interconnections with fan-out could improve its characteristic in a general point of view.

7.4 Photonic Networks

Although this is not the main scope of this chapter, the improvement of optical intra- and inter-chip communications also affects the advance of other complex systems as, for instance, networks-on-chip (NoC). For this reason, this section is devoted to the analysis of these networks.

A network-on-chip can be roughly defined as "a system made of carefully-engineered links representing a shared medium that is highly scalable and can provide enough bandwidth to replace many traditional bus-based and/or point to point links" [98].

These devices, as others systems, will suffer, in the near future, important limitations in bandwidth, latency, and power consumption due to the advances that determine the trend of systems on-chip. The potential advantages offered by optical devices for communication tasks, as it has been explained, have been considered as a feasible solution also for NoCs. However, a network-on-chip involves a signal routing much more complex than that of a direct connection. Then, more complex optical devices are necessary as, for instance, an optical switch. Fortunately, new advances in the field of photonics have allowed for the design and development of sources, waveguides, and photodetectors. Designs have also emerged for other devices such as switches, filters, etc. [2].

The recent designs in optical switches are based on micro-ring resonators [115], MZI modulators [16], or photonic crystals and located in the intersection of optical channels in such a way that the input light can be directed to one or other channels depending on the device conditions. For example, the model presented by Y. Vlasov and coauthors in reference [115] consists of an apodized fifth-order cascaded silicon micro-ring resonator that is able to transmit modulated signals at $> 100Gbp/s$. Tuning the resonances of the central micro-rings, by the injection of free carriers, the propagation through the drop-port is cut off and light is redirected to the through-port. However, these devices still present some challenges and constraints in their design and manufacturing as the size (footprints cannot be larger that $0.01 \ mm^2$) or the operation times that could be still too long due to the large spectral detuning necessary for switching, as occurs in [115].

Photonic NoC could present several advantages with respect to the conventional ones, mainly in terms of bandwidth and power consumption. However, some functions as buffering or processing of the signal cannot be implemented using optical devices. For this reason, a fully optical network-on-chip cannot be developed now. Typical optical NoCs described in the bibliography are, in fact, a hybrid system formed by an optical network and an electronic network with the same topology [8]. While the optical network is devoted to the transmission of the large-size packets or data, the electronic network takes over control signals and/or short message exchange. This configuration shows two main advantages with respect to conventional NoCs [17]:

1. Bit Transparency: unlike conventional NoC, in an optical network the power dissipation does not scale with the bit rate. Optical switches switch once per message independently of the message size.

2. Low-Loss in Optical Waveguide: power consumption in optical guides is almost independent of the transmission distance in such a way that the power requirements are quite similar if the message travels a few millimeters or several centimeters.

The step-by-step operation process on a hybrid optical network would be similar to that described as follows. First, an electric packet is routed through the electric network in order to manage the optical path of the forthcoming optical communication. When the electronic NoC has fixed the path, it sends a signal to the photonic one that starts sending the message through the determined path. If the message has to be buffered, this is made only in the electronic NoC.

In the last years, there have been several research groups and even high-technology companies such as IBM or Sun Technologies working on the development of these new networks. One of the first optimized designs is, for instance, described in reference [62].

7.5 New Optical Nanocircuits Based on Metamaterials

The future of the optics devices goes through the development of new materials and in particular the so-called metamaterials. These new materials can also be quite interesting for the performance improvement of OI and also for the implementation of optical nanocircuits. This section is a brief dissertation on the applications of metamaterials for systems-on-chip.

Metamaterials are internally structured medias that are made by packing or embedding subwavelengths structures [104, 102]. The details of the inner structure are much more smaller than the incident wavelength in such a way that the incident light only senses the "effective" materials characteristics [105, 78]. These effective optical properties can be modified changing the size, shape, distribution, density, or nature of the inclusions, inhomogeneities, or ensembles forming the metamaterial. This fact produces the values reached by the effective optical constants in an almost continuous range, allowing us to observe features that are impossible with conventional materials: negative refractive index [99], near-zero (ENZ) or very large (EVL) permittivities, or permeabilities [103].

Planar slabs with a negative refractive index present a very important and useful characteristic. A planar negative-refractive index metamaterial can reproduce the near-field as well as the far-field of a source with subwavelength resolutions, i.e., the metamaterial acts as a perfect lens [86]. A perfect lens is

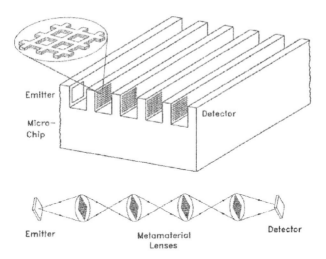

FIGURE 7.43
Scheme of a horizontal intrachip interconnect using a metamaterial for guiding light from the emitter to the receptor [39].

not subject to the diffraction limit that appears in conventional optics. Then, planar negative-refractive index or double-negative (DNG) metamaterials can potentially offer a lossless control of light propagation at sizes much smaller than the incident wavelength.

The main scope of optical interconnections in microelectronics is to focus and to guide light through a chip or some chips. Then, an optical device that integrates slabs made of a DNG metamaterial can improve the delay and energy characteristics of the link. In a recent U.S. patent, T. Gaylord and coworkers [39] showed the possible applications of this kind of optical system in the field of inter- and intra-chip optical communications. In Figure 7.43, an schematic illustration of the proposed optical on-chip interconnection is shown that uses the unconventional slabs as a perfect lens guiding light from the emitter to the detector. In this case, the metamaterial is composed of a multilayer structure that is called *fishnet metamaterial* [38]. A very interesting behavior of these interconnections is that the propagated beam can travel in both directions, i.e., this is a bidirectional system, provided a detector and a source are integrated simultaneously at the beginning and at the end of the communication.

Other futurist evolutions of optical communications goes through the implementation of optical devices with an electromagnetic response as that presented by other microelectronic elements. That is, the use of optical nanostructures whose interaction with light produces equivalent behaviors to that presented by, for instance, a resistance, an inductance, or a transistor [31]. If in the future, we are able to manufacture this kind of optical element, even the logic cells of the chip will be implemented with them, and light will be used to encode the information instead of electric signals [71]. The optical paths or links of these optical nano-circuits could also be implemented using nano-structures, like simple particles, made of those new metamaterials that will be able to control the direction of the light. The control of the optical properties of these materials means, as it was commented on previously, a control of their interaction with light, and in particular the direction at which the material scatters light. This item was first studied a few years ago in reference [49] and recently generalized in references [37, 36]. It has been even studied experimentally. For instance, A. Mirin and N. Halas [73] showed that light scattering by a metallic nanoparticle with a complex geometry (nano-cups) can be directed at certain directions. The idea of researchers is that in the near future composites of these kinds of nano-particles can transport light, in a similar way like waveguides, with much smaller sizes. However, the practical implementation of these optical links is still far from being real, and an intensive study and evolution of metamaterials are required.

7.6 Present and Future of the Intra-/Inter-Chip Optical Interconnections

During this chapter, the main advances in photonics and, in particular, in photonic devices with applications in architecture communications for system-on-chip have been analyzed. As has been shown, there has been an important evolution on this area of research, especially in silicon photonics, which has allowed that several designs and prototypes of fully-optical intra- and inter-chip optical interconnects have appeared [23, 20]. To finalize the chapter, a brief discussion about the present and future of optical interconnects in systems-on-chip is carried out in this section. The actual policies and the recent advances in this field are shown.

These trends have induced big manufacturers in microelectronics, such as *IBM* or *Sun Microelectronics*, to devote research groups on the analysis, development, and integration of new advanced optical devices on chips, as Dr. Vlasov's lab. Moreover, the U.S. government and the European Commission have dedicated a part of their budget to this field through the *DARPA Ultraperformance Nanophotonic Intrachip Communication* (UNIC) and the *Photonic Interconnect Layer on CMOS by Waferscale Integration* (PICMOS) programs, respectively. This shows that the scientific community really believes that this field has a promising future.

These future optical interconnects can be useful for several applications, in particular, for signaling or clocking. Recently, researchers have developed many optical interconnects on-chip. For instance, M. Kobrinsky and coworkers manufactured optimized optical links for these tasks [55]. They used a

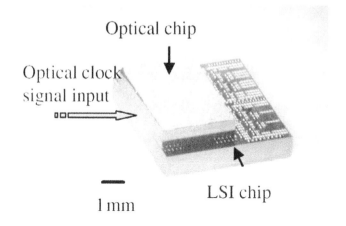

FIGURE 7.44
Photograph of the prototype of an optical clock module developed in [35].

pulsed light source for the signaling chip and an optical modulator and a continuous wave (CW) light source for the clocking one, both of them with silicon waveguides and a silicon photodetector. Their main conclusion is that although signaling applications offer greater challenges than signaling ones, the last one does not offer noticiable advantages with respect to its electrical counterpart. However, recent advantages in silicon photonics have improved the characteristics of a clocking optical chip and the first examples have appeared. For instance, J. Fujikata et al. presented in reference [35] a prototype of a Large-Scale Integration (LSI) on-chip fully optical clock system with a 4-branching H-tree structure. The optical chip (see Figure 7.44) is composed of an improved coupling and confinement architecture between a SiON waveguide and a silicon photodetector by a surface-plasmon-polariton (SPP) structure (see Figure 7.35). The input light is produced by a 850-nm CW light source and introduced in the chip with a lithium niobate modulator (not discussed in this chapter). A 5 GHz operation has been reported with this experimental and complete clock system, which is quite interesting and promising for OI.

In addition, many optical networks have been already manufactured for different applications. One example of these networks is called Iris [62]. This is a CMOS-compatible high-performance low-power nanophotonic on-chip network composed of two different subnetworks. While a linear-waveguide-based throughput-optimized circuit switched subnetwork supports large and throughput-sensitive messages, a planar-waveguide-based WDM broadcast-multicast nanophotonic subnetwork optimizes the transfer of short, latency-critical and often-multicast messages. Thus, a nano-photonic network, as this one, provides low-latency, high-throughput and low-power communications for many-core systems.

The last efforts have produced important results in a way that the last prototypes reach similar characteristics to that of their electric counterparts. The integration of optics in a chip have involved improvements in parameters as the bandwidth or latency. However, for other parameters, as cost or integration techniques, optical interconnects are still far from the desirable development. Then, although optical communications could be real and competitive, we still should wait for several years until those will be fully integrated on commercial microchips.

7.7 Glossary

ARC: Antireflection Coating

CMOS: Complementary Metal-Oxide-Semiconductor

CW: Continuous Wave

DARPA: Defense Advanced Research Projects Agency

DBR: Distributed Bragg Reflector

DNG: Double-Negative

EI: Electric Interconnection

ELO: Epitaxial Lateral Overgrowth

ENZ: Epsilon Near Zero

ELO: Epitaxial Lateral Overgrowth

EVL: Epsilon Very Large

FSOI: Free-Space Optical Interconnect

ICP: Induced-Coupled Plasma

LCOS-SLM: Liquid-Crystal on Silicon Spatial Light Modulator

LOCOS: Local Oxidation of Silicon

LED: Light Emitting Diode

LPCVD: Low-Pressure Chemical Vapor Deposition

LSI: Large-Scale Integration

LSPR: Localized Surface Plasmon Resonance

MD: Modulation Depth

MDR: Morphology-Dependent Resonance

MMI: Multi-Mode Interference

MRR: Micro-Ring Resonators

MOS: Metal-Oxide-Semiconductor

MSM: Metal-Semiconductor-Metal

MZI: Mach-Zhender Interferometer

NoC: Network-on-Chip

OI: Optical Interconnection

PECVD: Plasma-Enhance Chemical Vapor Deposition

PD: Photodetector

PICMOS: Photonic Interconnect Layer on CMOS by Waferscale Integration

Q-factor: Quality Factor

QW: Quantum Well

RF: Radio Frequency

SCH: Separate Confinement Heterostructure

SEM: Scanning Electron Microscope

SNR: Signal-to-Noise Ratio

SOI: Silicon-on-Insulator

SOS: Silicon-on-Sapphire

SPP: Surface Plasmon Polariton

SPR: Surface Plasmon Resonance

SRS: Stimulated Raman Scattering

SWAMI: Side-Wall Masked Isolation

TE: Transversal Electric

TIR: Total Internal Reflection

TM: Transversal Magnetic

TPA: Two-Photon-Absorption

UNIC: Ultraperformance Nanophotonic Intrachip Communication

VCSEL: Vertical-Cavity-Surface-Emitting Lasers

WDM: Wavelength Division Multiplexing

WGM: Whispering Gallery Mode

7.8 Bibliography

[1] *International Technology Roadmap for Semiconductors: 2007 Edition.* International SEMATECH, 2007.

[2] V. R. Almeida, C. A. Barrios, and R. R. Panepucci. All-optical control of light on a silicon chip. *Nature*, 431:1081–1084, 2004.

[3] D. K. Armani, T.J. Kippenberg, S. M. Spillane, and K. J. Vahala. Ultra-high-Q toroid microcavity on a chip. *Nature*, 421:925–928, 2003.

[4] S. Arnold. Microspheres, photonic atoms and the physics of nothing. *Am. Sci.*, 89:414–421, 2001.

[5] N. W. Ashcroft and N. D. Mermin. *Solid State Physics.* Thomson Learning Inc., New York, 1976.

[6] S. Assefa, F. Xia, S. W. Bedell, Y. Zhang, T. Topuria, P. M. Rice, and Y.A. Vlasov. CMOS-integrated high-speed MSM germanium waveguide photodetector. *Opt. Express*, 18:4986–4999, 2010.

[7] S. Assefa, F. Xia, and Y. A. Vlasov. Reinveniting germanium avalanche photodetector for nanophotonic on-chip optical interconnects. *Nature*, 464:80–85, 2010.

[8] S. Bahirat and S. Pasricha. Exploring hybrid photonic networks-on-chip for emerging chip multiprocessors. In *Proceedings of the 7th IEEE/ACM International Conference on Hardware/Software Codesign and System Synthesis*, 129–136, 2009.

[9] M. Balbi, V. Sorianello, L. Colace, and G. Assanto. Analysis of temperature dependence of Ge-on-Si p-i-n photodetectors. *Phys. E: Low-dimen. Sys. Nanostruc.*, 41:1086–1089, 2009.

[10] P. Barber and S. C. Hill. *Light Scattering by Particles: Computational Methods.* World Scientific, 1990.

[11] W. L. Barnes, A. Dereux, and T. W. Ebbesen. Surface plasmon subwavelength optics. *Nature*, 424:824–830, 2003.

[12] A. Biberman, B. G. Lee, K.Bergman, A. C. Turner-Foster, M. Lipson, M. A. Foster, and A. L. Gaeta. First Demonstration of On-Chip Wavelength Multicasting. In *Optical Fiber Communication Conference, OSA Technical Digest (CD)*, Oct. 13, 2009.

[13] R. Bicknell, L. King, C. E. Otis, J-S. Yeo, P. Kornilovitch, S. Lender, and L. Seals. Fabrication and characterization of hollow metal waveguides for optical interconnect applications. *Appl. Phys. A*, 95:1059–1066, 2009.

[14] C. F. Bohren and D. R. Huffman. *Absorption and Scattering of Light by Small Particles*. Wiley-Interscience, New York, 1983.

[15] O. Boyraz and B. Jalali. Demonstration of a silicon Raman laser. *Opt. Express*, 12:5269–5273, 2004.

[16] J. Van Campenhout, W. M. J. Green, and Y. A. Vlasov. Design of a digital, ultra-broadband electro-optic switch for reconfigurable optical networks-on-chip. *Opt Express*, 17:23793–23804, 2009.

[17] L. P. Carloni, P. Pande, and Y. Xie. Networks-on-chip in emerging interconnect paradigms: advantages and challenges. In *Proceedings of the 3rd ACM/IEEE International Symposium on Networks-on-Chip*, 93–102, 2009.

[18] J. Casas. *Óptica*. Lib. Pons, 1994.

[19] Y. C. Chang and L. A. Coldren. High effciency, high-speed VCSEL for optical interconnects. *Appl. Phys. A*, 95:1033–1037, 2009.

[20] L. Chen, K. Preston, S. Manipatruni, and M. Lipson. Integrated GHz silicon photonic interconnect with micrometer-scale modulators and detectors. *Opt. Express*, 17:15248–15256, 2009.

[21] Y-M. Chen, C-L. Yang, Y-L. Cheng, H-H. Chen, Y-C. Chen, Y. Chu, and T.E. Hsieh. 10 Gbps multi-mode waveguide for optical interconnect. In *Electronic Components and Technology Conference*, 1739–1743, 2005.

[22] Z. M. Chen and Y. P. Zhang. Inter-chip wireless communication channel: Measurament,characterization and modeling. *IEEE Trans Antennas Propag.*, 55:978–986, 2007.

[23] I-K. Cho, J-H. Ryu, and M-Y. Jeong. Interchip link using an optical wiring method. *Opt. Lett.*, 33:1881–1883, 2008.

[24] J. T. Clemens. Silicon microelectronics technology. *Bell Labs Tech. J.*, 2:76–102, 1997.

[25] L. Colace, G. Massini, and G. Assanto. Ge-on-Si approaches to the detection of near-infrared light. *IEEE J Quant. Electron.*, 35:1843–1852, 1999.

[26] J. H. Collet, F. Caignet, F. Sellaye, and D. Litaize. Performance constraints for Onchip optical interconnects. *IEEE J. Selec. Top. Quant. Elec.*, 9:425–432, 2003.

[27] J. A. Conway, S. Sahni, and T. Szkopek. Plasmonic interconnects versus conventional interconnects: a comparison of latency, crosstalk and energy cost. *Opt. Express*, 15:4474–4484, 2007.

[28] V. R. Dantham and P. B. Bisht. High-Q whispering gallery modes of doped and coated single microspheres and their efffect on radiative rate. *J. Opt. Soc. Am. B*, 26:290–300, 2009.

[29] D. Ding and D. Z. Pan. OLI: A nano-photonics optical interconnect library for a new photonic networks-on-chip architecture. In *Proceedings of the 11th International Workshop on System Level Interconnect Prediction*, 11–18, 2009.

[30] L. Eldada and L. W. Shacklette. Advances in polymer integrated optics. *IEEE J. Select. Topics Quantum Electron.*, 6:54–68, 2000.

[31] N. Engheta. Circuits with light at nanoscales: Optical nanocircuits inspired by metamaterials. *Science*, 317:1698–1702, 2007.

[32] M. R. Feldman, S. C. Esener, C. C. Guest, and S. H. Lee. Comparison between optical and electrical interconnections based on power and speed considerations. *Appl. Opt.*, 27:1742–1751, 1988.

[33] N-N. Fenga, S. Liao, D. Feng, P. Dong, D. Zheng, H. Liang, R. Shafiiha, G. Li, J. E. Cunningham, A.V. Krishnamoorthy, and M. Asghari. High speed carrier-depletion modulators with 1.4 V-cm integrated on 0.25 μm silicon-on-insulator waveguide. *Opt. Express*, 18:7994–7999, 2010.

[34] H. Fischer and O. J. F. Martin. Engineering the optical response of plasmonic nanoantennas. *Opt. Express*, 16:9144–9154, 2008.

[35] J. Fujikata, K. Nose, J. Ushida, K. Nishi ans M. Kinoshita, T. Shimizu, T. Ueno, D. Okamoto, A. Gomyo, M. Mizuno, T. Tsuchizawa, T. Watanabe, K. Yamata, S. Itabashi, and K. Ohashi. Waveguide-integrated Si nano-photodiode with surface-plasmon antenna and its application to on-chip clock distribution. *Appl. Phys. Exp.*, 1:022001, 2008.

[36] B. García-Cámara, J. M. Saiz, F. González, and F. Moreno. Distance limit of the directionality conditions for the scattering of nanoparticles. *Metamaterials*, 4:15–23, 2010.

[37] B. García-Cámara, J.M. Saiz, F. González, and F. Moreno. Nanoparticles with unconventional scattering properties. *Opt. Commun.*, 283:490–496, 2010.

[38] C. García-Meca, R. Ortuño, F. J. Rodriguez-Fortuño, J. Martí, and A. Martínez. Double-negative polarization-independent fishnet metamaterials in the visible spectrum. *Opt. Lett.*, 34:1603–1605, 2009.

[39] T. K. Gaylord, J. L. Stay, and J. D. Meindl. Optical interconnect devices and structures based on metamaterials. *US Patent*, US 2008/0212921 A1, 2008.

[40] J. W. Goodman, F.J. Leonberger, S.Y. Kung, and R. A. Athale. Optical interconnections for VLSI systems. *Proc. IEEE*, 72:850–866, 1984.

[41] M. Guillaumée, L. A. Dunbar, Ch. Santschi, E. Grenet, R. Eckert, O.J.F. Martin, and R. P. Stanley. Polarization sensitive silicon photodiode using nanostructured metallic grids. *Appl. Phys. Lett.*, 94:159–160, 2000.

[42] M. Haurylau, G. Chen, H. Chen, J. Zhang, N. A. Nelson ans D. A. Albonesi, E.G. Friedman, and P.M. Fauchet. On-chip optical interconnect roadmap: challenges and critical directions. *IEEE J. Selec. Top. Quant. Elec.*, 12:1699–1705, 2006.

[43] C. J. Henderson, D. G. Leyva, and T. D. Wilkinson. Free-space adaptive optical interconnect at 1.25 Gb/s with beam steering using a ferrolectric liquid-crystal SLM. *J. Lightwave Technol.*, 24:1989–1997, 2006.

[44] M. T. Hill, Y.-S. Oei, B. Smalbrugge, Y. Zhu, T. de Vries, P. J. Van Veldhoven, F. W. M. Van Otten, T. J. Eijkemans, J. P. Turkiemicz, H. De Waardt, E. J. Geluk, S.-H. Kwon, Y.-H. Lee, R. Nötzel, and M. K. Smit. Lasing in metallic-coated nanocavities. *Nature Phot.*, 1:589–594, 2007.

[45] T. Ishi, J. Fujikata and K. Makita, T. Baba, and K. Ohashi. Si nano-photodiode with a suface plasmon antenna. *Jpn. J. Appl. Phys.*, 44:L364–L366, 2005.

[46] B. Jalali and S. Fathpour. Silicon photonics. *J. Lightwave Technol.*, 24:4600–4615, 2006.

[47] M. Jarczynski, T. Seiler, and J. Jahns. Integrated three-dimensional optical multilayer using free-space optics. *Appl. Opt.*, 45:6335–6341, 2006.

[48] R. Jones, A. Liu, H. Rong, and M. Paniccia. Lossless optical modulation in a silicon waveguide using stimulated Raman scattering. *Opt. Express*, 13:1716–1723, 2005.

[49] M. Kerker, D. S. Wang, and C. L. Giles. Electromagnetic scattering by magnetic particles. *J. Opt. Soc. Am.*, 73:767–767, 1983.

[50] J. T. Kim, J. J. Ju, S. Park, M-S. Kim, S. K. Park, and M-H. Lee. Chip-to-chip optical interconnect using gold long-range surface plasmon polariton waveguides. *Opt. Express*, 16:13133–13138, 2008.

[51] K. Kim and K. K. O. Characteristics of integrated dipole antennas on bulk, SOI and SOS substrates for wireless communications. In *Proc. IITC, San Francisco*, 21–23, 1998.

[52] L. C. Kimberling. The economics of science: From photons to products. *Optics and Photonics News*, 9:19–52, 1998.

[53] L. C. Kimberling. Silicon microphotonics. *Appl. Surf. Sci.*, 159-160:8–13, 2000.

[54] N. Kirman, M. Kirman, R. K. Dokania, J. F. Martínez, A. B. Apsel, M. A. Watkins, and D.H. Albonesi. Leveraging optical technology in future bus-based chip multiprocessors. In *39th Annual IEEE/ACM International Symposium on Microarchitecture*, 2006.

[55] M. J. Kobrinsky et al. On-chip optical interconnects. *Intel Technol. J.*, 8:129–142, 2004.

[56] S. J. Koester, J. D. Schaub, G. Dehlinger, J. O Chu, Q. C. Ouyang, and A. Grill. High-efficieny Ge-on-SOI lateral PIN photodiodes with 29 GHz bandwidth. In *Proceedings of Device Research Conference*, 175–176, 2004.

[57] S. J. Koester, C. L. Schow, L. Schares, G. Dehlinger, J. D. Schaub, F. E. Doany, and R. A. John. Ge-on-SOI-detector/Si-CMOS-amplifier receivers for high performance optical-communications applications. *J. Light. Technol.*, 25:46–57, 2007.

[58] T. Kosugi and T. Ishigure. Polymer parallel optical waveguide with graded-index rectangular cores and its dispersion analysis. *Opt. Express*, 17:15959–15968, 2009.

[59] D. M. Kuchta, Y. H. Kwark, C. Schuster, C. Baks, C. Haymes, J. Schaub, P. Pepepljugoski, L. Shan, R. John, D. Kucharski, D. Rogers, M. Ritter, J. Jewell, L. A. Graham, K. Schrödinger, A. Schild, and H. M. Rein. 120-Gb/s VCESL-based parallel-optical interconnect and custom 120-Gb/s testing station. *J. Lightwave Technol.*, 22:2200–2212, 2004.

[60] T. A. Langdo, C. W. Leitz, M. T. Currie, E. A. Fitzgerald, A. Lochtefeld, and D. A. Antoniadis. High quality Ge on Si by epitaxial necking. *Appl. Phys. Lett.*, 76:3700–3702, 2000.

[61] J. S. Levy, A. Gondarenko, M. A. Foster, A. C. Turner-Foster, A. L. Gaeta, and M. Lipson. CMOS-compatible multiple wavelength oscillator for on-chip optical interconnects. *Nature Phot.*, 4:37–40, 2010.

[62] Z. Li et al. A high performance low-power nanophotonic on-chip network. In *Proceedings of the 14th ACM/IEEE International Symposium on Low Power Electronics and Design*, 291–294, 2009.

[63] T. K. Liang and H. K. Tsang. Role of free carriers from two-photon absorption in Raman amplification in silicon-on-insulator waveguides. *Appl. Phys. Lett.*, 84:2745–2747, 2004.

[64] L. Liao, D. Samara-Rubio, M. Morse, A. Liu, D. Hodge, D. Rubin, U. D. Keil, and T. Franck. High speed silicon Mach-Zhender modulator. *Opt. Express*, 13:3129–3135, 2005.

[65] A. Liu, R. Jones, L. Liao, D. Samara-Rubio, D. Rubin, O. Cohen, R. Nicolaescu, and M. Paniccia. A high-speed silicon optical modulator based on a metal-oxide-semiconductor capacitor. *Nature*, 427:615–618, 2004.

[66] A. Liu, L. Liao, D. Rubin, H. Nguyen, B. Ciftcioglu, Y. Chetrit, N. Izhaky, and M. Paniccia. High-speed optical modulation based on carrier depletion in a silicon waveguide. *Opt. Express*, 15:660–668, 2007.

[67] J. A. Lock. Morphology-dependent resonances of infinitely long circular cylinder illuminated by a diagonally incident plane wave or a focused Gaussian beam. *J. Opt. Soc. A*, 14:653–661, 1997.

[68] L. Martín-Moreno, F. J. García-Vidal, H. J. Lezec, A. Degiron, and T. W. Ebbesen. Theory of highly directional emission from a single subwavelength aperture surrounded by surface corrugation. *Phys. Rev. Lett.*, 90:167401, 2003.

[69] A. Matsko. *Practical Applications of Microresonantors in Optics and Photonics.* CRC Press, Buca Raton, FL, 2009.

[70] M. J. McFadden, M. Iqbal, T. Dillon, R. Nair, T. Gu, D. W. Prather, and M. W. Haney. Multiscale free-space optical interconnects for intrachip global communication: Motivation, analysis and experimental validation. *Appl. Opt.*, 45:6358–6366, 2006.

[71] D. A. Miller. Are optical transistors the next logical next step? *Nature Phot.*, 4:3–5, 2010.

[72] B. Min, E. Ostby, V. Sorger, E. Ulin-Avila, L. Yang, and X. Zhang an K. Vahala. High-Q surface-plasmon-polariton whispering-gallery microcavity. *Nature*, 457:455–459, 2009.

[73] N. A. Mirin and N. J. Halas. Light-bending nanoparticles. *Nano Lett.*, 9:1255–1259, 2009.

[74] M. I. Mishchenko and A. A. Lacis. Morphology-dependent resonances of nearly spherical particles in random orientation. *Appl. Opt.*, 42:5551–5556, 2003.

[75] H. T. Miyazaki and Y. Kurokawa. Squeezing visible light waves into a 3-nm-thick and 55-nm-long plasmon cavity. *Phys. Rev. Lett.*, 96:097401, 2006.

[76] G. Moore. Cramming more components onto integrated circuits. *Electronics*, 38:114–117, 1965.

[77] F. Moreno, B. García-Cámara, J. M. Saiz, and F. González. Interaction of nanoparticles with substrates: Effects on the dipolar behaviour of the particles. *Opt. Express*, 16:12487–12504, 2008.

[78] N. A. Mortensen, M. Yan, O. Sigmund, and O. Breinbjerg. On the unambiguous determination of effective optical properties of periodic metamaterials: A one-dimensional case study. *J. Europ. Opt. Soc. Rap. Public.*, 5:10010, 2010.

[79] K. Noguchi, O. Mitomi, and H. Miyazawa. Millimiter-wave Ti:LiNbO$_3$ optical modulators. *J. Lightwave Technol.*, 16:615–619, 1998.

[80] L. Novotny and B. Hecht. *Principles of Nano-Optics*. Cambridge University Press, New York, 2006.

[81] K. Ohashi and J. Fujikata. Photodetector using surface-plasmon antenna for optical interconnect. *Mater. Res. Soc. Symp. Proc.*, 1145:1145–MM01–05, 2009.

[82] G. R. Olbright and J. L. Jewell. Vertical-cavity surface emitting laser optical interconnect technology. US Patent, 5,266,794, 1993.

[83] J. Osmond, G. Isella, D. Chrastina, R. Kaufmann, M. Acciarri, and H. von Känel. Ultralow dark current Ge/Si(100) photodiodes with low thermal budget. *Appl. Phys. Lett.*, 94:201106, 2009.

[84] A. M. Pappu and A. B. Apsel. Analysis of intrachip electrical and optical fanout. *Appl. Opt.*, 44:6361–6372, 2005.

[85] L. Pavesi and D. J. Lockwood. *Silicon Photonics*. Springer-Verlag, Berlin Heidelberg, 2004.

[86] J. B. Pendry. Negative refraction makes a perfect lens. *Phys. Rev. Lett.*, 85:3966–3969, 2000.

[87] J. Piprek. *Semiconductor Optoelectronic Devices: Introduction to Physics and Simulation*. Elsevier Science, Amsterdam, NL, 2003.

[88] P. N. Prasad. *Nanophotonics*. Wiley-Interscience, Hoboken, NJ, 2004.

[89] A. Rahman, R. Eze, and S. Kumar. Novel optical sensor based on morphology-dependent resonances for measuring thermal deformation in microeletromechanical systems devices. *J. Micro/Nanolith. MEMS MOEMS*, 8:033071, 2009.

[90] C. V. Raman and K. S. Krishnan. A new type of secondary radiation. *Nature*, 121:501–502, 1928.

[91] G. C. Righini, S. Pelli, M. Ferrari, C. Armellini, L. Zampedri, C. Tosello, S. Ronchin, P. Rolli, E. Moser, M. Montagna, A. Chiasera, and S. J. L. Ribeiro. Er-doped silica-based waveguides prepared by different techniques: RF-sputtering, sol-gel and ion-exchange. *Opt. Quantum Electron*, 34:1151–1166, 2002.

[92] H. Rong, A. Liu, R. Jones, O. Cohen, D. Hak, R. Nicolaescu, A. Fang, and M. Paniccia. An all-silicon Raman laser. *Nature*, 433:292–294, 2005.

[93] T. Sakamoto, H. Tsuda, M. Hikita, T. Kagawa, K. Tateno, and C. Amano. Optical interconnection using VCSELs and polymeric waveguide circuits. *J. Lightwave Technol.*, 18:1487–1492, 2000.

[94] D. M. Schaadt, B. Feng, and E. T. Yu. Enhanced semiconductor optical absorption via surface plasmon excitation in metal nanoparticles. *Appl. Phys. Lett.*, 86:063106, 2005.

[95] G. Schweiger, R. Nett, and T. Weigel. Microresonantor array for high-resolution spectroscopy. *Opt. Lett.*, 32:2644–2646, 2007.

[96] A. Serpengüzel and A. Demir. Silicon microspheres for near-IR communication applications. *Semicond. Sci. Technol.*, 23:064009, 2008.

[97] A. Serpengüzel, A. Kurt, T. Bilici, S. Isçi, and Y. O. Yilmaz. Resonant channel-dropping filter with integrated detector system based on optical fiber coupler and microsphere. *Jpn. J. Appl. Phys.*, 43:5778–5881, 2004.

[98] A. Shacham, K. Bergman, and L. P. Carloni. On the design of a photonic network-on-chip. In *First International Symposium on Networks-on-Chip*, 2007.

[99] V. Shalaev. Optical negative-index metamaterial. *Nature Phot.*, 1:41–48, 2007.

[100] N. Sherwood-Droz, A. Gondarenko, and M. Lipson. Oxidized silicon-on-insulator (OxSOI) from bulk silicon: a new photonic plataform. *Opt. Express*, 18:5785–5790, 2010.

[101] SIA Semiconductor Industry Association. *The National Technology Roadmap for Semiconductors*. SIA Semiconductor Industry Association, 1997.

[102] A. Sihvola. Metamaterials in electromagnetics. *Metamaterials*, 1:2–11, 2007.

[103] A. Sihvola. Limiting response for extreme-parameter particles in composites. *Eur. Phys. J. Appl. Phys.*, 49:33001–33008, 2010.

[104] D. R. Smith, J. B. Pendry, and M. C. K. Wiltshire. Metamaterials and negative refractive index. *Science*, 305:788–792, 2004.

[105] D. R. Smith, D. C. Vier, N. Kroll, and S. Schultz. Direct calculation of the permeability and permittivity for a left-handed metamaterial. *Appl. Phys. Lett.*, 77:2246–2248, 2000.

[106] G. Solomon, S. Goetzinger, W. Fang, Z. Xie, and H. Cao. Ultra-low threshold lasing in a quantum dot microdisk cavity. In *Conference on Lasers and Electro-Optics/Quantum Electronics and Laser Science Conference and Photonic Applications Systems Technologies*, CMD4, 2007.

[107] R. A. Soref and B. R. Bennett. Electro-optical effects in silicon. *IEEE J. Quantum Electron.*, 23:123–129, 1986.

[108] E. M. Strzelecka, D. A. Louderback, B. J. Thibeault, G. B. Thompson, K. Bertilsson, and L. A. Coldren. Parallel free-space optical interconnect based on arrays of vertical-cavity lasers and detectors with monolithic microlenses. *Appl. Opt.*, 37:2811–2821, 1998.

[109] L. Tang, S. E. Kocabas, S. Latif, A. K. Okyay, D. S. Ly-Gagnon, K. C. Saraswat, and D. A. Miller. Nanometer-scale germanium photodetector enhanced by a near-infrared dipole antenna. *Nature Phot.*, 2:226–229, 2008.

[110] L. Tang and D. A. B. Miller. Metallic nanodevices for chip-scale optical interconnects. *J. Nanophot.*, 3:030302, 2009.

[111] L. Tang, D. A. B. Miller, A. K. Okyay, J. A. Matteo, Y. Yuen, K. C. Saraswat, and L. Hesselink. C-shaped nanoaperture-enhanced germanium photodetector. *Opt. Lett.*, 31:1519–1521, 2006.

[112] K. Tsuzuki, T. Ishibashi, T. Ito, S. Oku, Y.Shibata, R. Iga, Y. Kondo, and Y. Tohmori. A 40-Gb/s InGaAlAs-InAlAs MQW n-i-n Mach-Zhender modulator with a drive voltage of 2.3V. *IEEE Photon. Technol. Lett.*, 17:46–48, 2005.

[113] A. Tulek, S. Akbulut, and M. Bayindir. Ultralow threshold laser action from toroidal polymer microcavity. *Appl. Phys. Lett.*, 94:203302, 2009.

[114] J. T. Verdeyen. *Laser Electronics*. Prentice-Hall Inc., New York, 1995.

[115] Y. Vlasov, W. M. J. Green, and F. Xia. High-throughput silicon nanophotonic wavelength-insensitive switch for on-chip optical works. *Nature Phot.*, 2:242–246, 2008.

[116] A. C. Walker, T-Y. Yang, J. Gourlay, J. A. B. Dines, M. G. Forbes, S. M. Prince, D. A. Baillie, D. T. Neilson, R. Williams, L. C. Wilkinson, G. R. Smith, M. P. Y. Desmulliez, G. S. Buller, M. R. Taghizadeh, A. Waddie, C. R. Stanley I. Underwood, F. Pottier, B. Vögele, and W. Sibbet. Optoelectronic systems based on InGaAs-complementary-metal-oxide-semiconductor smart-pixel arrays and free-space optical interconnects. *Appl. Opt.*, 37:2822–2830, 1998.

[117] R. Wang, A. D. Rakić, and M. L. Majewski. Analyisis of lensless free-space optical interconnnects based on multi-transverse mode vertical-cavity-surface-emitting lasers. *Opt. Commun.*, 167:261–271, 1999.

[118] J. Witzens, T. Baehr-Jones, and M. Hochberg. Silicon photonics: On-chip OPOs. *Nature Phot.*, 4:10–12, 2010.

[119] Q. Xu, S. Manipatruni, B. Schmidt, J. Shakya, and M. Lipson. 12.5 Gbits/s carrier-injection-based silicon micro-ring silicon modulators. *Opt. Express*, 15:430–436, 2007.

[120] Q. Xu, B. Schmidt, S. Pradhan, and M. Lipson. Micrometer-scale silicon electro-optic modulator. *Nature*, 435:325–327, 2005.

[121] J. Xue, A. Garg, B. Ciftcioglu, S. Wang, J. Hu, I. Savidis, M. Jain, M. Huang, H. Wu, E. G. Friedman, G. W. Wicks, and D. Moore. An intra-chip free-space optical interconnect. In *37th International Symposium on Computer Architecture (ISCA)*, 2B, 2010.

[122] E. Yablonovitch. Can nano-photonic sicilon circuits become an intra-chip interconnect technology? In *Proceedings of the 2007 IEEE/ACM International Conference on Computer-Aided Design*, 309, 2007.

[123] L. Yan and G. W. Hanson. Wave propagation mechanisms for intra-chip communications. *IEEE Trans. Antennas Propag.*, 57:2715–2724, 2009.

[124] E. Yuce, U. O. Gurly, and A. Serpenguzel. Optical modulation with silicon microspheres. *IEEE Phot. Tech. Lett.*, 21:1481–1483, 2009.

8

Security Issues in SoC Communication

José M. Moya

Politecnica University of Madrid, Spain

Juan-Mariano de Goyeneche

Politecnica University of Madrid, Spain

Pedro Malagón

Politecnica University of Madrid, Spain

CONTENTS

8.1   Introduction

The power consumption and the electromagnetic emissions of any hardware circuit are functions of the switching activity at the wires inside it. Since the switching activity (and hence, power consumption) is data dependent, it is not surprising that special care should be taken when sensitive data has to be communicated between SoC components or to the outside.

A common approach to implementing tamper-resistance involves the use of a separate secure co-processor module [101], which is dedicated to processing all sensitive information in the system. Any sensitive information that needs to be sent out of the secure co-processor is encrypted.

Many embedded system architectures rely on designating and maintaining selected areas of their memory subsystem (volatile or nonvolatile, off-chip or on-chip) as secure storage locations. Physical isolation is often used to restrict the access of secure memory areas to trusted system components. When this is not possible, a memory protection mechanism adopted in many embedded SoCs involves the use of bus monitoring hardware that can distinguish between legal and illegal accesses to these locations. For example, the CrypoCell security solution from Discretix features BusWatcher, which performs this function. Ensuring privacy and integrity in the memory hierarchy of a processor is the focus of [114], which employs a hardware secure context manager, new instructions, and hash and encryption units within the processor.

ARM's TrustZone and Intel's Trusted Execution Technology (formerly La-Grande) are examples of commercial initiatives aimed at establishing a clear separation of access to sensitive information and other HW/SW portions of the chip. Those components, which share sensitive information, should be designed carefully to avoid attacks.

In traditional cryptanalysis, which views a cipher as a black box operation that transforms the plaintext into the ciphertext using a secret key, many ciphers have no practical known weaknesses, and the only way to unlock the secret key is to try all possible combinations. As a result, as long as the number of combinations is large enough such that a complete search becomes de facto impossible, the cipher is said to be secure. For instance, the Rivest-Shamic-Adleman (RSA) public key encryption algorithm using 2048 bits can be used at least until the year 2030 before the expected computing power will be available to do the integer factorization of a 2048-bit number.

However, the cipher has to be implemented on a real device, which will leak additional information that can be used to determine the secret key. Indeed, similarly that feeling or sound can help to find the combination of a padlock, the power consumption or time delay of the device can reveal the value of the secret key. Early smart card implementations, for instance, implemented the modular exponentiation of the RSA algorithm using the textbook version of the square-and-multiply algorithm, in which the multiplication is only executed if the exponent bit equals 1. Since a multiplication does not have the same power signature as a squaring operation, it was possible to find out by observing the power consumption when a multiplication took place and thus to read off the private key from a single power trace.

The attacks, that use additional information leaking from the practical implementation, are also known as side-channel attacks (SCAs). First proposed in 1996 [55], they have been used since then to extract cryptographic key material of symmetric and public key encryption algorithms running on microprocessors, DSPs, FPGAs, ASICs, and high-performance CPUs from variations in power consumption, time delay, or electromagnetic radiation. Note that SCAs are noninvasive attacks, which means that they observe the device in normal operation mode without any physical harm to the device, and making the device tamperproof does not protect against the attacks. For certain attacks, it is not even necessary to possess the device or be in close proximity as demonstrated with a remote attack that successfully found key material of an OpenSSL webserver from nonconstant execution time due to conditional branches in the algorithm [18].

A careful design can offer increased resistance against the vulnerabilities. As pointed out earlier [100], security adds a new dimension to a design in addition to area, performance, and power consumption optimization. Side-channel attack resistance, which can be a showstopper to achieve security, is no exception and the thus far prevailing strategies rarely come cheap. It is also not very well understood how to analyze the strength of a design and the precise cost of a mitigations strategy is seldom fully and clearly communicated.

This makes the resistance increase hard to quantify and the design trade-offs difficult to make.

The remainder of this chapter is organized as follows. This introductory section presents the basics of side-channel attacks and countermeasures. Section 8.2 provides a more in-depth description of the different attacks based on measurements of the total power consumption. Sections 8.3, 8.4, and 8.5 detail the available tools and techniques to avoid this kind of attacks at different levels (logic, architecture, and algorithm). Section 8.6 gives some notes about efficient validation of countermeasures against power analysis attacks. Finally, Section 8.7 provides some advice about miscellaneous design decisions, and Section 8.8 draws some conclusions.

8.1.1   Side-Channel Attacks

A cryptographic primitive can be considered from two points of view: on the one hand, it can be viewed as an abstract mathematical object or black box (i.e., a transformation, possibly parameterized by a key, turning some input into some output); on the other hand, this primitive will be finally implemented in hardware or in a program that will run on a given processor, in a given environment, and will therefore present specific characteristics. The first point of view is the one of classical cryptanalysis; the second one is the one of physical security. Physical attacks on cryptographic devices take advantage of implementation-specific characteristics to recover the secret parameters involved in the computation. They are therefore much less general (since they are specific to a given implementation) but often much more powerful than classical cryptanalysis, and are considered very seriously by cryptographic devices manufacturers.

Such physical attacks are numerous and can be classified in many ways. The literature usually sorts them among two orthogonal axes:

1. Invasive vs. noninvasive: invasive attacks require depackaging the chip to get direct access to its inside components; a typical example of this is the connection of a wire on a data bus to see the data transfers. A noninvasive attack only exploits externally available information (the emission of which is however often unintentional) such as running time, power consumption, electromagnetic emissions, etc.

2. Active vs. passive: active attacks try to tamper with the devices proper functioning; for example, fault-induction attacks will try to induce errors in the computation. As opposed, passive attacks will simply observe the devices' behavior during their processing, without disturbing it.

Side-channel attacks are closely related to the existence of physically observable phenomenons caused by the execution of computing tasks in present microelectronic devices. For example, microprocessors consume time and power to perform their assigned tasks. They also radiate an electromagnetic

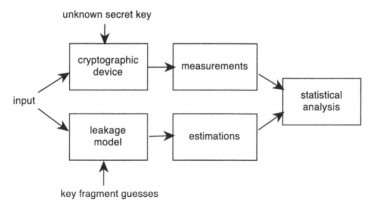

FIGURE 8.1
Structure of a typical side-channel attack.

field, dissipate heat, and even make some noise [108]. As a matter of fact, there are plenty of information sources leaking from actual computers that can consequently be exploited by malicious adversaries. In this chapter, we focus on power consumption and electromagnetic radiation that are two frequently considered side-channels in practical attacks. Since a large part of present digital circuits is based on CMOS gates, this introduction also only focuses on this technology.

Side-Channel Analysis (SCA) is one of the most promising approaches to reveal secret data, such as cryptographic keys, from black-box secure cryptographic algorithms implemented in embedded devices. Differential Side Channel Analysis (DSCA) exploits (small) differences in a set of measurements by means of statistics and is particularly well suited for the power analysis of block cipher implementations.

A side-channel attack works as follows (see Figure 8.1): it compares observations of the side-channel leakage (i.e., measurement samples of the supply current, execution time, or electromagnetic radiation) with estimations of the side-channel leakage. The leakage estimation comes from a leakage model of the device requiring a guess on the secret key. The correct key is found by identifying the best match between the measurements and the leakage estimations of the different key guesses. Furthermore, by limiting the leakage model to only a small piece of the algorithm, only a small part of the key must be guessed and the complete key can be found using a divide-and-conquer approach. For instance, an attack on the Advanced Encryption Standard (AES) generally estimates the leakage caused by a single key byte and as a result the 128-bit key can be found with a mere $16 \cdot 2^8$ tests. Finally, as the observations might be noisy and the model might be approximate, statistical methods are often used to derive the secret from many measurements.

Attacks that use a single or only a few observations are referred to as simple side-channel attacks. The "simple" refers to the number of measurements used and not to the simplicity of the attacks. In fact, they require a precise knowledge of the architecture and implementation of both the device and the algorithm and their effect on the observed measurement sample. As a result, they are relatively easy to protect from. For instance, the square-and-multiply algorithm can be implemented to perform the multiply independent of the exponent bit and only use the result if the exponent bit is actually one.

Attacks that use many observations are referred to as differential side-channel attacks. The timing attacks typically target variable instruction flow. The power attacks, and for that matter Electromagnetic Analysis (EMA) attacks, as electromagnetic fields are generated by the electric charge that flows, target data dependent supply current variations. These attacks are based on the fact that logic operations in standard static CMOS have power characteristics that depend on the input data. Power is only drawn from the power supply when a logic gate has a 0 to 1 output transition.

In the last decade, various attack methodologies have been put forward, such as Differential Power Analysis (Section 8.2.2) and Correlation Power Analysis (Section 8.2.3) as well as so called profiling attacks like the Stochastic Model [104] and Template Attacks [24]. As a consequence of the need for secure embedded devices such as smart cards, mobile phones, and PDAs research is also conducted in the field of DSCA prevention.

Early countermeasures include algorithmic masking schemes, noise generators, and random process interrupts. All of them have in common that they do not address the issue of side channel leakage directly, but aim at obfuscating the observables. Most of these countermeasures have been proven to be either insecure or circumventable, e.g., with higher-order attacks or with digital signal processing.

In recent years, research and industry have started to approach the issue of side-channel leakage right where it arises: at the gate level. There is a considerable body of research on gate level masking schemes, which again aim at obfuscating the leakage, and differential logic styles that focus on reducing the leakage. But also attacks against these secured logic styles have been published. Most of them exploit circuit "anomalies" as for example glitches and the early propagation effect.

In this chapter we focus on the power consumption side channel and hence on power analysis, which exploits the physical dependency of a device's power consumption and the data it is processing. Since they are noninvasive, passive and they can generally be performed using relatively cheap equipment, they pose a serious threat to the security of most devices requiring to handle sensitive information. Such devices range from personal computers to small embedded devices such as smart cards and Radio Frequency Identification Devices (RFIDs). Their proliferation in a continuously larger spectrum of applications has turned the physical security and side-channel issue into a real, practical concern. These attacks are better avoided by taking them into account during the whole design process.

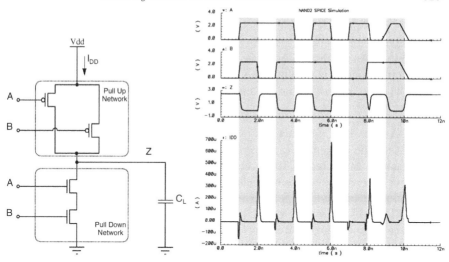

FIGURE 8.2
SPICE simulation result of a 2 input NAND gate in static CMOS logic. Figure
from [44]. Used with permission.

For this purpose, we start by covering the basics of side-channel attacks.
We discuss the origin of unintended leakages in recent microelectronic technologies and describe how simple measurement setups can be used to recover
and exploit these physical features. Then, we introduce some of the proposed
methods to avoid these leakages or to make them difficult to analyze.

8.1.2 Sources of Information Leakage

The single most important parameter used for in-chip or off-chip secret communications is the secret key used for encryption and decryption of data. While
the key stays constant for the duration of the encryption, the input values
for each subround of encryption are always changing. The input-dependent
characteristic of regular digital circuit will leak enough power consumption
information for a skilled adversary to successfully obtain the secret key.

Currently, the most widely used logic style to implement digital integrated
circuits is Complementary Metal-Oxide Semiconductor (CMOS). A main characteristic of CMOS logic is that it requires primarily dynamic power while its
static power consumption is almost zero (see Figure 8.2). The dynamic power
consumption is caused by transitions of logic signals that occur in the CMOS
circuit. The type and the probability of signal transitions depend on the logical function of a circuit and on the processed data. As a result, the power
consumption of a CMOS circuit depends on the data that is being processed
and hence, Differential Power Analysis (DPA) attacks as described in [54] are
possible.

For current CMOS technology, power consumption of a circuit is mainly contributed by dynamic power consumption and static power leakage consumption. Analyzing these types of power consumption helps us to find a solution on better controlling the total power consumption.

The sources of dynamic power information leakage come mainly from two categories:

- A digital circuit functions by evaluating the input voltage level and setting the output voltage level based on the input through a set of logic gates. In current CMOS technology, the logic value of a gate actually depends on the amount of charge stored on the parasitic capacitor at the output of the logic gate. A fully-charged capacitor represents logic-high (logic-1) whereas a depleted capacitor represents logic-low (logic-0). For each binary logic gate, there can only be four types of transitions on the gate output wire. Only one transition, from logic-0 to logic-1, actually draws current from the power supply to charge up the parasitic capacitor. By monitoring the amount of current consumed by the digital circuit at all times, we can get an idea on the relative amount of logic gates that are switching at any given time. This gives us some information about the circuit based on power consumption.

- Parasitic capacitances are not uniform for every gate. They depend on the type of gate, fanout of the gate, and also the length of the wire or net in between the current gate and its driven gates. Taking the length of wires as an example, even if two exact same gates with the same number of fanout are connected to the same set of successor gates, if the routing of the wires are different, the capacitance will differ. If both had a power-consuming transition, the amount of power consumed will be different, thus leaking important power information about the circuit.

Static power leakage consumption is a characteristic of the process used to manufacture the circuit. The exact amount of leakage for a given gate within a circuit is not controllable by a logic designer. Assuming the gates are manufactured exactly the same, then the static power leakage does not pose a threat. But this is not the case in the real world. Process variation plays an important role in the balancing of static power leakage. As process variation increases, the variation of the amount of charge leaked for every gate during a fixed period of time also increases. Unfortunately, the effect of the static power leakage due to process variations cannot be evaluated at design time, thus can only be seen through actual measurements on the finished product, whether it is an ASIC design or a design for FPGA.

Now we present a simple model of power consumption of cryptographic hardware in CMOS. This model is made to quickly simulate the contribution of one variable of the attacked algorithm in the power consumption of the chip and it also allows us to derive theoretical results.

We ignore coupling effects and create a linear model: the power consumption of the chip is the sum of the power consumption of its components. Hence, we can isolate the power consumption of the component (registers, memory cells, or bus lines) storing the temporary variable predicted by the selection function D from the power consumption of the rest of the device.

CMOS devices only consume power when logic states change after a clock pulse. Static power dissipation is negligible [23, 31]. Let $B = (\beta_m, \beta_{m-1}, ..., \beta_1)$ be the value of the relevant variables to store in the register. Let $A = (\alpha_m, \alpha_{m-1}, ..., \alpha_1)$ be the previous content of the register. Let c_{01} (resp. c_{10}) be the amount of power needed to flip a bit of the register from 0 to 1 (resp. from 1 to 0). If after the clock top the value B is stored in the register, the power consumption of the chip after the clock top is:

$$C(t) = \sum_{i=1}^{m}(1 - \alpha_i)\beta_i c_{01}(t) + \alpha_i(1 - \beta_i)c_{10}(t) + w(t) \qquad (8.1)$$

Where $w(t)$ is the power consumption of the rest of the chip, that can be modelled by a Gaussian noise whose mean represents the constant part of power consumption between two different executions of the algorithm. This assumes that the sequence of instructions executed by the chip does not depend on the data. This model can be easily simulated to verify the soundness of a new attack.

The data-dependencies in the dynamic power consumption causes four major sources of leakage:

1. *Energy imbalance.* Energy imbalance can be measured as the variation in energy consumed by a circuit processing different data. If e_1 and e_2 are the energy consumptions of two input patterns, then the numerical value of the imbalance is calculated as:

$$d = \frac{|e_1 - e_2|}{e_1 + e_2} \cdot 100\% \qquad (8.2)$$

2. *Exposure time.* The longer the imbalance is visible, the easier it is to measure. This is why the exposure time of the imbalance should be minimized alongside the imbalance reduction.

3. *Early propagation.* The early propagation is the ability of a gate to fire without waiting for all its inputs. Early propagation causes the data-dependent distribution of circuit switching events in time. The effect of early propagation is bounded by half of the clock cycle. One way to avoid the early propagation is to balance all paths by inserting buffers in such a way that all inputs of each gate arrive simultaneously.

4. *Memory effect.* The memory effect is the ability of a CMOS gate to remember its previous state, due to capacities not being completely discharged.

Note that early propagation and memory effect have much less impact on the security features of a circuit than imbalance and exposure time. So, it is essential to minimize the circuit imbalance and exposure time before optimizing the circuit for the early propagation and memory effect metrics.

8.1.3 Overview of Countermeasures

Countermeasures against side-channel attacks range among a large variety of solutions. However, in the present state-of-the-art, no single technique allows to provide perfect security. Protecting implementations against physical attacks consequently intends to make the attacks harder. In this context, the implementation cost of a countermeasure is of primary importance and must be evaluated with respect to the additional security obtained.

Side-channel attacks work because there exists some leakage that depends on intermediate values of the executed cryptographic algorithm. Therefore, the goal of a countermeasure is to avoid or at least to reduce these dependencies. Depending on the way these dependencies are avoided, current countermeasures against side-channel attacks can be classified into two main families: hiding techniques and masking techniques.

In the case of hiding, data-dependent leakages are avoided by breaking the link between the leaked magnitude and the processed data values. Hence, protected devices execute cryptographic algorithms in the same way as unprotected devices, but hiding countermeasures make it difficult to find exploitable information in the power traces.

The power consumption of a cryptographic device can be made independent of the processed data values in two different ways: by randomizing the power consumption in each clock cycle, or by equalizing the power consumption in every clock cycle. And both techniques can be applied in the time dimension (by shuffling operations or inserting dummy operations, for example) and the amplitude dimension (by increasing the non-data-dependent power consumption or by reducing the data-dependent power consumption). In most cases these countermeasures imply increasing significantly the required resources, but they are quite robust and attack-independent.

On the other hand, masking techniques attempt to remove the correlation between power consumption and secret data by randomizing the power consumption such that the correlation is destroyed. This is achieved by concealing intermediate values with a random number (mask). The operation performed with the mask depends on the cryptographic algorithm, but it is usually the Boolean exclusive-or function, the modular addition, or the modular multiplication.

An advantage of this approach is that it can be implemented at the algorithm level without changing the power consumption characteristics of the cryptographic device. It can also be implemented at the logic level.

Masking countermeasures are proved to be secure against first-order differential power analysis attacks, but they are not enough for higher-order attacks.

8.2 Power Analysis Attacks

Power Analysis attacks are performed by measuring the power consumption of a device as it operates, and then using these measurements to determine secret information (such as secret keys and/or user PINs). These robust attacks are often called "external monitoring attacks," as they are noninvasive and use observations of a device's power consumption during its operation.

Simple Power Analysis (SPA) attacks recover the secret keys from direct observation of individual power consumption measurements. They are most effective when there is a significant amount of sensitive information leakage.

Differential Power Analysis (DPA) attacks employ statistical techniques to extract information from multiple power consumption measurements. They are highly effective at extracting secrets even when the information available within any individual measurement is much smaller than unknown electrical activity, measurement error, and other noise sources.

With both SPA and DPA, the device under attack performs its ordinary cryptographic processing operations. As a result, the attacks generally cannot be stopped through traditional antitamper mechanisms such as intrusion sensors or other attack detectors.

SPA and DPA are effective against small single-chip devices, large SoCs, and multichip products. For systems where the cryptographic processing is only a small contributor to the overall variation in power consumption, DPA is typically required.

SPA and DPA attacks are normally classified as requiring a low to moderate degree of attacker sophistication. The hardware typically used for the process consists of a PC and a digital storage oscilloscope. Suitable oscilloscopes are widely available, and sell for under $500 used. Once automated, SPA attacks are virtually instantaneous, and typical DPA attacks on unprotected devices take a few minutes to a few hours to complete.

8.2.1 Simple Power Analysis

Simple Power Analysis (SPA) is a side-channel attack first introduced by Kocher et al. in [56] as "a technique that involves directly interpreting power consumption measurements collected during cryptographic operations." The goal of SPA attacks is to obtain information about the device under attack working from few power traces, even just one. The information revealed covers from the algorithm to the cryptographic key in a completely successful attack.

Let's suppose the attacker localizes an instant when an instruction that manipulates sensitive information is executed (e.g., load part of the secret key to the accumulator). Depending on the Hamming Weight (number of "1") of the key data manipulated, the amplitude of the power trace in that instant varies. If the attacker is expert and has a consumption reference model, he can estimate the HW of the key from the power amplitude.

Another scenario could be an implementation where the instructions executed depend on data (e.g., conditional branch depending on a bit value). If the attacker has information about the implementation and localizes the execution of the algorithm in the power trace, it can derive the data processed from the duration of the cycles. If the execution duration is different for different instructions, it is possible to assign sections of the power trace to concrete instructions executed.

We have two diffent sources for an attack, although authors typically refer SPA to the amplitude based ones.

SPA attacks require detailed knowledge about the implementation of the algorithm in the device. The attack process starts with a thorough analysis of the target device and its implementation of the algorithm. Useful information includes:

- Algorithm implementation: knowing the concrete implementation leads to points of interest in the power trace, where the sensitive data is manipulated. Sometimes it is well known and published and there is no obstacle. On the other hand, the implementation or the algorithm might be unknown to the attacker, who needs to analyze power traces, search patterns and conclude which algorithm is used and where data is manipulated. Given a power trace, a target device and the characteristics of the different encryption algorithms, the attacker infers which algorithm is used. In [56], the authors show an example of power trace with a pattern repeated 16 times, the rounds of DES algorithm.

- Points of interest: once the algorithm is known, it is mandatory to map the power trace to the operations of the algorithm. It can be a high-level mapping, assigning sections of the power trace to blocks of the algorithm, or a precise mapping, assigning concrete instructions to the power trace.

- Target device: the hardware implementation affects the power trace. In an 8-bit microcontroller, there is typically a high correlation between data and consumption. Data bus lines are the main source of leakage. However, the buses can be precharged to "1" (which means that there is more consumption as long as are more "0" in data) or to "0," data and code can share buses, a hardware module can have special consumption (barrel shifter, hardware multiplier, etc.). These assumptions should be confirmed with a device similar to the target of the attack and the attacker must create its own model. The experience of the attacker must include these knowledges to perform a profitable attack.

We can find descriptions of profiling methods since 1999, when Biham and Shamir [11] described a method to map parts of a power trace to the key scheduling operation of the AES algorithm. Power traces of a device executing the AES algorithm were analyzed in a device similar to the target of the attack. Profiling does not require a lot of traces of the device under attack, which is one of the limitations that SPA overcomes, but traces of similar devices available are needed to get experience and create a model. In [36], Fahn and Pearson describe the profiling stage of their attack Inferential Power Analysis (IPA). In the process, they included in the profiling stage every round of the DES algorithm, as it is the same code with different arguments.

In [76] and [2], the authors describe the process followed to extract a model of an 8-bit microcontroller before attacking a device. In [5], authors obtain their own model of smart-cards from experience after realizing that stated models were not suitable for their devices.

Simple Power Analysis includes three major attack families: Visual Inspection, Template, and Collision Attacks.

Visual Inspection requires great personal knowledge from the attacker about the implementation of the algorithm and the device. An example is [56], where the authors highlight the visual recognition of the DES algorithm and its 16 rounds. In higher resolution views authors point out different rotations of the key based on the repetition of a concrete pattern inside the round power trace. Moreover, they distinguish between instructions, so they can conclude if a conditional branch skips a jump instruction. All these results can lead to useful information for other attacks, even if they fail in the extraction of the key.

Template attacks were introduced by Chari et al. [24]. In a template-based power analysis attack, the attacker is assumed to know the power consumption characteristics of some instructions of a device. This characterization is stored and called a *template*. Templates are then used as follows. In a template-based DPA attack, the attacker matches the templates based on different key hypotheses with the recorded power traces. The templates that match best indicate the key. This type of attack is the best attack in an information theoretic sense, see [24].

Template attacks compare the power trace of the attack with templates created from previous analysis following the maximum-likelihood decision rule. In the profiling stage, the characterization phase, a template is created as a multivariate normal distribution, defined by its mean vector and its covariance matrix, from power traces. We can have templates for a pair instruction-operand, or for a pair data-key. Once every possible value has its template (e.g., every pair key-value with its template) the comparison with the power trace of the device under attack can be done. This second stage, the matching phase, involves calculating the probability density function of the multivariate normal distribution with every template. The template with the highest probability indicates the correct key.

Some difficulties emerge when considering the practical characterization of a device. Power traces from the same source data (instruction-operand, data-key) are grouped, the points of interest are set and the mean vector and covariance matrix are calculated. More interesing points involve more information. On the other hand, they grow the covariance matrix quadratically. The attacker must arrive at a compromise solution depending on the device under attack. These attacks were first described in [24].

Collision attacks exploit the coincidence of an intermediate value in two different encryption runs. If two different plain-texts (or cipher-texts) have a common intermediate value detected through SPA, the collection of possible key values is reduced to a subset. There are more than one point to detect the collision of the two encryptions, because intermediate values are manipulated in more than one point: load into accumulator, operate, save in memory, ... To detect a collision, the previous attacks (mainly template attack). Collision attacks were first applied by Wiemers and Schramm et al. [106], identifying collisions in a power trace of a DES implementation, following its application to AES [105]. These attacks were enhanced by [65] including almost-collisions in the attack (much more points of interest reducing possible key values) for Feistel ciphers.

8.2.2 Differential Power Analysis Attacks

Differential Power Analysis (DPA) attacks are the most popular type of power analysis attacks. The main advantage of DPA attacks is that a detailed knowledge about the attacked device is not required. Moreover, DPA attacks can reveal the secret key even if the recorded power traces are extremely noisy. However, DPA presents some drawbacks. The first one is that the attacker needs several power traces of the execution of the algorithm in the device under attack. The concrete number of traces depends on the noise of the signal and the leakage of the implementation, but it usually requires physical access to the device for a long time to obtain the needed traces. Another drawback is that the traces need to be perfectly synchronized.

It was first introduced by P. Kocher et al. [54, 56]. There are two main stages in a DPA attack: data acquisition and data analysis.

In data acquisition stage, the attacker executes several times the algorithm in the device under attack. For each execution, it must save data related to the execution (input or output of the algorithm) and power traces of the execution, with a common reference that enables the synchronization of the traces. We define the following: data collection involves running the encryption algorithm for N random values of plain-text input. For each of the N inputs (PTI_i), an output cipher-text is obtained (CTO_i). During the execution a discrete time power signal (S_{ij}) is collected, which is a sampled version of the power consumed during the portion of the algorithm that is being attacked. i corresponds to the plain-text input PTI_i that produced the signal and j corresponds to the instant of the sample.

The idea of the second stage, data analysis, is to try different keys and check which matches the traces collected. This operation is not done to the whole trace at once, but in a concrete instant of the trace, an instant where the power consumption depends on an intermediate value of data (v) and part of the key (k). Consequently, the first step is to choose an instant (concrete j) for the attack that corresponds to a known intermediate state of the algorithm. For each possible k, associated v can be calculated from known data input (PTI_i) applying the algorithm or from known data output (CTO_i) applying the inverse algorithm. At this point, the attacker has a vector of real power values (S_{ij}) and, for each possible k, a vector of intermediate values (v_{ik}), $i \in 1, ..., N$, j instant chosen, and k the possible key. The next step is to find the key with greater correlation between calculated intermediate values and real power values. In the DPA attack the difference of means method is used.

The S_{ij} vector is split into two vectors (S_0 and S_1) using a *selection function*, D. D assigns S_{ij} to S_0 when v_{ik} is 0 and to S_1 when v_{ik} is 1. For every possible part of the key we have two subsets S_0 and S_1. For each subset the average power signal is:

$$A_0[j] = \frac{1}{|S_0|} \sum_{S_{ij} \in S_0} S_{ij} \qquad (8.3)$$

$$A_1[j] = \frac{1}{|S_1|} \sum_{S_{ij} \in S_1} S_{ij} \qquad (8.4)$$

where $|S_0| + |S_1| = N$. The difference is DT_j. If we had this DT_j for every j in the power trace, we would have a function $DT[j]$, denoted as the differential trace. The differential trace is:

$$DT[j] = A_0[j] - A_1[j] \qquad (8.5)$$

In $DT[j]$, peaks appear in the instants where there is a correlation. For a key guess, peaks appear when there is a relationship between traces associated to the same subset in that instant and it is different than the other subset (data dependant instant). If the point is not data dependent, an almost-zero value can be supposed: relationship between traces associated to the same subset exists but it should be similar in both subsets. If the instant is data dependent but the key guess is not correct, the relationship in each subset should be smaller and similar in both of them. P. Kocher [56] assumed that if the key guess is incorrect, $DT[j]$ should approach a flat trace but researchers have noticed (like T. Messerges in [80]) that "the actual trace may not be completely flat, as D with K_k incorrect may have a weak correlation to D with K_k correct." However, it has been shown [9] that this correlation even can be strong and depends on the attacked algorithm and the selection function D.

Consequently, the partial key guess with highest value of the difference DT_j is the best candidate. If possible, the selection function D and the instant j should correspond to intermediate values mutually independent in order to be sure that the greatest value of DT_j corresponds to the correct hypothesis [9].

Although we have seen a selection function D for one instant and one data bit, multiple bits and multiple instants can be used in order to increase the difference between the correct key guess and the incorrect ones. For instance, Messerges et al. [80] use d-bit data and two sets, and they assign those with greater Hamming Weight to S_1 ($H(V_{ij}) \geq d/2$) and the rest to S_0 ($H(V_{ij}) < d/2$). In [9] Bevan improves the DPA attack using 4-bit D function. Instead of deciding the key value when the four selection functions agree, they sum the four differences of means to reach a solution faster. This solution is possible because every bit influences the power consumption at the same time. Instead of always having a binary division, in [5] they use d-bit attacks and they divide the set of traces into $d+1$ subsets. In [64] there is a formal definition of the differential trace that includes the mentioned attacks as particular cases, assigning values to a_j in

$$DT = \sum_{c=0}^{d} a_c \frac{\sum_{S_c} S_{ij}}{|S_c|} \tag{8.6}$$

8.2.3 Correlation Power Analysis

Correlation Power Analysis (CPA) attack is the name for a DPA attack that uses the correlation factor method instead of the difference of means to solve the problem of the "ghost peaks," which are peaks obtained with erronous key guesses whose intermediate values partially correlate to the correct ones. It was first proposed in [16]. Brier states that the chosen intermediate value is a uniform random variable. If values stand between 0 and $2^d - 1$ (d-bit value) the average of the Hamming Weight is $\mu_H = d/2$ and its variance is $\sigma_H^2 = d/4$. Moreover, they generalize the model, as stated in [29] and [31], using the Hamming Distance from a previous value (constant) referred as R, which also has the same distribution. They model the power consumption at one point (S), which is a linear function of the Hamming Distance $HW(D \oplus R)$:

$$S = aH(D \oplus R) + b \tag{8.7}$$

where a is a scalar gain and b represents power consumptions independent from data (noise, time dependent components, ...). The correct key guess is the one that maximizes the correlation factor ρ_{SH}.

The estimated value of the correlation factor, $\hat{\rho}_{SH}$, is obtained from the traces stored and data known with the formula:

$$\hat{\rho}_{SH}(i, R) = \frac{N \sum S_{ij} H_{j,R} - \sum S_{ij} \sum H_{j,R}}{\sqrt{N \sum S_{ij}^2 - (\sum S_{ij})^2} \sqrt{N \sum H_{j,R}^2 - (\sum H_{j,R})^2}} \tag{8.8}$$

According to [64] CPA requires 15% of traces required by DPA, and it solves the problem of "ghost peaks." [64] relates the correlation factor to the

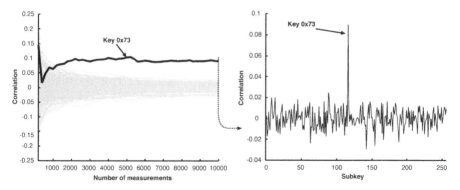

FIGURE 8.3
CPA attack results of the *Fastcore* AES chip. Figure from [44]. Used with permission.

general expression of DPA attacks, dividing traces into $d+1$ classes depending on the Hamming Distance in the intermediate value chosen. The result, adapting nomenclature to CPA, is

$$\rho_{SH}(i, R) = \frac{cov(S_i, H)}{\sigma_{S_i} \sigma_H} = \frac{\sum_{j=0}^{d} a_j \frac{\sum_{G_j} S_{ij}}{|G_j|}}{\sigma_{S_i} \sigma_H} \quad (8.9)$$

where G_j is the subset of traces whose Hamming Distance between the actual state of the message and the reference state R is equal to j, and $|G_j|$ is the cardinal of the subset. a_j are the coefficients, similar to those in DPA; and for CPA, if data is uniformly distributed and the number N of traces is large, these coefficients are calculated as

$$a_j = \frac{1}{2^d} \binom{j}{d} \left(j - \sum_{k=0}^{d} \frac{k}{2^d} \binom{k}{d} \right) \quad (8.10)$$

Figure 8.3 shows the results of a sample CPA attack to an ASIC AES implementation. The graph on the left shows the correlation of all $K = 256$ subkey permutations to the measurement results as a function of the number of measured samples S. On the right, the correlation of all $K = 256$ subkey permutations is given for 10,000 measurements.

8.2.4 Stochastic Methods

Advanced stochastic methods have turned out to be efficient tools to optimize pure timing and combined timing and power attacks. Using such methods, the efficiency of some known attacks could be increased considerably (up to a factor of fifty) [104].

The underlying working hypothesis for side-channel cryptanalysis assumes that computations of a cryptographic device have an impact on instantaneous physical observables in the (immediate) vicinity of the device, e.g., power consumption or electromagnetic radiation. The dependency of the measurable observables on the internal state of a cryptographic algorithm is specific for each implementation and represents the side channel. This relationship can be predicted, e.g., by applying a (standard) power consumption model of the implementation such as the Hamming weight or Hamming distance model. Alternatively, the probability density of the observables can be profiled in advance for every key dependent internal state of the implementation.

The Stochastic Model [104] assumes that the physical observable $I_t(x, k)$ at time t is composed of two parts, a data-dependent part $h_t(x, k)$ as a function of known data x and subkey k and a noise term R_t with zero mean: $I_t(x, k) = h_t(x, k) + R_t$. $I_t(x, k)$ and R_t are seen as stochastic variables.

The attack consists of two main phases: a profiling phase and a key extraction phase. The profiling phase processes $N = N_1 + N_2$ samples representing a known subkey k and known data $x_1, x_2, ..., x_N$ and consists of two parts. The first part yields an approximation of $h_t(\cdot, \cdot)$, denoted as $\tilde{h}_t^*(\cdot, \cdot)$, i.e., the data-dependent part of the side channel leakage, in a suitable u-dimensional chosen vector subspace $\mathcal{F}_{u;t}$ for each instant t. The second part then computes a multivariate density of the noise at relevant instants. For the computation of $\tilde{h}_t^*(\cdot, \cdot)$, an overdetermined system of linear equations has to be solved for each instant t. The $(N1 \times u)$ design matrix is made up by the representation of the outcome of a selection function combining k and $x_n (1 \le n \le N1)$ in $\mathcal{F}_{u;t}$ and the corresponding N_1-dimensional vector includes the instantiations i_{t_n} of the observable. As a preparation step for the computation of the multivariate density, p side-channel relevant time instants have to be chosen based on $\tilde{h}_t^*(\cdot, \cdot)$. The complementary subset of N_2 measurements is then used to compute the covariance matrix C. For this, p-dimensional noise vectors have to be extracted from all N_2 measurements at the p instants by subtracting the corresponding data-dependent part. Given the covariance matrix C, this leads to a Gaussian multivariate density $\tilde{f}_0 : \mathbb{R}^p \to \mathbb{R}$.

Key extraction applies the "maximum likelihood principle." Given N_3 measurements at key extraction, one decides for key hypothesis $k \in 1, ..., K$ that maximizes

$$\alpha(x_1, ..., x_{N_3}; k) = \prod_{j=1}^{N_3} \tilde{f}_0 \left(i_t(x_j, k^o) - \tilde{h}_t^*(x_j, k) \right) \tag{8.11}$$

Herein, k^o is the unknown correct key value.

The "minimum principle" could also be applied as shown in [104]. Although it is less efficient for key extraction, it requires less measurements.

Gierlichs et al. [40] show that towards a low number of profiling measurements stochastic methods are more efficient than Template attacks, whereas towards a high number of profiling samples Templates achieve superior performance results.

8.2.5   Higher Order DPA

The commonly suggested way to fight against first-order power analysis is random masking [29] wherein intermediate computations are handled under a probabilistic form to defeat statistical correlation. In fact, Boolean and arithmetic maskings are certainly the most intensively used approach to protect power-sensitive cryptographic software as it appears that data randomization usually works well in practice, even when hardware countermeasures are not available. It is known, however, that masking can be defeated if the attacker knows how to correlate power consumption more than once per computation. This is known as second-order, or more generally higher-order, power analysis and was originally suggested by Messerges in [79]. These attacks are known to be more complex and delicate to carry out because they usually require the attacker to have a deeper knowledge of the device, although this might be alleviated in particular cases [129].

k-order DPA attacks generalize (first-order) DPA attacks by considering simultaneously k samples, within the same power consumption trace, that correspond to k different intermediate values.

The main application of higher-order DPA attacks is to attack systems protected against first-order DPA [79]. A method commonly used to fight (first-order) DPA attacks is masking. Each intermediate sensitive value w is XOR-ed with a random value r, unknown to the adversary. Therefore, the adversary no longer sees a DPA peak in the differential trace.

However, if the adversary knows the time periods, j and k, when the values of r and of $w \oplus r$ are manipulated, respectively, then he can evaluate

$$\overline{DT}_2 = \langle S_{ik} - S_{ij} \rangle_{S_{ij} \in S_1} - \langle S_{ik} - S_{ij} \rangle_{S_{ij} \in S_0} \qquad (8.12)$$

In case the adversary only knows the offset $\delta = k - j$ (but not j nor k), the previous attack can be extended as a "known-offset second-order DPA attack" [129]. The adversary evaluates the second-order differential trace:

$$DT_2[j] = \langle S_{i(j+\delta)} - S_{ij} \rangle_{S_{ij} \in S_1} - \langle S_{i(j+\delta)} - S_{ij} \rangle_{S_{ij} \in S_0} \qquad (8.13)$$

Again, under certain assumptions, the second-order DPA trace exhibiting the highest DPA peak will likely uncover the value of the intermediate result.

8.2.6   Attacks on Hiding

Hiding data-dependent information of a cryptographic device can be achieved by two different approaches. The first approach blurs the information by varying the power-consumption characteristic in its amplitude. The second approach randomizes the execution of operations in the time dimension. However, hiding can also occur in an unintended manner. There, misaligned traces in the amplitude and also in the time make the analysis of side channels largely infeasible.

The common goal of hiding countermeasures is to make the power consumption independent of the data values and the operations. However, in practice this goal can only be achieved to a certain degree. Therefore, attacks to protected devices are still possible [72].

There exist techniques that increase the performance of attacks on hiding through trace preprocessing. The most obvious and commonly used preprocessing technique is filtering. By applying different filters it is possible to reduce noise that originates from narrowband interferers such as RFID readers. Filtering of these perturbing signals helps to evade hiding in the amplitude dimension. Hiding in the time dimension, in contrast, can be obviated by for example integration of power or Electromagnetic (EM) traces. Specific points in time are thereby summed up before performing the attack. In practice, only points are chosen that exhibit a high side-channel leakage. These points form a kind of comb or window that can be swept through the trace in order to obtain the highest correlation. This technique is therefore often referred as to *windowing*. However, it is evident that this technique implies the knowledge of certain points in time where the leakage of information is high. If no knowledge of this leakage is available, it shows that the performance of this attack is rather low due to the integration of unimportant points.

Hiding techniques reduce the signal-noise ratio (SNR) in the leaking side channel. Therefore, apart from trying to undo the hiding process, the way to attack the protected device is just to analyze more power traces.

DPA attacks on misaligned power traces are significantly less effective than attacks on aligned traces [73]. Therefore, the goal of attackers is always to align the traces before performing DPA attacks. However, this is not always possible. For example, the power traces can be too noisy to be aligned successfully. In this case, attackers usually resort to other preprocessing techniques, such as integration. The misalignment of traces reduces the correlation linearly with the number of clock cycles inserted, while the integration reduces the correlation only with the square root of the number of integrated clock cycles [73].

Chari et al. [23] analyzed the effectiveness of the randomization of the execution of cryptographic algorithms. They pointed out that the effect of the randomization can be undone by aligning the power traces with the help of signal processing techniques, and the number of needed power traces grows quadratically with the number of shuffled opperations. With the sliding-window DPA attack, introduced by Clavier et al. [29], the required power traces grow only linearly with the number of shuffled opperations.

Another related technique uses Fast Fourier Transform (FFT) to transform the traces into the frequency domain. Instead of performing differential analysis in the time domain (such as done in standard DPA and Differential Electromagnetic Analysis (DEMA) attacks), the analysis is performed in the frequency domain. This allows a time-invariant analysis of side-channel leakages across the overall signal spectrum. This analysis is also referred as Differential Frequency Analysis [38].

8.2.7 Attacks on Masking

In a masked implementation, all intermediate values that are computed during the execution of a cryptographic algorithm are concealed by a random value, which is called *mask*. This ensures, if correctly implemented, a pairwise independence between masked values, unmasked values, and masks. Hence, only if pairwise independence does not hold for some reason, a masking scheme is vulnerable to DPA attacks. DPA attacks based on one intermediate value that is predicted are referred to as first-order DPA attacks, and they do not work to attack perfect mask schemes.

As previously seen, higher-order DPA attacks exploit the joint leakage of several intermediate values. In practice, second-order DPA attacks are enough for real masking implementations, as masks are reused for efficiency [73].

Second-order DPA attacks are very similar to first-order DPA attacks except that they require preprocessing the power traces, in order to combine the effect of the two intermediate values in a single point of the DPA trace. If we choose two intermediate values u and v, these values will not occur as such in the device, but their masked versions $u_m = u \oplus m$ and $v_m = v \oplus m$. In attacks on Boolean masking, the combination function for the two intermediate values is typically the exclusive-or function:

$$w = u \oplus v = u_m \oplus v_m \qquad (8.14)$$

Note that we can calculate the value of the combination of two masked intermediate values without having to know the mask.

However, second-order DPA attacks are not always required to break a masking scheme. There are some cases where the independence property is not fulfilled, and therefore first-order attacks may suffice (see [73]):

- *Multiplicative masking.* Multiplicative masking, that is often used in asymmetric ciphers, does not satisfy the independence condition because if $v = 0$, then $v \times m = 0$. The masked value has certain dependence on the unmasked value, and therefore it is vulnerable to first-order DPA attacks. In particular, it is vulnerable to zero-value DPA attacks [4].

- *Reused masks.* If masks are reused, the result of an exclusive-or of two intermediate values concealed with the same Boolean mask would be unmasked. Also, if a device leaks the Hamming distance of the intermediate values, and two values concealed with the same mask pass consecutively through the bus, the Hamming distance of the unmasked values would be leaked.

- *Biased masks.* Uniformly distributed masks are essential for the security of a masking scheme. Hence, the attacker could try to force a kind of bias into the masks, either by manipulating the device (e.g., fault attacks), or by selecting a subset of the measured traces that correspond to a subset of masks.

8.2.8   ECC-Specific Attacks

Elliptic curve cryptosystems (ECC) where introduced in the late 80s [81, 53] and they constitute attractive alternatives to other public-key cryptosystems, such as RSA, due to their comparable grade of security achieved at higher speeds with shorter keys. That characteristic constitutes a benefit for constrained devices, such as smart-cards, with scarce memory and computational resources.

These kinds of devices are a primary target for DPA attacks, and several simulations and experimental results show that they are effective on the ECC on such devices [30, 46].

In essence, an elliptic curve defines a set of points with some properties, including scalar multiplication. Given a point of the curve P and a scalar integer d, it is easy to compute the scalar multiplication dP, but there is no known efficient algorithm to compute d given dP and P. The problem is called *the elliptic curve discrete logarithm problem*, and it assures the security of ECC. The point P and the scalar multiplied point dP are public, while the scalar d is secret.

DPA and SPA attacks concentrate on the scalar multiplication (which is by far the most time consuming part in ECC) to extract d, so several countermeasures were proposed against them.

The *Montgomery method* [82], the *add-and-double always* [30] method and the *unified add/double formula* [17] where countermeasures accepted for some time against SPA, with the Montgomery method also avoiding timing attacks [89, 90].

However, these countermeasures were unable to avoid DPA attacks, and the *random projective coordinates*, the *random elliptic curve isomorphisms* and the *random field isomorphisms* complementary countermeasures were proposed [30, 50] to prevent them, so a combination of one of the above SPA and DPA countermeasures seemed secure.

This changed in 2003 when some ECC-specific DPA variations were developed: Goubin proposed the *refined power analysis* (RPA) attack [41] and Akishita and Takagi extended it to the *zero value analysis* (ZVA) attack [4], both being resistant to the above combination of countermeasures.

The RPA attack is based on the exploitation of *special points* of the elliptic curve with one of their coordinates being zero: $(x, 0)$ or $(0, y)$. One characteristic of these points is that, even if they are randomized, one of their coordinates always remains zero, and the attacker can detect, using DPA, when such a point is used in the scalar multiplication because the power consumption of 0 is quite characteristic. Thus, the attacker chooses values of P based on the guess of the bit-values of d, one bit at a time, so a special point is reached (and detected) when the guess is correct. The attack depends on the existence of such 0-coordinate points in the curve but, as shown by Goubin, that is the case with most standardized elliptic curves.

In the same year the RPA attack was published, it was extended by means of the *zero-value point attacks* [4], which tries to observe that characteristic power consumption trace of 0 in the outputs of the elliptic curve addition and elliptic curve doubling blocks rather than in intermediate points.

8.2.9 Power Analysis Attacks on Faulty Devices

Even if the device is designed carefully, avoiding any data-dependent leakage during its normal operation, it still may leak information in presence of fault. In September 1996 Boneh, Demillo, and Lipton from Bellcore announced an ingenious new type of cryptanalytic attack that received widespread attention [12]. This attack is applicable only to public key cryptosystems such as RSA, and not to secret key algorithms such as the Data Encryption Standard (DES). According to Boneh, "The algorithm that we apply to the device's faulty computations works against the algebraic structure used in public key cryptography, and another algorithm will have to be devised to work against the non-algebraic operations that are used in secret key techniques." In particular, the original Bellcore attack is based on specific algebraic properties of modular arithmetic, and cannot handle the complex bit manipulations that underly most secret key algorithms.

In [10], Biham and Shamir describe a new attack, called Differential Fault Analysis, or DFA, and showed that the occurrence of faults is a serious threat to cryptographical devices. Their attacks exploit faults within the computation of a cryptographic algorithm to reveal secret information. They showed that it is applicable to almost any secret key cryptosystem. They also implemented DFA in the case of DES, and demonstrated that under the same hardware fault model used by the Bellcore researchers, the full DES key could be extracted from a sealed tamper-resistant DES encryptor by analyzing between 50 and 200 cipher-texts generated from unknown but related plaintexts. In more specialized cases, as few as five ciphertexts are sufficient to completely reveal the key. The power of Differential Fault Analysis is demonstrated by the fact that even if DES is replaced by triple DES (whose 168 bits of key were assumed to make it practically invulnerable), essentially the same attack can break it with essentially the same number of given cipher-texts.

Such faults can be provoked by an adversary in several ways. One method to inject faults is to insert peaks into the clock supply. These peaks may corrupt data transferred between registers and memory [57]. This is called a *glitch attack*. As no modification of the device is needed, glitch attacks belong to the noninvasive attacks. In contrast to noninvasive methods, invasive and semi-invasive ones are also targeting data storage and not only its transfer. Both attack methods require direct access to the chip. Mostly, a decapsulation procedure has to be applied to expose it. Semi-invasive attacks inject faults without electrical contact to the chip surface.

A conventional semi-invasive attack uses light [110], generating transient faults, or electromagnetic fields [98], resulting in transient or permanent faults. Invasive attacks establish direct contact to the chip. The behavior of the device can be permanently changed in this way. Using a probing needle, the content of Electrically-Erasable Programmable Read-Only Memory (EEPROM) cells can be manipulated [6]. It is also possible to injected permanent faults by modifying the device itself and cutting some wires using a laser cutter or a focused ion beam (FIB) [57].

Differential fault analysis is not limited to finding the keys of known ciphers. An asymmetric fault model makes it possible to find the secret key stored in a tamper-resistant cryptographic device even when nothing is known about the structure and operation of the cryptosystem.

Moreover, it is possible to use DFA to extract the exact structure of an unknown DES-like cipher sealed in the tamper-resistant device, including the identification of its round functions, S-boxes, and subkeys.

8.2.10   Multichannel Attacks

While it seems plausible that side-channel attacks can be significantly improved by capturing multiple side-channel signals such as the various EM channels and possibly the power channel, a number of questions remain. Which side-channel signals should be collected? How should information from various channels be combined? How can one quantify the advantage of using multiple channels? These issues are especially relevant to an attacker since a significant equipment cost is associated with capturing each additional side-channel signal. Furthermore, in some situations, the detection risk associated with the additional equipment/probes required to capture a particular side channel has to be weighed against the benefit provided by that channel.

To address these issues, Agrawal et al. [1] present a formal adversarial model for multichannel analysis using the power and various EM channels, and based on a leakage model for CMOS devices and concepts from the Signal Detection and Estimation Theory. This formal model can be used to assess how an adversary can best exploit the wide array of signals available to him. In theory, this model can also deal with the problem of optimal channel selection and data analysis. However, in practice, a straightforward application of this model can sometimes be infeasible. But they show a judicious choice of approximations that renders the model useful for most practical applications.

Multichannel DPA attack is a generalization of the single-channel DPA attack. In this case, the adversary collects N signals, $\mathbf{S}_i, i = 1, ..., N$. In turn, each of the signals $\mathbf{S}_i$ is a collection of L signals collected from L side channels. Thus, $\mathbf{O}_i = [\mathbf{O}_i^1, ..., \mathbf{O}_i^L]^T$ where $\mathbf{O}_i^l$ represents the i-th signal from the l-th channel. Note that all DPA style attacks treat each time instant independently and leakages from multiple channels can only be pooled together if they occur at the same time. Thus, in order for multichannel DPA attacks to be effective, the selected channels must have very similar leakage characteristics.

The formulae for computing the metric for multichannel DPA attack are generalizations of those for the single channel. The main difference is that the expected value of sample mean difference at time j under hypothesis H is a vector of length L, with the l-th entry being the sample mean difference of the l-th channel. Furthermore, the variance of the b-bin under hypothesis H at time j, is a covariance matrix of size $L \times L$ with the i, j-th entry being the correlation between signals from the i-th and j-th channels.

8.3 Logic-Level DPA-Aware Techniques

During the last years, many logic styles that counteract side-channel analysis (SCA) attacks have been proposed. The big advantage of counteracting SCA attacks at the logic level is that this approach treats the problem right where it arises. If the basic building blocks, i.e., the logic cells, are resistant against this kind of attacks, a designer can build a digital circuit with an arbitrary functionality and it will also be resistant against SCA attacks. Having SCA-resistant cells means that hardware as well as software designers do not need to care about side-channel leakages any more. This greatly simplifies the design flow of a cryptographic device. Only the designers of the logic cells themselves need to be aware of these risks.

In recent years, research and industry have started to approach the issue of side-channel leakage at the gate level. There is a considerable body of research on gate-level masking schemes, which again aim at obfuscating the leakage and differential logic styles, which focus on reducing the leakage. Tiri and Verbauwhede introduced Wave Dynamic Differential Logic (WDDL) [125] where they use the concept of Fat Wires for the balanced routing of the complementary wire pairs. As a result, a WDDL circuit ideally has a constant power consumption and hence no side channel leakage. Popp and Mangard introduced MDPL [96], which applies both aforementioned concepts: it does not use special differential routing but instead randomizes the signals on the complementary wire pairs. As a result, the remaining leakage of an MDPL circuit is assumed to be randomized to the quality of the random numbers provided. On the other hand, also attacks against these secured logic styles have been published. Most of them exploit circuit "anomalies," as, for example, glitches and the early propagation effect.

In this section we review some of the most important SCA-aware logic families, and the security concerns that have been pointed out in the last years.

During the last years, several proposals to counteract DPA attacks at the logic level have been published. The basic idea of these proposals is to design logic cells with a power consumption that is independent of the data they process. Essentially, there exist two approaches to build such cells. The first approach is to design these cells from scratch. This implies that a completely new cell library needs to be designed for every process technology. Examples of such logic styles are SABL [121], RSL [117], DRSL [27], and TDPL [111].

The alternative to this approach is to build secure logic cells based on existing standard cells. In this case, the design effort for new cell libraries is minimal. This is the motivation for logic styles like WDDL [122] or MDPL [96].

Of course, each of the proposed logic styles also has other pros and cons besides the design effort for the cells. Dual-rail precharge (DRP) logic styles (e.g., SABL, TDPL, WDDL), which belong to the group of hiding logic styles, are for example smaller than masked logic styles (e.g., MDPL, RSL, DRSL). However, the security of DRP logic styles strongly depends on the balancing of complementary wires in the circuit, while this is not the case for masked logic styles. Design methods to balance complementary wires can be found in [123], [43], and [125].

8.3.1 Dual-Rail Precharge Logic

So far, the most promising logic styles to make devices resistant against SCA attacks are dual-rail precharge (DRP) logic styles that consume an equal amount of power for every transition of a node in a circuit. The most relevant logic styles of this kind are SABL [59, 121], WDDL [122], and Dual-Spacer DRP [111]. In these DRP logic styles, the signals are represented by two complementary wires. The constant power consumption is achieved by guaranteeing that in every clock cycle one of these two wires is charged and discharged again. Which one of the two wires performs this charge and discharge operation depends on the logical value that the wires represent.

Obviously, a constant power consumption can only be achieved, if the complementary wires have the same capacitive load. Otherwise, the amount of energy needed per clock cycle would depend on which of the two nodes is switched and therefore would be correlated to the logical value. Unfortunately, the requirement to balance the capacitive load of two wires is hard to fulfill in a semi-custom design flow.

Bucci et al. [20] proposed random precharging of combinational logic and registers in order to randomize the power consumption, instead of making it constant.

In a semi-custom design flow, so-called EDA tools place and route a digital circuit automatically. There exist only suboptimal mechanisms to tailor the place and route operation such that the capacitive load of two wires is equal. Such a partial solution is, for example, parallel routing as introduced by Tiri and Verbauwhede [123]. However, integrating such mechanisms becomes more and more difficult for deep submicron process technologies where the transistor sizes and wiring widths continuously shrink. Hence, the capacitance of a wire more and more depends on the state of adjacent wires rather than on the length of the wire and its capacitance to VDD or GND. Therefore, it is very hard to guarantee a certain resistance against SCA attacks, if a DRP circuit is placed and routed automatically. Placing and routing a circuit manually, i.e., doing a full-custom design, significantly increases the design costs.

8.3.1.1 Sense-Amplifier Based Logic

Sense-Amplifier Based Logic (SABL) [121] is a logic style that uses a fixed amount of charge for every transition, including the degenerated events in which a gate does not change state. In every cycle, a SABL gate charges a total capacitance with a constant value.

SABL is based on two principles. First, it is a Dynamic and Differential Logic (DDL) and therefore has exactly one switching event per cycle and this independently of the input value and sequence. Second, during a switching event, it discharges and charges the sum of all the internal node capacitances together with one of the balanced output capacitances. Hence, it discharges and charges a constant capacitance value.

SABL controls exactly the contribution of the internal parasitic capacitances of the gate into the power consumption by charging and discharging each one of them in every cycle. It also has symmetric intrinsic input and output capacitances at the differential signals such that it has balanced output capacitances.

In addition to the fact that every cycle the same amount of charge is switched, the charge goes through very similar charge and discharge paths during the precharge phase and during the evaluation phase respectively. As a result, the gate is subject to only minor variations in the input-output delay and in the instantaneous current. This is important since the attacker is not so much interested in the total charge per switching event, as in the instantaneous current and will sample several times per clock cycle in order to capture the instantaneous current.

The resulting differential power traces are more than an order of magnitude smaller than standard CMOS implementation. The resulting increased security will as always come in a trade-off with some cost. Here, the cost will be an increase in power, area, and an initial design time for a perfect symmetric standard cell.

8.3.1.2 Wave Dynamic Differential Logic

One of the most prominent of constant-power technologies is Wave Dynamic Differential Logic (WDDL) [122]. It is logic style designed first for ASICs, based on SABL (Sense-Amplifier Based Logic) technology. To adopt to standard cell ASIC design, it has been improved to form the current WDDL logic style.

Creating a compound standard cell, which has a dynamic and differential behavior, is done with the help of (1) the De-Morgan's Law, which allows expressing the false output of any logic function, using the false inputs of the original logic function and (2) AND-ing the differential output with a precharge signal. Because of the AND-ing with the precharge signal, whenever the precharge signal is 1, the outputs are predischarged to 0 independently of the inputvalues. On the other hand, whenever the precharge signal is 0, exactly one output, which is specified by the inputs, will evaluate to 1.

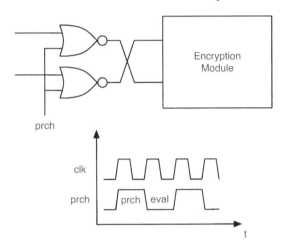

FIGURE 8.4
Precharge wave generation in WDDL.

However, this methodology does not guarantee that each compound gate has only one switching event per cycle. Both timing and value of the inputs influence the number of switching events. WDDL resolves this problem by conceiving a secure version of the AND- and OR-operator. Any other logic function in Boolean algebra can be expressed with these two differential operators. The differential inverter is not needed, as it can be implemented by simply exchanging the outputs of the previous gate.

Contrary to simple dynamic differential logic, WDDL gates remove the logic cells required to precharge the outputs to 0, and so they do not precharge simultaneously. As the input signals are usually connected to the outputs of other dynamic gates, whenever the inputs of any AND- or OR-gate are precharged to 0, the outputs are automatically at 0. The precharged 0s ripple through the combinatorial logic. Instead of a precharge signal that resets the logic, there is a precharge wave: hence the name "Wave Dynamic Differential Logic."

In order to launch the precharge wave, a precharge operator is inserted at the inputs of the encryption module, and Master-Slave DDL registers should be used, as shown in Figure 8.4. The registers store the precharged 0s, sampled at the end of the preceding precharge phase, during the evaluation phase, and they launch the precharge wave again.

WDDL can be characterized by the following properties:

- Consistent Switching Activities: In order to keep power consumption constant, dual-rail differential logic is used. WDDL replaces the standard gates with their WDDL versions. Each WDDL gate has four inputs and two outputs. Two of the inputs correspond to the original gate inputs and the other two are their logical complements. The two outputs also complement each other. These logic gates guarantee an opposite switching behavior for the two complementary outputs. As can be easily seen, when the direct logic block switches high (consuming current from power supply), the complementary logic block will switch low (discharging the capacitive load).

- Precharge Wave Generation: The gates introduced previously will not provide a transition if the inputs are not changing. This behavior renders the gates data-dependent. To fix this problem, in the absence of input data changes, a transition is provided by means of a precharge circuit. When the clock is high, the signal connected to the precharge circuit enter a precharge phase, where the connected net is driven to logic 0. When the clock becomes low, the circuit enters the evaluation phase, where actual computation is done. The operation can be more accurately described as predischarge since the logics are driven to logic low. We will keep the terminology to precharge since the idea behind it is the same. The reason to assign the clock as high at the precharge phase as opposed to the evaluation phase is because regular flip-flops latch in data value at clock up-edge; this means that, when the clock is low, the circuit should be in the evaluation phase.

WDDL implements the precharge circuit only at the register outputs and system inputs [122], and allows a logic-0-wave ripple through the whole module. It can even implement precharge only at the system inputs, by using Master-Slave DDL registers to propagate the precharge wave.

The dual-rail logic gates of WDDL and precharge logic together forms the most important feature of WDDL logic style: during every clock cycle, there is *exactly one* switching activity per WDDL gate. This feature is the basis for implementing constant power consumption secure circuits.

- Negative Logic: Notice that, all the WDDL gates have no inverters in them, even though they still implement negative logic, for example, WDDL NAND gate. Inverters disrupt the precharge wave propagation because the wavefront is inverted. Therefore, WDDL uses only positive logic and implements inverters by cross-coupling wires from the direct gate with those from the complementary gate.

Note that the above three properties address the first source of information leakage from power consumption. Rather than altering the probability of power-consuming transitions like masking-based logic, WDDL makes such transitions happen on every clock cycle. This effectively deflects any attempt to correlate the number of transitions to the secret data.

- Routing Procedure: One major hurdle for WDDL to be overcome is the routing of complementary nets. As discussed previously, any routing asymmetry between complementary nets results in unbalanced parasitic capacitive loading and in a residual power variation between direct and complementary transitions. Current techniques to control routing keep the direct and complementary gates close to each other, so that resulting nets are as symmetrical as possible [123].

A major advantage of WDDL is that it can be incorporated by the common EDA tool flow. In synchronous logic, the design of a module can be done with a standard hardware description language, such as VHDL. Next, synthesis is done with a subset of the standard cell library. The subset consists of the inverter, all AND- and OR-gates and a register. Subsequently, a script transforms the resulting synthesized code at the gate level to a code that reflects the differential gates. The script replaces the single-ended gates with WDDL gates, removes the inverters, and establishes the right connections. Next, placement and routing should be performed as usual.

The automated design flow generates a secure design from the VHDL netlist. The digital designer does not need specialized understanding of the methodology. He can write the code for a crypto coprocessor like for every other design. In the resulting encryption module, each gate will have constant power consumption independently of the input signals and thus idependently of which and how the operation has been coded.

One interesting property of WDDL is that the implementation of any combinational function consists of two dual parts. One can be derived from the other by inverting the inputs and by replacing single-ended AND-gates by single-ended OR-gates and vice versa. One generates the true outputs and the other the false outputs.

As a result, it is possible to place and route the original gate-level netlist and subsequently take the layout and interchange the AND- and OR-gates. This Divided Wave Dynamic Differential Logic has the same properties of the original WDDL and avoids the requirement of the differential signals to be matched to guarantee constant load capacitance.

As the AND-gate and the OR-gate, which make up a compound standard cell, are not identical, the internal and the intrinsic capacitances cannot be identical between both single-ended gates. However, as the channel-length decreases, the interconnect capacitance becomes dominant. If the differential signals travel in the same environment, the interconnect capacitances should be equivalent.

As shown in [122], WDDL can guarantee a 100% switching factor.

Security Concerns about WDDL

Two main factors have been pointed out as leakage sources that can be exploited by DPA attacks [123, 116]:

1. Leakage caused by the difference of loading capacitance between two complemetary logic gates in a WDDL gate.

2. Leakage caused by the difference of delay time between the input signals of WDDL gates.

The impact of these leakages has been studied by Suzuki and Saeki [115]. The power consumption at the CMOS gate can be generally evaluated by:

$$P_{total} = p_t \cdot C_L \cdot V_{dd}^2 \cdot f_{clk} + p_t \cdot I_{sc} \cdot V_{dd} \cdot f_{clk} + I_{leakage} \cdot V_{dd} \qquad (8.15)$$

where C_L is the loading capacitance, f_{clk} is the clock frequency, V_{dd} is the supply voltage, p_t is the transition probability of the signal, I_{sc} is the direct-path short circuit current, and $I_{leakage}$ is the leakage current. As realized from the formula 8.15, the power consumption at the first term is different between the gates if there is a difference in the loading capacitance between each complementary logic gate. Since the existence of a transition at each complementary logic gate is determined by the values of the input signals, the total power consumption differs in dependence of the signal values even if the total number of transitions is equal between the gates.

As described previously, the transition probability during an operation cycle at the WDDL gates is guaranteed to be $p_t = 1$ independently of the input signals. However, the operation timing of each complementary logic gate is generally different mainly due to the delay time of the input signals during an operation cycle. Therefore, since the average power traces specified by the predictable signal values have different phases, a spike can be detected after the DPA operation.

Avoiding Leakages in WDDL

The two complementary parts of a WDDL circuit have exactly the same number of gates, and the same structure, but AND- and OR- gates are interchanged. Therefore, the difference of load capacitances in the WDDL circuit arises due to the difference of capacitance of the AND/OR gates themselves and the difference due to placement and routing. This place-and-route difference of loading capacitances is independent of the logic function and is usually more important than the difference of the AND/OR capacitances.

Suzuki et al. [116] analyzed the leakage due to timing differences between the signals of a WDDL AND gate. When designing a cryptographic component, unless a special design is made as described in [85], the input signals at the gates, a and b (or $\bar{a}$ and $\bar{b}$), have a different number of logic steps and are easy to cause differences in the total delay time. On the contrary, since the number of gates connected to each complementary output of WDDL is equal as described above, the difference of place-and-route is predominant over a difference in the delay time between a and $\bar{a}$ (or b and $\bar{b}$).

A main factor of the place-and-route leakage is the process automatization that is generally carried out in the VLSI design. Therefore, this leakage is likely to improve with manual placement and routing or semi-automatic placement and routing using special constraints. Actually, Tiri and Verbauwhede. [123]

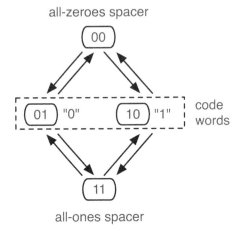

FIGURE 8.5
State machine for the dual spacer protocol.

proposed "fat wire," and Guilley et al. [43] proposed "backend duplication" as countermeasures in the placement and routing to improve the DPA-resistance.

However, we are not aware of any study of a countermeasure against the inevitable leakage due to the differences in the delay of two different signals. The S-box design method for low power consumption proposed by Morioka and Safoh [85] is recommended as one technique to reduce this leakage. In general, adjusting the delay time between the input signals at each gate requires a high effort.

8.3.1.3 Dual-Spacer Dual-Rail Precharge Logic

Dual-rail code uses two rails with only two valid signal combinations 01, 10, which encode values 0 and 1, respectively. Dual-rail code is widely used to represent data in self-timed circuits, where a specific protocol of switching helps to avoid hazards. The protocol allows only transitions from all-zeroes 00, which is a noncode word, to a code word and back to all-zeroes; this means the switching is monotonic. The all-zeroes state is used to indicate the absence of data, which separates one code word from another. Such a state is often called a *spacer*.

Dual-rail techniques certainly help to balance switching activity at the level of dual-rail nodes. Assuming that the power consumed by one rail in a pair is the same as in the other rail, the overall power consumption is invariant to the data bits propagating through the dual-rail circuit. However, the physical realization of the rails at the gate level is not symmetric, and experiments with these dual-rail implementations show that power source current leaks the data values.

In order to balance the power signature, Sokolov et al. [112, 111] proposed to use two spacers (i.e., two spacer states, 00 for all-zeroes spacer, and 11 for all-ones spacer), resulting in a dual spacer protocol (see Figure 8.5). It defines the switching as follows: *spacer → code word rightarrow spacer rightarrow code word*. The polarity of the spacer can be arbitrary and possibly random. A possible refinement for this protocol is the alternating spacer protocol. The advantage of the latter is that all bits are switched in each cycle of operation, thus opening a possibility for perfect energy balancing between cycles of operation.

Single-rail circuits can be converted automatically into dual-spacer dual-rail circuit with a software tool named the "Verimap design kit," from Sokolov et al., and it successfully interfaces to the Cadence CAD tools. It takes as input a structural Verilog netlist file, created by Cadence Ambit (or another logic synthesis tool), and converts it into dual-rail netlist. The resulting netlist can then be processed by Cadence or other EDA tools.

8.3.1.4 Three-Phase Dual-Rail Precharge Logic

Bucci et al. [19] propose Three-Phase Dual-Rail Precharge Logic (TDPL), a dual-rail precharge logic family whose power consumption is insensitive to unbalanced load conditions thus allowing adopting a semi-custom design flow (automatic place and route) without any constraint on the routing of the complementary wires.

The proposed concept is based on a three-phase operation where an additional discharge phase is performed after the precharge/evaluation steps typical of any dynamic logic style. Although the concept is general, it can be implemented as an improvement of the SABL logic with a limited increase in circuit complexity.

The circuit operation is as follows (see Figure 8.6):

1. *charge phase*: at the beginning of each cycle, signals *charge* and *discharge* go low, and both output lines are precharged to V_{DD}.

2. *evaluation phase*: during the charge phase new input data are presented to the circuit. On the raising edge of signal *eval* one of the output lines discharges according to the input data.

3. *discharge phase*: at the end of each operating cycle, input *discharge* is activated in order to pull down the output line, which has not been discharged during the evaluation phase.

The results show [111] a reduction of one order of magnitude in the normalized energy deviation, defined as $(\max(E) - \min(E))/\max(E)$, and the normalized standard deviation $(\sigma_E/\overline{E})$ with respect to SABL.

8.3.2 Charge Recovery Logic

The charging through the DC voltage source causes enormous energy dissipation because the charge (the charging current) experiences a potential drop on

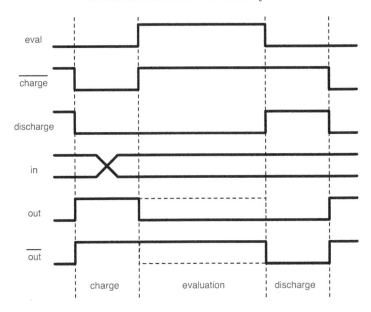

FIGURE 8.6
Timing diagram for the TDPL inverter.

its way from the supply node to the load. In contrast, in charge recovery circuits each capacitance node is charged steadily, and the voltage drops across the resistive elements are made small in order to reduce the energy dissipation during the charge or discharge of the capacitive loads via a power clock signal.

Basically, charge recovery logic styles have been devised for low-power purposes. However, they have some other characteristics such as inherent pipelining mechanism, low data-dependent power consumption, and low electromagnetic radiations that are usually neglected by researchers. These properties can be useful in other application areas such as side-channel attack-resistant cryptographic hardware. Moradi et al. [84] have recently proposed a charge recovery logic style, called 2N-2N2P, as a side-channel attack countermeasure. The observed results show that the usage of this logic style leads to improve DPA-resistance as well as energy saving.

Several charge recovery styles have been proposed, each one with its own characteristic and efficiency, but their fundamental structure does not differ much from each other. Due to its simplicity, Moradi et al. choose 2N-2N2P [58] to examine the DPA-resistance of charge recovery logics [84].

A 2N-2N2P gate consists of two main parts:

- two functional blocks whose duty is to construct the gate outputs q and $\bar{q}$

- two cross-coupled inverters, which are formed by two p-MOS and two n-MOS transistors and allow maintaining the right (high or low) voltage levels at the outputs.

All 2N-2N2P gates operate at four different phases: input phase, evaluation phase, hold phase, and reset phase. During the input phase, the inputs reach their own valid values. During the evaluation phase the outputs are calculated and reach their valid values. During the hold phase the inputs discharge to 0 and the output values remain valid. Finally, during the reset phase the outputs are discharged to 0.

The 2N-2N2P is a full-custom logic style. The transistor cost of all 2N-2N2P logic cells is less than the corresponding SABL and TDPL cells. From the power consumption point of view, the peak of power consumption traces in DPA-resistant logic styles does not depend on the frequency but in charge recovery logic families it does. But the average energy consumption per cycle of a 2N-2N2P gate is much smaller than the SABL and TDPL ones.

A sudden current pulse in a CMOS circuit causes a sudden variation of electromagnetic field surrounding the device. According to Faraday's law of induction, any change in the magnetic environment of a coil of wire will cause a voltage to be induced in the coil. In this way the data-dependent electromagnetic variation can be captured by inductive sensors. The peak of power consumption traces and the slope of supply current changes are also much smaller than SABL, especially for low power clock frequencies. Therefore, 2N-2N2P has also less electromagnetic emanations than the other static DPA-resistance logic styles.

The pipelining structure of 2N-2N2P (and other charge recovery logic families) causes the circuit to process multiple data simultaneously. Therefore, the power consumption at each cycle depends on several data that are being processed. Obviously, a pipeline does not provide an effective countermeasure against DPA attacks, and it can be viewed as a noise generator that has the advantage of decreasing the correlation between predictions and measurements. However, by evaluating an information theoretic metric, mutual information, Moradi et al. [84] conclude that the information leakage is much smaller than in SABL, specially at low frequencies.

Detailed experimental results about charge recovery logics as DPA-countermeasures are still missing.

8.3.3 Masked Logic Styles

Besides hiding, the second way to hide the data-dependent power consumption was to randomize the power consumption. We have already presented masking techniques at the algorithm level (see Section 8.1.3), but they can also be applied to the logic level.

Masking-based logic attempts to remove the correlation between power consumption and secret data by randomizing the power consumption such that the correlation is destroyed. Masking-logic works by eXclusive OR-ing signals with a masking-bit, then later removes the mask by doing another XOR operation. In this approach, each probably attacked signal b is represented by $b_m = b \oplus m_b$, where m_b is a uniformly distributed random variable (i.e., $p(mb = 0) = p(mb = 1) = 1/2$) and is independent of b. Consequently, the b_m also is a uniformly distributed random variable. In the masking-approach, a circuit is replaced with a masked implementation. For example, a 2-input XOR function $g = a \oplus b$ is replaced with $g_m = a_m \oplus b_m$ and $m_g = m_a \oplus m_b$. We refer to the implementation g_m as M-XOR gate. The m_g signal is called its *correction mask.*

Masking on the gate level was first considered in the U.S. patent 6295606 of Messerges et al. in 2001. However, the described masked gates are extremely big because they are built based on multiplexors. A different approach has been pursued later on by Gammel et al. in patent DE 10201449. This patent shows how to mask complex circuits such as crypto cores, arithmetic-logic units, and complete microcontroller systems.

The problem with those masked logic styles is that glitches occur in these circuits. As shown in [117], glitches in masked CMOS circuits reduce the SCA resistance significantly. Therefore, glitches must be considered when introducing an SCA countermeasure based on masking.

One widely used masking-based logic style that avoids glitches is Random Switching Logic (RSL) [117].

Popp and Mangard [96] proposed a new logic style called *masked dual-rail precharge logic (MDPL)* that applies random data masking into WDDL gates. There are no constraints for the place and route process. All MDPL cells can be built from standard CMOS cells that are commonly available in standard cell libraries.

Figure 8.7 shows the basic components of MDPL. The logic AND- and OR-gates in WDDL were implemented with a pair of standard two-input AND- and OR-gates. In MDPL, both compound gates apply a pair of standard 3-input majority (MAJ) gates.

The architecture of cryptographic circuits using MDPL is shown in Figure 8.8. The signals $(a_m, b_m, \overline{a_m}, \overline{b_m})$, which are masked with the random data m and $\overline{m}$, and these random values are the input signals of the combinational circuit.

When examining the security of MDPL against DPA, we assume that an attacker can predict the architecture of the combinational circuit and the pre-masking signals $(a, b, \overline{a}, \overline{b})$ corresponding to the masked counterparts $(a_m, b_m, \overline{a_m}, \overline{b_m})$. And the random numbers m and $\overline{m}$ can be predicted only with a probability of $1/2$.

MDPL circuits have the following features:

1. MDPL gates have complementary outputs $(q, \overline{q})$.

2. The precharge signal controls the precharge phase to transmit $(0,0)$ and the evaluation phase to transmit $(0,1)$ or $(1,0)$.

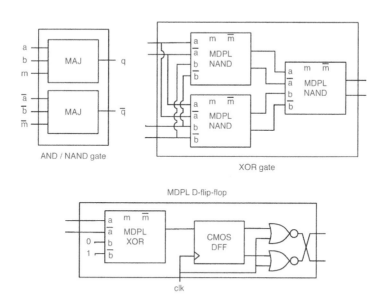

FIGURE 8.7
Basic components of MDPL.

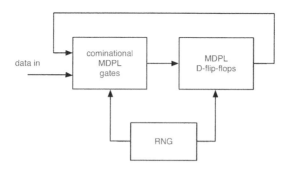

FIGURE 8.8
Architecture of a MDPL-based cryptographic circuit.

3. The precharge operation is performed at the first step in the combinational circuit, and the components are limited to AND- and OR-gates. Inverters are implemented by rewiring.

4. The number of transitions in all circuits generated during an operation cycle is constant without depending on the values of the input signals.

5. Even if the correct signal values (a, b) are predictable, the random transition occurs at the MAJ gate according to the value of random data m.

The first four properties are shared with WDDL. The last one guarantees that the power consumption is made uniform even if there is a difference in the load capacitance of each complementary logic gate. Thus, MDPL is claimed to not require constraints on the place-and-route to adjust the load capacitance.

8.3.3.1 Random Switching Logic

The countermeasures that equalize the signal transition frequency by complementary operations are dependent on wire length or fanout. This often makes the design very difficult. To solve this problem, Suzuki et al. [117] propose a new countermeasure against DPA called *Random Switching Logic (RSL)*. RSL does not require complementary operations and it avoids glitches in the circuit. Yet, RSL needs a careful timing of enable signals. Furthermore, a new standard cell library must be compiled where all combinational gates have enable inputs.

RSL replaces traditional logic gates such as NAND, NOR, and XOR gates with their RSL version respectively. Each RSL gate has four inputs as opposed to two of their traditional counterparts. The two extra inputs are *enable* and *RandomBit*. The *RandomBit* input is used to alter transition probabilities of the circuit and achieve the randomized switching property. The enable signal on the other hand is used to suppress spurious transitions. The circuit starts operating when enable is asserted, otherwise, the circuit is driven to logic-0.

One critical condition for the RSL circuit to be secure is that the enable signal is to be asserted only after *all* input signals are fixed. One question that comes to mind is that since enable is connected to every gate, how can the signal be coordinated such that the circuit still operates? Imagine the following scenario: *RSLGate1* is driving *RSLGate2*, see Figure 8.9. Assuming they use the same enable signal, one of the input of *RSLGate2*, the one driven by *RSLGate1*, will not arrive until *RSLGate1* is enabled. This means *RSLGate2* must also be enabled. Following the condition for security, *RSLGate2* should not be enabled because output of *RSLGate1* has not arrived yet. The implication of this conflict is that every gate or level of gates must have a different enable signal. If so, the cost of enabling every gate could be prohibitive.

Assuming the RSL enable signal can be created and routed, it is shown through simulated attacks in [120], that RSL can still be attacked fairly effectively. It is shown that, indeed, when RSL is in normal operation, the power

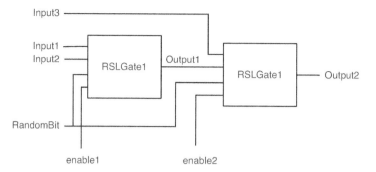

FIGURE 8.9
Random Switching Logic example.

consumption correlation with the secret key is removed. However, further analysis by using a threshold filter can remove the randomizing effect caused by the random bits used. After the random bits are removed, the stripped RSL circuit becomes just like the original circuit with no countermeasure and can be attacked using power analysis to single out the secret key.

Logic styles that are secure against DPA attacks must avoid early propagation. Otherwise, a power consumption occurs that depends on the unmasked data values due to data-dependent evaluation moments. In [27], the logic style Dual-rail Random Switching Logic (DRSL) is presented. In DRSL, a cell avoids early propagation by delaying the evaluation moment until all input signals of a cell are in a valid differential state.

Popp et al. [95] point out that DRSL does not completely avoid an early propagation effect in the precharge phase. The reason is that the input signals, which arrive at different moments, can still directly precharge the DRSL cell. The propagation delay of the evaluation-precharge detection unit (EPDU) leads to a time frame in which this can happen. Only after that time frame, the EPDU unconditionally precharges the DRSL cell.

8.3.3.2 Masked Dual-Rail Precharge Logic

In case of standard CMOS logic ($E_{00} \simeq E_{11} \ll E_{10} \neq E_{01}$), the differential trace in a DPA attack is different from zero and can therefore be detected by an attacker. At all other moments of time, except for t, the partitioning of the traces according to DT is meaningless. Consequently, the expected value for the difference of the means at this time is zero. Furthermore, if a wrong key value is used to calculate DT, the partitioning of the traces is again meaningless and also leads to an expected value of zero. As a result, the attacker in general gets a significant peak for a single key hypothesis, which is then the correct key.

The straightforward method to prevent an attacker from seeing such a peak in the differential trace is to use cells with the property that $E_{00} =$

$E_{01} = E_{10} = E_{11}$. This is in fact the motivation for using dual-rail precharge (DRP) logic styles such as SABL [59] or WDDL [122]. DRP logic styles have the property that transitions need the same amount of energy, if all pairs of complementary wires are perfectly balanced, i.e., have the same capacitive load. However, as already discussed, this requirement is very hard or even impossible to guarantee. This is the motivation for Masked Dual-rail Precharge Logic (MDPL) [96]. MDPL is based on a completely different approach to prevent DPA attacks.

MDPL is a masked logic style that prevents glitches by using the DRP principle. Hence, for each signal d_m also the complementary signal $\overline{d_m}$ is present in the circuit. Every signal in an MDPL circuit is masked with the same mask m. The actual data value d of a node n in the circuit results from the signal value d_m that is physically present at the node and the mask m: $d = d_m \oplus m$.

Figure 8.7 show the basic elements of the MDPL logic style. An MDPL AND gate takes six dual-rail inputs $(a_m, \overline{a_m}, b_m, \overline{b_m}, m, \overline{m})$ and produces two output values $(q_m, \overline{q_m})$. As shown in [96], q_m and $\overline{q_m}$ can be calculated by the so-called majority (MAJ) function. The output of this function is 1, if more inputs are 1 than 0. Otherwise, the output is 0: $q_m = MAJ(a_m, b_m, m)$ and $\overline{q_m} = MAJ(\overline{a_m}, \overline{b_m}, \overline{m})$. A majority gate is a commonly used gate and it is available in a typical CMOS standard cell library.

In an MDPL circuit, all signals are precharged to 0 before the next evaluation phase occurs. A so-called precharge wave is started from the MDPL D-flip-flops, similar to WDDL [122]. First, the outputs of the MDPL D-flip-flops are switched to 0. This causes the combinational MDPL cells directly connected to the outputs of the D-flip-flops to precharge. Then, the combinational gates in the next logic level are switched into the precharge phase and so on. Note that also the mask signals are precharged. The output signals of the MDPL AND gate are precharged if all inputs are precharged. All combinational MDPL gates are implemented in that way. Therefore, in the precharge phase, the precharge wave can propagate through the whole combinational MDPL circuitry and all signals are precharged correctly.

A majority gate in a precharge circuit switches its output at most once per precharge phase and at most once per evaluation phase, i.e., there occur no glitches. In a precharge circuit, all signals perform monotonic transitions in the evaluation phase (0 to 1 only) and in the precharge phase (1 to 0 only), respectively. Furthermore, the majority function is a so-called monotonic increasing (positive) function. Monotonic transitions at the inputs of such a gate lead to an identically oriented transition at its output. Hence, a majority gate performs at most one (0 to 1) during the evaluation phase and at most one (1 to 0) during the precharge phase. Since an MDPL AND gate is built from majority gates, an MDPL AND gate will produce no glitches.

Other combinational MDPL gates are based in the AND gate, as shown in [96].

However, the use of MDPL gates has a significant cost. It implies an average area increment by 4.5, the maximum speed is also reduced by 0.58, and the power consumption is increased by 4 to 6.

Security Concerns about Masked Logic Styles

As previously seen when discussing WDDL, there are two main factors that can be exploited by DPA attacks [123, 116]:

1. Leakage caused by the difference of loading capacitance between two complemetary logic gates in a MDPL gate.

2. Leakage caused by the difference of delay time between the input signals of MDPL gates (early propagation effect, [60]).

As MDPL uses the same operator for both parts of the compound gate, it can improve in principle the leakage caused by the difference of loading capacitance between a signal and its complementary, so the most important leakage source is the delay differences between signals.

Differences of delay time between independent signals (e.g., a_m and b_m) are more likely to occur than those between complementary signals (e.g., a_m and $\overline{a_m}$) in the design of a dual-rail circuit. In the case of the MDPL gate, if there are differences in the delay time between the signals a_m, b_m, and m (or $\overline{a_m}$, $\overline{b_m}$, and $\overline{m}$), there are six possible delay conditions depending on the order of arrival of the three signals. As shown in [115], the leakage occurs under any of these delay conditions. In short, there is no secure delay condition in MDPL on the single input change model. Therefore, in order to implement the secure logic circuits using MDPL gates, it is required to adjust differences in the delay time between the input signals.

Gierlichs [39] points out another possible source of leakage. Studying the switching behavior of a majority gate in detail, he discovered a potential problem. There are "internal nodes" in the pull-up and pull-down networks, which cannot always be fully charged (resp. discharged), depending on the input signals' values and delays. This fact induces a kind of memory effect. Possibly, there exist combinations of delays in the input signals, for which a (small) bias in the distribution of the random masks leads to a data-dependent memory effect. In that case, the side channel leakage of an AND gate would be increased. Note that such delay combinations need not necessarily lead to early propagation.

To avoid leakages, mask signals need to be unbiased, and the circuit must consume the same power for each value of the unmasked input in a statistical sense. However, real circuits have a large number of higher-order effects that can cause the power consumption to become dependent on the circuit state. Two of them have been studied by Chen et al. [25]: glitches and interwire capacitance. When intermediate values contain both the information of the mask and the corresponding masked value (not necessary to be exactly m and a_m), an attack based on the probability density function (PDF) is possible [120, 103].

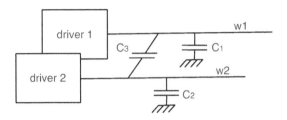

FIGURE 8.10
Interwire capacitance model.

- **Glitches.** First, consider again the effect of glitches. A glitch results in additional net transitions in a circuit, so that the switching factor appears to be increasing. Glitches are also state-dependent. For example, a glitch may appear when $a_m = 1$ and $m = 1$ ($a = 0$) but not in any other combination. Hence, glitches may cause an imbalance which in turn results in a violation of the perfect masking condition.

- **Interwire Capacitance.** Second, consider the effect of the interwire capacitance on the average power consumption. The total capacitance of a circuit has two components: gate capacitance and wire capacitance. In deep submicron technology, the interwire capacitance accounts for more than 50% of the total capacitance for narrow wires [131]. Modeling of the wire capacitance in an integrated circuit is a nontrivial task. For simplicity, the following discussion is on the basis of the simplified model as shown In Figure 8.10. This circuit shows two wires w_1 and w_2, each with a different driver. There are two parts for the capacitance seen by each wire, one part between the wire and the semiconductor substrate (C_1 and C_2), and the other part between adjacent wires (C_3). Wire w_1 sees a single, fixed capacitor C_1 and a second floating capacitor C_3. The effective value of C_3 as seen from wire w_1 changes in function of w_2's voltage level. For example, when w_1 carries a logic-1 while w_2 carries a logic-0, then the effective capacitance on w_1 is $C_1 + C_3$. However, when both w_1 and w_2 carry a logic-1, then the effective capacitance on w_1 is only C_1, since there is no current required to charge C_3.

The conditions for perfect masking can be easily achieved at the logic-level, which abstracts voltage into discrete logic-levels and which abstracts time into clock cycles. The results in [25] confirm that a logic-level simulation of a masked circuit indeed remains side-channel-free. However, the conditions for secure masking are much harder to achieve in real circuits, in which we have to allow for various electrical and analog effects. Chen et al. showed that glitches are not the only cause of side-channel leakage in masked circuits. As an example, the effect of interwire capacitance is elaborated. These second-order effects were used to successfully mount a first-order attack, using HSPICE simulations and measurements, on a glitch-free masked AES S-Box in a FPGA.

Although theoretically possible, Popp et al. have recently shown that the masked logic style MDPL is not completely broken because of early propagation. In regular designs where the signal-delay differences are small, MDPL still provides an acceptable level of protection against DPA attacks. With a PDF attack [103], no key byte of the attacked MDPL AES coprocessor was revealed with 3,511,000 power measurements. This is also clear from the following perspective. With the settings chosen for the PDF-attack in theory (no electronic and other forms of noise, no perfect balancing of dual-rail wire pairs, ...), it can easily be shown that all DPA-resistant logic styles that try to achieve a constant power consumption are also completely broken. However, various publications draw opposite conclusions from experimental results.

Avoiding Leakages in MDPL

As it clearly turned out in the last section, logic styles that are secure against DPA attacks must avoid early propagation. Otherwise, a power consumption occurs that depends on the unmasked data values due to data-dependent evaluation moments.

The differential encoding of the signals in MDPL circuits detects the point in time in the evaluation phase where all input signals of a cell are in a valid differential state. A cell that avoids early propagation must delay the evaluation moment until this point in time. In [27], the logic style DRSL is presented, which implements such a behavior in the evaluation phase.

DRSL does not completely avoid an early propagation effect in the precharge phase. The reason is that the input signals, which arrive at different moments, can still directly precharge the DRSL cell. The propagation delay of the evaluation-precharge detection unit (EPDU) leads to a time frame in which this can happen. Only after that time frame, the EPDU unconditionally precharges the DRSL cell. Simulations with an intermediate version of an improved MDPL cell confirmed this [95], there still occurred correlation peaks in the precharge phase. Thus, the input signals of a cell must be maintained until the EPDU generates the signal to precharge the cell.

Figure 8.11 shows the schematic of an improved MDPL (iMDPL, see [95]) cell with respect to the early propagation effect. The three OR and the NAND cell on the left side implement the EPDU, which generates 0 at its output only if all input signals a_m, b_m, and m are in a differential state. The following three set-reset latches, each consisting of two cross-coupled 3-input NORs, work as gate elements. As long as the EPDU provides a 1, each NOR produces a 0 at its output. Thus, the outputs of both MAJ cells are 0 and the iMDPL cell is in the precharge state.

When the EPDU provides a 0 because all input signals have been set to a differential state, the set-reset latches evaluate accordingly and the MAJ cells produce the intended output according to the masked AND function. Note that this evaluation only happens after all input signals have arrived differentially, i.e., no early propagation occurs. However, this is only true if

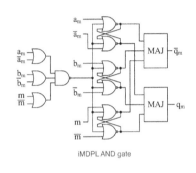

iMDPL AND gate

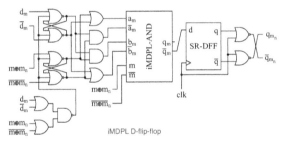

iMDPL D-flip-flop

FIGURE 8.11
Basic components of iMDPL.

the input signals reach the inputs of the three latches before the EPDU sets its output to 0. Fortunately, this timing constraint is usually fulfilled because of the propagation delay of the EPDU.

Finally, if the first input signal is set back to the precharge value, the EPDU again produces a 1 and all six outputs of the set-reset latches switch to 0. Note that the set-reset latches are only set to this state by the EPDU and not by an input signal that switches back to the precharge value. Thus, also an early propagation effect at the onset of the precharge phase is prevented. An iMDPL-OR cell can be derived from an iMDPL-AND cell by simply swapping (i.e., inverting) the mask signals m and $\overline{m}$.

Obviously, the price that has to be paid for the improvements in terms of early propagation is a further significant increase of the area requirements of iMDPL cells compared to MDPL. Since the iMDPL cells are already quite complex, exact figures for the area increase cannot be given in general because it depends significantly on the particular standard cell library that is used to implement an iMDPL circuit. However, one can expect an increase of the area by a factor of up to 3 compared to the original MDPL. This makes it clear that carefully finding out which parts of a design really need to be implemented in DPA-resistant logic is essential to save the chip area.

A significant reduction of the cell size can be achieved by designing new standard cells that implement the functionality of iMDPL. Of course, that has the well-known disadvantages of a greatly increased design and verification effort. Furthermore, a change of the process technology would then mean spending all the effort to design an iMDPL standard cell library again.

8.3.3.3   Precharge Masked Reed-Muller Logic

In [67], Lin et al. propose a logic design style, called *Precharge Masked Reed-Muller Logic (PMRML)* to overcome the glitch and Dissipation Timing Skew (DTS) problems in the design of DPA-resistant cryptographic hardware. Both problems can significantly reduce the DPA-resistance, but WDDL and MDPL still leaks information due to the DTS problem. The PMRML design can be fully realized using common CMOS standard cell libraries. It can be used to implement universal functions since any Boolean function can be represented as the Reed-Muller form (also known as "algebraic normal form" or "fixed polarity Reed-Muller form") [127].

The Fixed Polarity Reed-Muller (FPRM) form of a Boolean function is the XOR sum of products (ANDs) in which every variable has either positive or negative polarity. In the PMRML, a combinational logic function is realized as the FPRM form, as shown in Figure 8.12. To provide DPA resistance, each AND gate is one-to-one replaced with its masked implementation, a 4X1 MUX with Dual-rail Selection signals (MUX-DS). Lin et al. prove that the masked circuit is glitch-safe and DTS-safe under the scheme. In the XOR-part, each XOR gate is one-to-one replaced with the corresponding M-XOR circuit. The corresponding masks to recover the plain data are manipulated

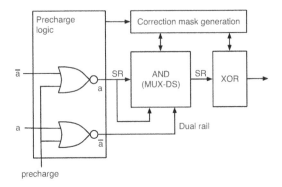

FIGURE 8.12
Basic components of PMRML.

in the correction masks generator. Initial masks should come from a Random Number Generator (RNG), which is assumed to be already available to the design.

The precharge logic is used to ensure at most one transition at an AND (NAND) gate during a cycle. This makes the gates glitch-free. Though the precharge method has been used in many other logic styles, there are two main differences. First, only a subset of data is conveyed by dual-rail signals. Specifically, only the selection signals of MUX-DSs need them. Second, a multistage precharge scheme is used to reduce the performance penalty caused by the precharging time.

The proposed masked circuit for an AND gate ($g = c \cdot d$) is a 4X1 MUX-DS, which is implemented by a 2-level NAND network, and the selection variable is encoded with a dual rail. In this work, a NAND gate is assumed to be atomic, but the whole MUX-DS circuit is not. Lin et al. [67] prove that the masked implementation of an AND function $g = c \cdot d$ shown in Figure 8.12 is glitch-safe and DTS-safe when it is used in a PMRML design.

In a multistage PMRML structure, each stage is controlled by separate PE (precharge enable) signals. All stages start precharge ($PE_k = 1$) simultaneously, and start evaluation ($PE_k = 0$), at different time. The duration of the precharge pulse of each stage should be enough to ensure that all the nodes in the stage have been reset. Also, no other evaluation phase should be started until all input signals from the stage currently evaluating have been set stable.

AES encryption hardware design was successfully synthesized by Synopsys DC with conservative wire load model under UMC $0.18 \, \mu m$ technology. Compared to the unprotected design, as shown in Table 8.1, the area is increased by 100% and the speed is decreased by 29% in the PMRML design, what is not bad compared to other alternatives. Although a precharge scheme is used in the PMAXL design, the multistage schemes make the speed not halved.

TABLE 8.1
Relative cost of different anti-DPA logic styles

Protection	Area	Performance
Unprotected	1	1
WDDL	3	0.26
MDPLtextsuperscript*	4.54	0.58
PMRMLtextsuperscript*	2	0.71

*Does not include RNG circuit

8.3.4 Asynchronous Circuits

An asynchronous logic style that makes devices more resistant against SCA attacks has been presented, for example, in [32]. However, it has been shown in [37] that this logic style has some weaknesses.

Additionally to their absence of a clock signal that demonstrates the practical way to eliminate a global synchronization signal, asynchronous logic is well-known for its ability to decreasing the consumption and to shape the circuits' current. In [32], Cunningham et al. demonstrate how 1-of-n encoded speed-independent circuits improve security by eliminating data-dependent power consumption. Symmetry in data communication and data processing, persistent storage, timing information leakage, and propagating alarm signals (to defend chip against fault induction) are design aspects addressed by those papers for increasing chip resistance. The countermeasures that used the self-timed circuit properties are all focused on balancing the operation through special delay insensitive coding scheme.

Moreover, Paul Kocher also developed some countermeasures based on the same properties (European Patent No. EP1088295/WO9967766); and a new design concept has been presented in [112] by Danil Sokolov et al., who used standard dual-rail logic with a two-spacer protocol working in a synchronous environment. The results obtained by exploiting self-timed logic have been reported in several papers. J. Fournier et al. evaluated and demonstrated in [37] that Speed Independent asynchronous circuits increase resistance against side channel attack and the concrete results of the effectiveness of the Quesi-Delay Insensitire (QDI) asynchronous logic against DPA has been reported in [13].

However, all these papers concluded in terms of DPA that there still exists some residual sources of leakage that can be used to succeed an attack. These residual sources of leakage that are still observable when implementing a DPA attack on balanced QDI asynchronous circuits are addressed by G.F. Bouesse et al. in [13]. They show that the residual sources of leakage of balanced QDI circuits come from the back-end steps that introduce some electrical asymmetries, especially through the routing capacitances. The solutions implemented in [13] in order to remove electrical asymmetries, consist in constraining the placement and routing. They defined a place and route methodology, which enables the designer to control the net capacitances.

8.3.4.1 Path Swapping

The path swapping (PS) method [14] takes advantage of the structural symmetries that exist in QDI asynchronous circuits or those proposed in [83]. In fact, in such circuits many identical structures called *paths* exist that can be alternatively used to compute a given function. Therefore, the idea is to randomly choose one of the possible paths to compute the function which, hence, averages the electrical signature over all the paths. The issue lies in succeeding to do so with minimum overhead.

The goal of this design approach is to eliminate the electrical effects that succeed the DPA attack on QDI circuits. To do so, Bouesse et al. propose to exploit the circuit structure that exhibits a lot of symmetries. Indeed, in the blocks of such circuits there exist many identical physical paths from their primary inputs to their primary outputs. The idea is to randomly choose one of the possible paths to compute the function.

This method is called *path swapping* because interchanging the inputs and/or outputs leads to swap the execution from logical paths to other logical paths inside the circuit. The realization of this technique requires the use of multiplexers/demultiplexers and a random number generator (RNG). Multiplexers/Demultiplexers are used to permute inputs/outputs and are controlled by the random number generator. The use of a random number generator guarantees an equiprobable and unpredictable distribution function of inputs/outputs.

The path-swapping method can only be efficiently implemented with design logic that offers an opportunity to implement symmetrical and balanced circuits as it is the case with QDI asynchronous circuits. This type of logic enables to implement the PS method with a minimum area overhead and by slightly changing the performance of the circuit.

8.4 Architecture-Level DPA-Aware Techniques

Architecture-level countermeasures can be classified as time-oriented hiding techniques, amplitude-oriented hiding techniques, and masking techniques.

There are essentially two groups of proposals of time-oriented hiding countermeasures at the architecture-level. The first group aims at randomly insert dummy operations or cycles and to shuffle the performed operations:

- Random insertion of dummy operations and suffling.

- Random insertion of dummy cycles. This usually requires the registers to be duplicated. The original registers are used to store intermediate values of the cryptographic algorithm, and the others are used to store random values.

The second group of time-oriented hiding techniques affects the clock signal. These proposals randomly change the clock signal to make the alignment of power traces more difficult:

- Randomly skipping clock pulses, by filtering the clock signal.

- Randomly changing the clock frequency.

- Randomly switching between multiple clock domains.

For all these countermeasures, if the attacker is able to identify the countermeasure, its effect can be undone.

Amplitude-oriented hiding countermeasures try to make the power consumption equal for all operations and all data values, or, alternatively, to add noise to the power traces to counteract power analysis attacks.

- Filtering the power suply.

- Noise generators.

It is important to note that the signal-noise ratio (SNR) not only depends on the cryptographic device, but also on the measurement setup. If a countermeasure reduces the SNR for one measurement setup, it does not necessarily reduce the SNR for all setups. For example, filtering the power consumption makes attacks more difficult if the power consumption is measured via a resistor, but it does not affect measurements of electro-magnetic emissions significantly.

Architecture-level masking countermeasures usually apply Boolean masking schemes to the different circuit elements. Other types of masking are not worth to design unless Boolean masking is not applicable. Some of the most used techniques include:

- Masking adders and multipliers.

- Random precharging. The duplicates of the registers contain random values and they are connected randomly to the combinational cells.

- Masking buses. Buses are particularly vulnerable to power analysis attacks because of their large capacitance. Basic bus encryption prevents eavesdropping on the bus.

Next, some interesting architecture-level countermeasures will be discussed in more detail.

8.4.1 Current Flattening

The real-time power analysis resistant architecture, called *PAAR architecture* [87], targets a dynamically controlled current level with a very low detectable current variation. It is composed of 2 modules: feedback current flattening module (FCFM) and the pipeline current flattening module (PCFM).

The FCFM is responsible for measuring the instantaneous current consumption at the processor supply pin and generating two feedback signals to the PCFM. The PCFM is responsible for inserting nonfunctional instructions into the pipeline in order to bring the current consumption of the processor to a value that is within two programmable current variation limits (L_1 to L_2).

The FCFM has software programmable capabilities and can be activated only when needed. After reset, the block's signals are inactive until the current level registers are programmed. The FCFM has 6 control lines and one 8-bits output data bus. A current resistor R_p is connected between the power supply line V_{dd} and the power supply entry to the processor core. Two voltage signals, V_1 and V_2, are used to measure the voltage drop across the current resistor R_p. The voltage drop across the current resistor is proportional to the processor's current at all times. A differential amplifier is used to bring the small signal value of the voltage drop to a comparable signal level CL_p that is proportional to the instantaneous current consumption of the processor. The architecture assumes that the two current level registers are loaded with the binary values L_1 and L_2. These values can be loaded every time a secure communication is executing. The binary values L_1 and L_2 are converted by digital to analog (D/A) converters into the current levels CL_1 and CL_2, respectively. The current levels CL_1 and CL_2 represent the minimum and the maximum current level allowed, respectively. The current levels CL_1 and CL_2 are compared with the instantaneous current level CL_p by two comparators, which produce the feedback digital output signals of the FCFM to the PCFM. The feedback signals are "Cut Current Signal" (CCS) and "Increase Current Signal" (ICS). The CCS signals that the processor's instantaneous current is above the allowed maximum value and the ICS signals that the processor's instantaneous current is below the allowed minimum value.

8.4.2　Double-Data-Rate Computation

In [70], Maistri and Leveugle present a detection scheme based on temporal redundancy in order to avoid fault analysis: each encryption round is computed twice and the results are compared. The novel contribution is the exploitation of a computation approach based on the Double-Data-Rate (DDR) technique: at each clock cycle, both clock edges are used to sample data within the registers. This approach allows doubling the throughput of some operations at a given frequency; as side effect, it implies that the overall encryption latency can be halved and a second computation can be initiated. Under certain assumptions, the main and the verifying computation can be completed in the same (or comparable) time of the original computation process. The detection mechanism has been validated against fault injection by means of hardware emulated faults.

The DDR design brings some additional costs, which must be evaluated against comparable solutions. In particular, it has higher design costs than pipeline redundancy; on the other hand, it may operate in different modes that provide higher security against certain fault models. The area overhead for a 32-bit AES prototype is 36%, and the power consumption is increased by 55%.

8.4.3 Dynamic Reconfiguration

Dynamically reconfigurable systems are known to have many advantages such as area and power reduction. The drawbacks of these systems are the reconfiguration delay and the overhead needed to provide reconfigurability. Mentens, Gierlichs, and Verbauwhede [78] show that dynamic reconfiguration can also improve the resistance of cryptographic systems against physical attacks.

Some advantages of dynamic reconfiguration for cryptosystems have been explored before. In such systems, the main goal of dynamic reconfigurability is to use the available hardware resources in an optimal way. Mentens et al. consider to use a coarse-grained partially dynamically reconfigurable architecture in cryptosystems to prevent physical attacks by introducing temporal and/or spatial jitter. Note that the proposed countermeasures do not represent an all-embracing security solution and should be complemented by other countermeasures.

This type of countermeasure provides increased resistance, in particular against fault attacks, by randomly changing the physical location of functional blocks on the chip area at run-time. Besides, fault detection can be provided on certain devices with negligible area-overhead: the partial bitstreams can be read back from the reconfigurable areas and compared to a reference version at run-time and inside the device. These countermeasures do not change the device's input-output behavior, thus they are transparent to upper-level protocols. Moreover, they can be implemented jointly and complemented by other countermeasures on algorithm-, circuit-, and gate-level. Also, dynamic reconfiguration can realize a range of countermeasures that are standard for software implementations and that were practically not portable to hardware so far.

To port the idea of temporal jitter to hardware implementations, many registers could be foreseen in combination with multiplexors deciding whether to bypass a register or not. Because this would create a large overhead in resources, this option is highly impractical. Mentens et al. [78] propose an architecture with a dynamically reconfigurable switch matrix to avoid such a problem. The matrix determines the position of one or more registers in between functional blocks. Since a register causes a delay of one clock cycle, randomly positioning registers in between subfunctions desynchronizes the observations.

The number of possible configurations depends on the number m of registers and the number n of blocks. The value n depends on the number of reasonable subfunctions in the algorithm, which may depend on the width of the datapath. The number of options increases if we allow cascaded registers in between functional blocks.

Note, however, that if cascaded registers are allowed, there exist several configuration options that lead to identical sequences of combinatorial and sequential logic. Concerning this matter and allowing up to m cascaded registers, the number c of distinct configurations is $\binom{n+m-1}{m}$, i.e., the number of combinations of m elements out of n, where the order does not matter and repetition is allowed. The probability to observe the same configuration twice is $1/c$. However, the number of possible configurations determines the size of the memory needed to store the configuration data and is therefore bounded. Further, an increasing number of intermediate registers increases the number of cycles needed for one encryption. The number of registers, however, does not affect the maximal clock frequency, because more than one register can be cascaded. In general, the number of options for the temporal shift is determined by the number m of registers and the number n of blocks, and is bounded above by c.

It is important to remember that the number of needed power traces for a successful DPA attack grows quadratically with the number of operations shuffled [23]. And in this case, a preprocessing phase to properly align the traces can be very difficult.

8.4.4   Stream-Oriented Reconfigurable Unit (SORU)

Another way to take advantage of reconfigurability as SCA countermeasure is by adding reconfigurable coprocessors into the SoC design. SORU2 [35] is a dynamically reconfigurable coprocessor especially designed to support run-time self-adaptation in environments with tight resource and power restrictions. Besides providing a simple mechanism to dynamically change the power and timing profiles, it also allows to accelerate computationally intensive multimedia processing on portable/embedded devices maintaining low energy consumption.

The overall structure is depicted in Figure 8.13. Memory accesses are controlled by an external arbiter that prioritizes SORU2 memory accesses over the main processor. A vector load/store unit decouples operand fetch of SORU2 operations from the normal processor operation. It allows to feed the SORU2 RFU with up to four independent in-memory vectors, and store the results in data memory with no further intervention from the main datapath. This feature lets the processor execute any non memory-dependent data processing at the same time the SORU2 unit is executing a complex vector operation.

Most SORU2 operations are SIMD, and they will not take advantage from using a cache memory. The main processor accesses to memory through a cache but the SORU2 coprocessor accesses directly. This way, by exploiting the reference locality of most well-written programs, the main processor will be able to continue its operation almost normally.

As shown in Figure 8.14, the SORU2 execution stage is divided into four cascade-connected basic reconfigurable units (BRU). Each BRU gets three 32-

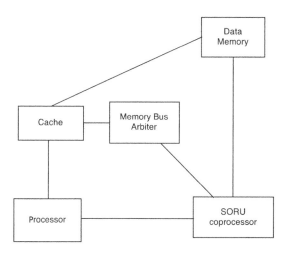

FIGURE 8.13
Overview of a SORU2 system.

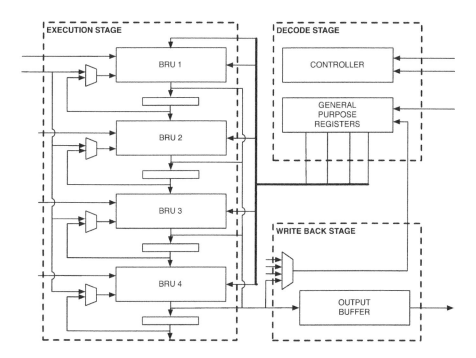

FIGURE 8.14
SORU2 datapath.

bit data operators: 1) the result from the previous BRU, 2) a new data item from the SORU2 register file, and 3) the last result computed by itself.

The result of a BRU operation is stored in the pipeline register, so it can be used by the next BRU at the next clock cycle. Moreover, that result can be sent to the register file where it will be stored.

Different configurations for one-cycle operations can be prepared off-line and stored in an external memory. The compiler is in charge of inserting configuration code to write the required contexts while the program is loaded. Additionally, the run-time support system can use dynamic information to re-optimize parts of the program by using a different set of configuration contexts.

One of the main advantages of the SORU2 architecture is the ease of the compilation process. We have ported Low-Level Virtual Machine (LLVM) [63] to our prototype processor, adding a new vectorization pass to implement loops as SORU2 SIMD operations.

SORU2 has many characteristics that can be used effectively to avoid side-channel attacks in an embedded system. We can classify them into low-power characteristics and nondeterministic behavior.

Unlike most reconfigurable architectures, SORU2 does not include long lines connecting many gates all over the device. Data flow inside the SORU2 execution unit is very directional, and therefore the load capacitances are usually very low.

Memory buses usually have a high power consumption, and therefore, whenever a cryptographic key goes through them, it is leaking significant information that can be analized with DPA techniques. However, in a SORU2 implementation of the algorithm, the key would only go through the memory bus once. Then, it would be stored in a SORU2 internal register, and operated only inside the coprocessor. As power traces of the SORU2 execution unit are much smaller, the signal-noise ratio would be much smaller, and any statistical analysis would require many more traces to succeed.

The multiple SORU2 configuration contexts can be preloaded before starting to execute the cipher algorithm, and it can change the active configuration every clock cycle.

The most straightforward approach to take advantage of the proposed architecture for avoiding DPA attacks is by generating multiple SORU2 implementations of the program loops, and randomly change between them in run-time. Non-deterministic changes between functionally-equivalent implementation of loop bodies would increase the noise level significantly at no cost for the embedded system. An attack based on power analysis would become very difficult.

The compiler finds the loops that can be extracted to SORU2 SIMD operations, and uses standard compilation techniques to generate a first implementation. Additionally, a simulated annealing pass in the compiler generates many different implementations of these loops starting from the first implementation.

One of the most interesting aspects of SORU as a DPA-countermeasure is that it also helps to improve the global performance and reduce the power consumption at the same time. Benchmarks from MiBench saw a speedup ranging from 2 to 4.5, while reducing the energy consumption up to 80%. Also, the main datapath is mostly idle during the execution of the SORU2 SIMD operations. This idle time can be used to schedule other independent operations (leading to a better global speedup), or alternatively the main datapath can be left in a hibernation state until the SORU2 unit finishes (leading to a better energy reduction).

The experimental results show at least three orders of magnitude improvement in DPA resistance for a custom Keeloq implementation, by using only static techniques. The performance results on a DSP benchmark show a mean speedup of 4 with only doubling the total area when connected to a simple RISC processor, and reducing the instruction fetches also significantly.

The advantages become more obvious when considering energy consumption in battery-powered embedded systems. As opposed to typical FPGA devices, where interconnect power consumption is dominant, SORU2 units force a predefined pipelined data flow, reducing significantly the length of interconnections and the load capacitance. The loss in flexibility is highly compensated by the ease of mapping complete loops into a single SORU2 instruction, enabling the use of dynamic context-aware optimization techniques.

Moreover, our SORU2-based approach to avoid side-channel attacks allows a high degree of decoupling between the application development and the security-aware implementation, taking into account architecture, compilation, and run-time issues.

8.5 Algorithm-Level DPA-Aware Techniques

We can classify the algorithm-level countermeasures into three main categories: time-oriented hiding techniques, amplitude-oriented hiding techniques, and masking techniques.

The most commonly used time-oriented hiding countermeasure at the algorithm-level is to randomize the execution of the algorithms. The random insertion of dummy operations as well as shuffling can also be implemented easily at this level. However, it is important to note that these countermeasures require random numbers that need to be generated on the cryptographic device.

Amplitude-oriented hiding techniques aim at choosing the algorithm implementation that minimizes the leakages. Some of these techniques include:

- Choosing the operations that leak the minimum amount of information about their operands.

- Changes in the program flow, although they are usually easy to detect by visual inspection of the power traces.

- Key-independent memory addresses.

- Parallel activity to increase the noise.

Some of the most used techniques between the algorithm-level masking techniques include:

- Masking table look-ups, in order to implement masking in a simple and efficient way. However, table initialization needs to be done for all the masks involved in the operations, and the computational effort and memory requirements are high.

- Random precharging to perform implicit masking. If the device leaks the Hamming distance, loading or storing a random value before the actual intermediate value occurs, works as if the intermediate value were masked.

Next, some interesting algorithm-level countermeasures will be discussed in more detail.

8.5.1   Randomized Initial Point

The Randomized Initial Point (RIP) is a countermeasure proposed by Itoh et al. [47] (later extended by the same authors [48]) against DPA and the RPA/ZVA ECC-specific attacks.

Basically, when computing the scalar multiplication, a register is initially loaded with a randomly generated point R, so $R + dP$ is computed instead of dP. To give the expected result, R is subtracted afterwards. This way, the attacker cannot force special points by carefully choosing P so information on d is leaked.

It is important to note that the initial point R must be properly chosen to be random: RIP would be vulnerable to DPA if R were fixed or reused.

Another characteristic of RIP, aside from being resistant to RPA and ZVA, is its small performance impact: according to its creators, RIP has almost no penalty from scalar multiplication algorithms without countermeasures.

Randomized exponentiation algorithms were recently considered as effective countermeasures against DPA by introducing randomization onto the input message or into the computational process of the algorithm in order to remove correlation between the private key and the collected power traces. One of such countermeasures is the Binary-Based Randomized Initial Point (BRIP) algorithm [71] (it means binary expansion with random initial point/value) in which the input RSA message is blinded by multiplying with a random integer $R - 1 \mod n$. It was claimed in [71] that the BRIP algorithm can be secure against SPA. However, Yen et al. [133] published a SPA attack that breaks BRIP and other well-known anti-SPA countermeasures.

BRIP was originally proposed for ECC context to protect against RPA, which requires an inversion for the computation. However, BRIP's authors say that their algorithm can also be applied in $\mathbb{Z}^n$ for cryptosystems based on integer factorization or discrete logarithm.

8.5.2 Avoiding Cold Boot Attacks

Contrary to popular assumption, DRAMs used in most modern computers retain their contents for several seconds after power is lost, even at room temperature and even if removed from a motherboard. Although DRAMs become less reliable when they are not refreshed, they are not immediately erased, and their contents persist sufficiently for malicious (or forensic) acquisition of usable full-system memory images [45]. This phenomenon limits the ability of an operating system to protect cryptographic key material from an attacker with physical access. Moreover, remanence times can be increased dramatically with simple cooling techniques.

Although not a classical one, these *cold boot attacks* are certainly another kind of side-channel attack.

Memory imaging attacks are difficult to defend against because cryptographic keys that are in active use need to be stored somewhere. Halderman et al. [45] suggest to focus on discarding or obscuring encryption keys before an adversary might gain physical access, preventing memory-dumping software from being executed on the machine, physically protecting DRAM chips, and possibly making the contents of memory decay more readily.

Countermeasures begin with efforts to avoid storing keys in memory. Software should overwrite keys when they are no longer needed, and it should attempt to prevent keys from being paged to disk. Run-time libraries and operating systems should clear memory proactively; Chow et al. show that this precaution does not need to be expensive [28]. Of course, these precautions cannot protect keys that must be kept in memory because they are still in use, such as the keys used by encrypted disks or secure web servers.

8.6 Validation

In principle, the design flow of secure cryptosystems can take advantage of the concept of power analysis simulation for early assessing the susceptibility of a given system to selected PA attacks, but the ever-increasing complexity of SoC architectures makes it difficult to quantitatively understand the degree of vulnerability of a system at the design time. In fact, although circuit-level, gate-level, and even register transfer level (RTL) simulations of a whole SoC are nearly unfeasible for the average performance and power consumption, they are even more time consuming for SCA simulation, where a much higher detail is needed than in the average power consumption simulation. System-level performance and average power simulations have been in use for some years now based on modeling languages such as SystemC, C/C++, and Java [75, 69, 88, 126, 119], but there is presently a significant lack of consolidated tools and results in system-level SCA simulation, which may assess the potential failure points of a SoC architecture with respect to SCA attacks.

Several previous works have addressed the issue of accelerating power analysis (PA) simulation at several abstraction levels. In [20], a countermeasure against PA attacks is evaluated by analyzing power traces generated from accelerated transistor-level simulations (Synopsys Nanosim). In [125], a specialized CMOS logic style is evaluated by executing a PA attack on power traces produced by SPICE simulations. In [37], simulations of PA attacks are performed on power traces generated by Synopsys Primepower from a postlayout gate-level model of the design. A sophisticated extension of this approach is presented in [66], where a methodology for simulating EMA attacks is presented, which takes into account both the circuit and the package of the given system yet relies on postlayout power traces.

In [91], a PA attack is executed by RTL power simulations accounting for logic-value transitions in CMOS registers, and the results are interestingly compared with the corresponding ones based on real-power traces collected on the given physical system.

The first instance of system-level simulation is shown in [34, 33], which present a Java tool, i.e., PINPAS for simulating side-channel effects in smart card microprocessors. The simulator models processor instructions and accounts for the Hamming weight (HW) of the bits in the logical registers of the processor architecture. Another interesting example of the instruction-level simulation of PA attacks is shown in [68], which uses cycle-accurate power traces generated by the SimplePower processor simulator.

In [77], Menichelli et al. present a methodology for system-level modeling and simulation of a whole SoC architecture executing an industrial standard cryptography algorithm. The approach is based on the SystemC 2.0 language, which is a de facto standard in complex digital system simulations because of its straightforward and efficient hardware-software cosimulation capabilities. The flexibility inherent to SystemC allows to model and evaluate both microprocessor-based designs (i.e., relying on software-implemented algorithms) and hardwired designs (i.e., relying on some applicationspecific hardware units), including the presence of hardware and/or software countermeasures against SCA.

8.7 Traps and Pitfalls

Typical targets of side-channel attacks are security integrated circuits (ICs) used in embedded devices and smart cards. They are not only an attractive target as they are dedicated to performing secure operations, they are also rather easy to analyze.

Yet the interaction between software and a micro-architecture optimized for performance leads to similar vulnerabilities and in recent times, SCAs have been demonstrated that successfully attack software implementations of symmetric and public key encryption algorithms running on 32-bit microprocessors [128]. Without proper attention, the pitfalls are prone to contribute side-channel leakage and hence they can be used to identify potential problems. Given that a sidechannel is in fact a covert channel without conspiracy or consent, it should not be a surprise that both share facilitators and consequently mitigation strategies [62, 109].

Tiri [118] provides some important advice to avoid traps and pitfalls. The book [73] also includes an important set of tips to enhance the security of electronic systems.

Resource Sharing May Make Things Worse

Resource sharing, which reduces the amount of hardware needed to implement a certain functionality, can both create side-channel information and facilitate its observation.

For example, the cache is shared between the different processes running on the same CPU and since the cache has a finite size, they compete for the shared resource. This means that one process can evict another process's data from the cache if some of their data is mapped to the same cache line, where a cache line is the smallest unit of memory that can be transferred between the main memory and the cache. This has two consequences. First, a spy process is able to observe which cache lines are used by the crypto process and thus determine its cache footprint. Second, the spy process is able to clean the cache. It can remove all of crypto's data from the cache, which will increase the number of cache misses of the crypto process as the cache does not contain any of its data. This also has the advantage that the attacker can readily hypothesize a known initial state of the cache, which is required to estimate the leakage.

Optimization May Generate Leakage

Optimization features, which improve a system's performance or cost, can create side-channel information. In general, only the typical case is being optimized, and hence the corner cases leak information.

For example, in the timing attack from Brumley and Boneh [18], the compiler optimizations can make a huge difference in order to increase the leakage. And moreover, the leakage found by Brumley and Boneh in the OpenSSL code was due to algorithm-level optimizations.

Increased Visibility/Functionality

Increased visibility, or for that matter increased functionality, can facilitate the observation of side-channel information. In general, new functionality increases the complexity and hence introduces new interactions that might ease the difficulty of mounting the measurements.

For example, special performance counters have been added to modern microprocessors to count a wide array of events that affect a program's performance. They are able to count the number of cache accesses, as the cache behavior is an important factor in a program's performance. Compared with the time stamp counter, which is currently used in the cache attacks, the performance counters increase the visibility. The counters can be programmed to solely count the events of interest. For instance, it is possible to specify to only measure the cache read misses and not the cache write misses. Time measurements, on the other hand, measure all events that influence the time delay. The performance counters paint a more accurate picture and they could enable better and faster attacks than the timestamp counter.

Masking Does Not Avoid Attacks

Many of the mitigations that are intuitively put forward, such as the randomization of the execution sequence or the addition of a random power-consuming module or a current sink, hardly improve the resistance against the power attacks [29, 56, 107]. In the present state-of-the-art, the countermeasures try to make the power consumption of the cryptographic device independent of the signal values at the internal circuit nodes by either randomizing or flattening the power consumption. None of the techniques, however, provides perfect security; on the contrary, they increase the required number of measurements.

Randomizing the power consumption is done with masking techniques that randomize the signal values at the internal circuit nodes while still producing the correct cipher-text. This can be done at the algorithmic level where a random mask is added to the data prior to the encryption and removed afterwards without changing the encryption result (e.g., [97]) or at the circuit level where a random mask-bit equalizes the output transition probabilities of each logic gate [117, 96]. Flattening the power consumption is usually done at the circuit level, such that each individual gate has a quasi data-independent power dissipation. This is done with dynamic differential logic, sometimes also referred to as dual rail with precharge logic to assure that every logic gate has a single charging event per cycle [59]. In self-timed asynchronous logic [83], the terminology refers to dual rail encoded data interleaved with spacers. As an example, Tiri et al. [119] show the measurements and the attack result of an AES ASIC protected using WDDL to flatten out the power consumption. 1.5M measurements are not sufficient to find the key byte under attack.

Algorithmic masking was conceived as a mathematically proven countermeasure, but recent developments show that practical implementation issues can often break perfect theoretical security schemes. Earlier work has already shown that it was vulnerable against higher order attacks. These attacks can combine multiple samples to remove an unknown mask because of the Hamming weight or distance leakage estimation model used [93]. But now, using template attacks, in which the leakage model is built from the measurements, the authors of [92] conclude that masking has zero improvement on the security of an implementation.

Besides, the power dissipation of masked hardware circuits is uncorrelated to the unmasked data values, and therefore cannot be used for DPA. However, the power dissipation of a masked hardware circuit may still be correlated to the mask. Because of this correlation, it is possible to bias the mask by selecting only a small slice over the entire power probability density function (PDF). This technique has been successfully applied using an AES S-Box with perfect masking [26]. Using logic-level simulation, Chen and Schaumont have demonstrated the dependency between the power dissipation and the mask value. By slicing the power PDF before mounting a DPA, each bit can be biased from the mask. Therefore, hardware masking remains susceptible to direct DPA by making clever use of the power probability density function.

Countermeasures Are Not Free

The increased power attack resistance does not come free. The algorithmic level masking has a factor 1.5 overhead when compared with a regular (unprotected) design [97]. The masked logic styles have a factor 2 and 5 area overhead [117, 96]. The dual rail logic styles have a factor 3 area overhead [119]. Yet, the figures for the algorithmic and logic masking do not include the random number generator. It is thus important that the full implementation cost of a countermeasure is clearly communicated and taken into account for evaluation. Several techniques have been proposed to reduce the area overhead. For instance, custom logic cells can be made more compact than compound standard logic cells, and security partitioning reduces the part of the design that has to be protected [119].

Security Evaluation Is Hard

Yet, one has to be careful to declare a (new) mitigation as secure. A visual inspection, or even the standard deviation of the power consumption, does not provide any indication [124]. Thus far, the best figure of merit is probably the required number of measurements for a successful attack on a realistic circuit. The success of an attack, however, depends both on the information in the power consumption and on the strength of the attacker, which encompasses the measurement setup but also the leakage estimation and the statistical technique used. Indeed, if the power estimation is more accurate the attack will be more successful. The statistical analysis technique to compare the measurements with the estimations is also important: the difference of means test requires more measurements than the correlation test, which requires more measurements than the Bayesian classification. Some work has been done to distinguish the quality of an implementation from the strength of a side-channel adversary [113], but it is not clear how in a practical way a design can be evaluated without an attack and with abstraction of the statistical tool or distinguisher. Can an expression be found that based on design parameters such as the activity factor or a power-consumption profile indicates the strength of a design?

Matching the Interconnect Capacitances

For the dual-rail circuit styles to be effective it is crucial that the load capacitances at the differential output are matched. Matching the interconnect capacitances of the signal wires is essential. This can be done with differential routing [125] or back-end duplication [43]. Yet with shrinking deep submicron technology, and with the fast development of new SCA attacks, it is difficult to make sure that the nets are matched sufficiently.

A Global Approach to Security Is a Must

Aside from algorithmic techniques for software running on microprocessors, resistance cannot be added as an afterthought. Some design flows have been developed to automatically create more secure designs [125, 43, 3, 61]. Yet, design time security assessment remains a crucial design phase [66]. The quality of the assessment, however, is only as good as the power-consumption simulation model used. It is important to note that side-channel resistance cannot be isolated at one abstraction level: a technique that can be proven secure at a high abstraction level is not necessarily secure when gate delays or load capacitances are taken into account. This has been proven true both for the masking techniques and for the current flattening techniques. Glitches make attacks on the former possible [74]. Early propagation of data enables attacks on the latter [60]. These issues could be addressed with strict control over all the signal arrival times. And this become more difficult to assess as the complexity of the designs grow. Building a correct simulation model of the side-channel leaks for use in the design time security assessment is not easy. Minor differences, even second-order effects, in the power model can have a big influence in the resistance assessment [124].

Three factors contribute to make security in distributed embedded systems a very difficult problem: 1) many nodes in the network have very limited resources; 2) pervasiveness implies that some nodes will be in noncontrolled areas and are accessible to potential intruders; and 3) all these computers are usually interconnected, allowing attacks to be propagated step by step from the more resource-constrained devices to the more secure servers with lots of private data.

Current ciphers and countermeasures often imply a need for more resources (more computation requirements, more power consumption, specific integrated circuits with careful physical design, etc.), but usually this is not affordable for this kind of application. But even if we impose strong requirements for any individual node to be connected to our network, it is virtually impossible to update hardware and software whenever a security flaw is found. The need to consider security as a new dimension during the whole design process of embedded systems has already been stressed [100], and there have been some initial efforts towards design methodologies to support security [102, 7, 8], but to the best of our knowledge no attempt has been made to exploit the special characteristics of wireless sensor networks.

Other Side-Channel Attacks

Power analysis attacks are not the only ones. There are many other side-channel attacks, based on leakages on timing, EM emissions, system behavior in presence of faults, sound, etc.

For example, a cache attack works as follows. The cache is used to store recently used data in a fast memory block close to the microprocessor. Whenever this data is used subsequently, it can be delivered quickly. On the other hand, whenever the microprocessor requires data that is not in the cache, it has to be fetched from another memory with a larger latency. The time difference between both events is measurable and provides the attacker with sufficient data on the state and the execution of the algorithm to extract some or even all secret key bits [128]. Note that numerous mitigations have been put forward to limit the information leakage (e.g., [15]), some of which have been incorporated into cryptographic tools and libraries (e.g., OpenSSL).

Therefore, security-critical parts of the system should be kept to a minimum and their design, including the interface and the protocols, should be left to experienced designers.

Secure Applications Based on Untrusted Elements

Even when all kinds of precautions have been taken, a new attack may compromise our design. There are numerous examples of this.

Applications built on distributed embedded systems have to live with the fact that privacy and integrity cannot be preserved in every node of the network. This poses restrictions on the information that a single node can manage, and also in the way the applications are designed and distributed in the network.

Of course, the inherent insecurity of embedded systems should not lead us to not try hard to avoid compromises. We should guarantee that a massive attack cannot be fast enough to avoid the detection and recovery measures to be effective. Therefore, we should design the nodes as secure as the available resources allow.

In spite of the disadvantages of distributed embedded systems from the security point of view, they provide two advantages for fighting against attacks:

- Redundancy. A wireless sensor network usually has a high degree of spatial redundancy (many sensors that should provide coherent data), and temporal redundancy (habits, periodic behaviors, causal dependencies), and both can be used to detect and isolate faulty or compromised nodes in a very effective manner.

- Continuous adaptation. Wireless sensor networks are evolving continuously, there are continuous changes of functional requirements (data requests, service requests, user commands...), nodes appear and disappear continuously and therefore routing schemes change, low batteries force some functionality to be migrated to other nodes, etc.

These properties have been used to build an application framework for the development of secure applications using sensor networks [86].

8.8   Conclusions

Very often, SoCs need to communicate sensitive information inside the chip or to the outside. To avoid eavesdropping on the communications, it is common to use encryption schemes at different levels. However, using side-channel information, it can be very easy to gain the secret information from the system. Protecting against these attacks, however, can be a challenge, it is costly and must be done with care.

Every SoC component involved in the communication of sensitive information (emitter, receiver, and communication channels) should be carefully designed, avoiding leakages of side-channel information at every design level. And these security concerns should be considered globally, as one design level may interact with others. For example, compiler optimizations may create data-dependent asymmetries in a perfectly designed algorithm, but also the small variance in the capacitance of signal lines in the chip may be used to discover the cryptographic key. These considerations should be also taken into account when reusing cores.

Therefore, to keep the design time reasonable, the communication of sensitive information within a SoC should be limited to a minimum set of components, that should be designed carefully, trying to avoid any kind of information leakage at every design level. Reuse is not possible unless these issues were taken into account during the design of the reused component. And even in that case, it is important to consider the impact in the leakages of any change from previous designs (technology, logic style, components connected to the bus, protocols usage, etc.).

There is a huge catalog of countermeasures against side-channel attacks, ranging from algorithmic transformations, to custom logic styles, but no one by itself completely avoids these attacks. Each countermeasure has its weaknesses. Hence, a reasonable compromise needs to be found between the resistance against side-channel attacks and the implementation costs of the countermeasures (performance, area, power consumption, design time, etc.).

8.9 Glossary

AES: Advanced Encryption Standard

ASIC: Application-Specific Integrated Circuit

BBRIP: Binary-Based Randomized Initial Point

CMOS: Complementary Metal-Oxide-Semiconductor

CPA: Correlation Power Analysis

DDL: Dynamic and Differential Logic

DDR: Double-Data-Rate

DEMA: Differential Electromagnetic Analysis

DES: Data Encryption Standard

DFA: Differential Fault Analysis

DPA: Differential Power Analysis

DRP: Dual-Rail Precharge logic styles

DSCA: Differential Side Channel Attacks

DSP: Digital Signal Processor

DT: Differential Time

DTS: Dissipation Timing Skew

ECC: Elliptic Curve Cryptosystem

EEPROM: Electrically-Erasable Programmable Read-Only Memory

EM: Electromagnetic

EMA: Electromagnetic Analysis

EPDU: Evaluation-Precharge Detection Unit

FFT: Fast Fourier Transform

FIB: Focused Ion Beam

FPGA: Field Programmable Gate Array

FPRM: Fixed Polarity Reed-Muller canonical form

HW/SW: Hardware/Software

iMDPL: Improved Masked Dual-rail Precharge Logic

IPA: Inferential Power Analysis

LLVM: Low-Level Virtual Machine

MDPL: Masked Dual-rail Precharge Logic

PA: Power Analysis

PDA: Personal Digital Assistant

PDF: Probability Density Function

PMRML: Precharge Masked Reed-Muller Logic

QDI: Quasi-Delay Insensitive

RFID: Radio Frequency Identification Device

RIP: Randamized Initial Point

RNG: Random Number Generator

RPA: Refined Power Analysis

RSA: Rivest-Shamir-Adleman public key encryption algorithm

RSL: Random Switching Logic

SABL: Sense Amplifier Based Logic

SCA: Side Channel Analysis

SNR: Signal-Noise Ratio

SORU: Stream-Oriented Reconfigurable Unit

SPA: Simple Power Analysis

TDPL: Three-phase Dual-rail Precharge Logic

VHDL: VHSIC Hardware Description Language

VHSIC: Very-High-Speed Integrated Circuit

WDDL: Wave Dynamic Differential Logic

ZVA: Zero Value Analysis

8.10 Bibliography

[1] D. Agrawal, J. R. Rao, and P. Rohatgi. Multi-channel attacks. In Walter et al. [130], 2–16.

[2] M. Aigner and E. Oswald. Power analysis tutorial. Technical report, University of Technology Graz, 2008.

[3] M. Josef Aigner, S. Mangard, F. Menichelli, R. Menicocci, M. Olivieri, T. Popp, G. Scotti, and A. Trifiletti. Side channel analysis resistant design flow. In *ISCAS*. IEEE, 2006.

[4] T. Akishita and T. Takagi. Zero-value point attacks on elliptic curve cryptosystem. In Colin Boyd and Wenbo Mao, editors, *ISC*, volume 2851 of *Lecture Notes in Computer Science*, 218–233. Springer, Berlin Heidelberg, 2003.

[5] M.-L. Akkar, R. Bevan, P. Dischamp, and D. Moyart. Power analysis, what is now possible... In Tatsuaki Okamoto, editor, *ASIACRYPT*, volume 1976 of *Lecture Notes in Computer Science*, 489–502. Springer, Berlin Heidelberg, 2000.

[6] R. J. Anderson and M. G. Kuhn. Low cost attacks on tamper resistant devices. In Bruce Christianson, Bruno Crispo, T. Mark, A. Lomas, and Michael Roe, editors, *Security Protocols Workshop*, volume 1361 of *Lecture Notes in Computer Science*, 125–136. Springer, Berlin Heidelberg, 1997.

[7] D. Arora, A. Raghunathan, S. Ravi, M. Sankaradass, N. K. Jha, and S. T. Chakradhar. Software architecture exploration for high-performance security processing on a multiprocessor mobile SoC. In *Proceedings of the 43rd Annual Conference on Design Automation*, 496–501, San Francisco, CA, USA, 2006. ACM.

[8] L. Benini, A. Macii, E. Macii, E. Omerbegovic, F. Pro, and M. Poncino. Energy-aware design techniques for differential power analysis protection. In *Proceedings of the 40th Conference on Design Automation*, 36–41, Anaheim, CA, USA, 2003. ACM.

[9] R. Bevan and E. Knudsen. Ways to enhance differential power analysis. In *ICISC'02: Proceedings of the 5th International Conference on Information Security and Cryptology*, 327–342, Springer, Berlin Heidelberg, 2003.

[10] E. Biham and A. Shamir. Differential fault analysis of secret key cryptosystems. In Burton S. Kaliski Jr., editor, *CRYPTO*, volume 1294 of *Lecture Notes in Computer Science*, 513–525. Springer, 1997.

[11] E. Biham and A. Shamir. Power analysis of the key scheduling of the AES candidates. In *Proceedings of the Second AES Candidate Conference*, 115–121. Addison-Wesley, New York, 1999.

[12] D. Boneh, R. A. DeMillo, and R. J. Lipton. On the importance of checking cryptographic protocols for faults (extended abstract). In *EUROCRYPT*, 37–51, 1997.

[13] G. F. Bouesse, M. Renaudin, S. Dumont, and F. Germain. DPA on quasi delay insensitive asynchronous circuits: Formalization and improvement. In *DATE '05: Proceedings of the Conference on Design, Automation and Test in Europe*, 424–429, Washington, DC, USA, 2005. IEEE Computer Society.

[14] G. F. Bouesse, G. Sicard, and M. Renaudin. Path swapping method to improve DPA resistance of quasi delay insensitive asynchronous circuits. In Goubin and Matsui [42], 384–398.

[15] E. Brickell, G. Graunke, M. Neve, and J. P. Seifert. Software mitigations to hedge AES against cache-based software side channel vulnerabilities. iacr eprint archive, report 2006/052, 2006.

[16] E. Brier, C. Clavier, and F. Olivier. Correlation power analysis with a leakage model. In Joye and Quisquater [49], 16–29.

[17] E. Brier and M. Joye. Weierstraß elliptic curves and side-channel attacks. In David Naccache and Pascal Paillier, editors, *Public Key Cryptography*, volume 2274 of *Lecture Notes in Computer Science*, 335–345. Springer, Berlin Heidelberg, 2002.

[18] D. Brumley and D. Boneh. Remote timing attacks are practical. *Computer Networks*, 48(5):701–716, 2005.

[19] M. Bucci, L. Giancane, R. Luzzi, and A. Trifiletti. Three-phase dual-rail pre-charge logic. In Goubin and Matsui [42], 232–241.

[20] M. Bucci, M. Guglielmo, R. Luzzi, and A. Trifiletti. A power consumption randomization countermeasure for DPA-resistant cryptographic processors. In Enrico Macii, Odysseas G. Koufopavlou, and Vassilis Paliouras, editors, *PATMOS*, volume 3254 of *Lecture Notes in Computer Science*, 481–490. Springer, Berlin Heidelberg, 2004.

[21] Ç. K. Koç and C. Paar, editors. *Cryptographic Hardware and Embedded Systems, First International Workshop, CHES'99, Worcester, MA, USA, August 12-13, 1999, Proceedings*, volume 1717 of *Lecture Notes in Computer Science*. Springer, Berlin Heidelberg, 1999.

[22] Ç. K. Koç and C. Paar, editors. *Cryptographic Hardware and Embedded Systems - CHES 2000, Second International Workshop, Worcester, MA,*

USA, August 17-18, 2000, Proceedings, volume 1965 of *Lecture Notes in Computer Science*. Springer, Berlin Heidelberg, 2000.

[23] S. Chari, C. S. Jutla, J. R. Rao, and P. Rohatgi. Towards sound approaches to counteract power-analysis attacks. In Wiener [132], 398–412.

[24] S. Chari, J. R. Rao, and P. Rohatgi. Template attacks. In Kaliski Jr. et al. [51], 13–28.

[25] Z. Chen, S. Haider, and P. Schaumont. Side-channel leakage in masked circuits caused by higher-order circuit effects. In J. H. Park, H.-H. Chen, M. Atiquzzaman, C. h. Lee, T.-H. Kim, and S.-S. Yeo, editors, *ISA*, volume 5576 of *Lecture Notes in Computer Science*, 327–336. Springer, Berlin Heidelberg, 2009.

[26] Z. Chen and P. Schaumont. Slicing up a perfect hardware masking scheme. In Mohammad Tehranipoor and Jim Plusquellic, editors, *HOST*, 21–25. IEEE Computer Society, Anaheim, CA, 2008.

[27] Z. Chen and Y. Zhou. Dual-rail random switching logic: A countermeasure to reduce side channel leakage. In Goubin and Matsui [42], 242–254.

[28] J. Chow, B. Pfaff, T. Garfinkel, and M. Rosenblum. Shredding your garbage: reducing data lifetime through secure deallocation. In *SSYM'05: Proceedings of the 14th Conference on USENIX Security Symposium*, 22–22, Berkeley, CA, USA, 2005. USENIX Association.

[29] C. Clavier, J.-S. Coron, and N. Dabbous. Differential power analysis in the presence of hardware countermeasures. In C. K. Koç and C. Paar [22], 252–263.

[30] J.-S. Coron. Resistance against differential power analysis for elliptic curve cryptosystems. In C. K. Koç and C. Paar [21], 292–302.

[31] J.-S. Coron, D. Naccache, and P. Kocher. Statistics and secret leakage. *ACM Trans. Embed. Comput. Syst.*, 3(3):492–508, 2004.

[32] P. Cunningham, R. Anderson, R. Mullins, G. Taylor, and S. Moore. Improving smart card security using self-timed circuits. In *ASYNC '02: Proceedings of the 8th International Symposium on Asynchronus Circuits and Systems*, 211, Washington, DC, 2002. IEEE Computer Society.

[33] J. den Hartog and E. P. de Vink. Virtual analysis and reduction of side-channel vulnerabilities of smartcards. In T. Dimitrakos and F. Martinelli, editors, *Formal Aspects in Security and Trust*, 85–98. Springer, New York, 2004.

[34] J. den Hartog, J. Verschuren, E. P. de Vink, J. de Vos, and W. W. Pinpas: A tool for power analysis of smartcards. In D. Gritzalis, S. De Capitani di Vimercati, P. Samarati, and S. K. Katsikas, editors, *SEC*, volume 250 of *IFIP Conference Proceedings*, 453–457. Kluwer, Norwell, MA, 2003.

[35] J. M. Moya et al. SORU: A reconfigurable vector unit for adaptable embedded systems. In *ARC '09: Proceedings of the 5th International Workshop on Reconfigurable Computing: Architectures, Tools and Applications*, 255–260, Springer-Verlag, Berlin Heidelberg, 2009.

[36] P. N. Fahn and P. K. Pearson. IPA: A new class of power attacks. In C. K. Koç and C. Paar [21], 173–186.

[37] J. J. A. Fournier, S. Moore, H. L. R. Mullins, and G. Taylor. Security evaluation of asynchronous circuits. In *In Proceedings of Cryptographic Hardware and Embedded Systems - CHES2003*, 137–151. Springer-Verlag, Berlin Heidelberg, 2003.

[38] C. H. G., S. Ho, and C. C. Tiu. Em analysis of rijndael and ecc on a wireless java-based pda. In Rao and Sunar [99], 250–264.

[39] B. Gierlichs. DPA-resistance without routing constraints? In Paillier and Verbauwhede [94], 107–120.

[40] B. Gierlichs, K. Lemke-Rust, and C. Paar. Templates vs. stochastic methods. In Goubin and Matsui [42], 15–29.

[41] L. Goubin. A refined power-analysis attack on elliptic curve cryptosystems. In Y. Desmedt, editor, *Public Key Cryptography*, volume 2567 of *Lecture Notes in Computer Science*, 199–210. Springer, Berlin Heidelberg, 2003.

[42] L. Goubin and M. Matsui, editors. *Cryptographic Hardware and Embedded Systems - CHES 2006, 8th International Workshop, Yokohama, Japan, October 10-13, 2006, Proceedings*, volume 4249 of *Lecture Notes in Computer Science*. Springer, Berlin Heidelberg, 2006.

[43] S. Guilley, P. Hoogvorst, Y. Mathieu, and R. Pacalet. The "backend duplication" method. In Rao and Sunar [99], 383–397.

[44] F. K. Gürkaynak. *GALS System Design: Side Channel Attack Secure Cryptographic Accelerators*. PhD thesis, ETH, Zürich, 2006.

[45] J. A. Halderman, S. D. Schoen, N. Heninger, W. Clarkson, W. Paul, J. A. Calandrino, A. J. Feldman, J. Appelbaum, and E. W. Felten. Lest we remember: Cold-boot attacks on encryption keys. *Commun. ACM*, 52(5):91–98, 2009.

[46] K. Itoh, T. Izu, and M. Takenaka. Address-bit differential power analysis of cryptographic schemes OK-ECDH and OK-ECDSA. In Kaliski Jr. et al. [51], 129–143.

[47] K. Itoh, T. Izu, and M. Takenaka. Efficient countermeasures against power analysis for elliptic curve cryptosystems. In J.-J. Quisquater, P. Paradinas, Y. Deswarte, and A. A. El Kalam, editors, *CARDIS*, 99–114. Kluwer, Norwell, MA, 2004.

[48] K. Itoh, T. Izu, and M. Takenaka. Improving the randomized initial point countermeasure against DPA. In J. Zhou, M. Yung, and F. Bao, eds, *ACNS*, volume 3989 of *Lecture Notes in Computer Science*, 459–469, 2006.

[49] M. Joye and J.-Jacques Quisquater, editors. *Cryptographic Hardware and Embedded Systems - CHES 2004: 6th International Workshop Cambridge, MA, USA, August 11-13, 2004. Proceedings*, volume 3156 of *Lecture Notes in Computer Science*. Springer, Berlin Heidelberg, 2004.

[50] M. Joye and C. Tymen. Protections against differential analysis for elliptic curve cryptography. In C. K. Koç, D. Naccache, and C. Paar, editors, *CHES*, volume 2162 of *Lecture Notes in Computer Science*, 377–390. Springer, Berlin Heidelberg, 2001.

[51] B. S. Kaliski Jr., Ç. K. Koç, and C. Paar, editors. *Cryptographic Hardware and Embedded Systems - CHES 2002, 4th International Workshop, Redwood Shores, CA, USA, August 13-15, 2002, Revised Papers*, volume 2523 of *Lecture Notes in Computer Science*. Springer, Berlin Heidelberg, 2003.

[52] W. H. Joyner Jr., G. Martin, and A. B. Kahng, editors. *Proceedings of the 42nd Design Automation Conference, DAC 2005, San Diego, CA, USA, June 13-17, 2005*. ACM, 2005.

[53] N. Koblitz. Elliptic curve cryptosystems. *Mathematics of Computation*, 48:203–209, 1987.

[54] P. Kocher, J. Jaffe, and B. Jun. Introduction to differential power analysis and related attacks. Whitepaper, Cryptography Research, San Francisco, CA, USA, 1998.

[55] P. C. Kocher. Timing attacks on implementations of Diffie-Hellman, RSA, DSS, and other systems. In N. Koblitz, editor, *CRYPTO*, volume 1109 of *Lecture Notes in Computer Science*, 104–113. Springer, Berlin Heidelberg, 1996.

[56] P. C. Kocher, J. Jaffe, and B. Jun. Differential power analysis. In Wiener [132], 388–397.

[57] O. Kömmerling and M. G. Kuhn. Design principles for tamper-resistant smartcard processors. In *USENIX Workshop on Smartcard Technology*, 1999.

[58] A. Kramer, J. S. Denker, B. Flower, and J. Moroney. 2nd order adiabatic computation with 2n-2p and 2n-2n2p logic circuits. In M. Pedram, R. W. Brodersen, and K. Keutzer, editors, *ISLPD*, 191–196. ACM, 1995.

[59] M. A. K. Tiri and I. Verbauwhede. A dynamic and differential CMOS logic with signal independent power consumption to withstand differential power analysis on smart cards. In *Proceeding of the 28th European Solid-State Circuits Conf. (ESSCIRC '02)*, 2002.

[60] K. J. Kulikowski, M. G. Karpovsky, and A. Taubin. Power attacks on secure hardware based on early propagation of data. In *IOLTS*, 131–138. IEEE Computer Society, 2006.

[61] K. J. Kulikowski, A. B. Smirnov, and A. Taubin. Automated design of cryptographic devices resistant to multiple side-channel attacks. In Goubin and Matsui [42], 399–413.

[62] B. W. Lampson. A note on the confinement problem. *Commun. ACM*, 16(10):613–615, 1973.

[63] C. Lattner and V. Adve. LLVM: a compilation framework for lifelong program analysis & transformation. In *Code Generation and Optimization, 2004. CGO 2004. International Symposium on Code Generation and Optimization*, 75–86, 2004.

[64] T.-H. Le, J. Clédière, C. Canovas, B. Robisson, C. Servière, and J.-L. Lacoume. A proposition for correlation power analysis enhancement. In Goubin and Matsui [42], 174–186.

[65] H. Ledig, F. Muller, and F. Valette. Enhancing collision attacks. In Joye and Quisquater [49], 176–190.

[66] H. Li, A. T. Markettos, and S. W. Moore. Security evaluation against electromagnetic analysis at design time. In Rao and Sunar [99], 280–292.

[67] K. J. Lin, S. C. Fang, S. H. Yang, and C. Chia Lo. Overcoming glitches and dissipation timing skews in design of DPA-resistant cryptographic hardware. In R. Lauwereins and J. Madsen, editors, *DATE*, 1265–1270. ACM, 2007.

[68] M. Liskov, R. L. Rivest, and D. Wagner. Tweakable block ciphers. In M. Yung, editor, *CRYPTO*, volume 2442 of *Lecture Notes in Computer Science*, 31–46. Springer, Berlin Heidelberg, 2002.

[69] E. Macii, M. Pedram, and F. Somenzi. High-level power modeling, estimation, and optimization. *IEEE Trans. on CAD of Integrated Circuits and Systems*, 17(11):1061–1079, 1998.

[70] P. Maistri and R. Leveugle. Double-data-rate computation as a countermeasure against fault analysis. *IEEE Trans. Computers*, 57(11):1528–1539, 2008.

[71] H. Mamiya, A. Miyaji, and H. Morimoto. Efficient countermeasures against RPA, DPA, and SPA. In Joye and Quisquater [49], 343–356.

[72] S. Mangard. Hardware countermeasures against DPA? a statistical analysis of their effectiveness. In T. Okamoto, editor, *CT-RSA*, volume 2964 of *Lecture Notes in Computer Science*, 222–235. Springer, Berlin Heidelberg, 2004.

[73] S. Mangard, E. Oswald, and T. Popp. *Power Analysis Attacks: Revealing the Secrets of Smart Cards (Advances in Information Security)*. Springer-Verlag New York, Inc., Secaucus, NJ, USA, 2007.

[74] S. Mangard and K. Schramm. Pinpointing the side-channel leakage of masked AES hardware implementations. In Goubin and Matsui [42], 76–90.

[75] D. Marculescu, R. Marculescu, and M. Pedram. Information theoretic measures for power analysis [logic design]. *IEEE Trans. on CAD of Integrated Circuits and Systems*, 15(6):599–610, 1996.

[76] R. Mayer-Sommer. Smartly analyzing the simplicity and the power of simple power analysis on smartcards. In C. K. Koç and C. Paar [22], 78–92.

[77] F. Menichelli, R. Menicocci, M. Olivieri, and A. Trifiletti. High-level side-channel attack modeling and simulation for security-critical systems on chips. *IEEE Trans. Dependable Sec. Comput.*, 5(3):164–176, 2008.

[78] N. Mentens, B. Gierlichs, and I. Verbauwhede. Power and fault analysis resistance in hardware through dynamic reconfiguration. In Elisabeth Oswald and Pankaj Rohatgi, editors, *CHES*, volume 5154 of *Lecture Notes in Computer Science*, 346–362. Springer, Berlin Heidelberg, 2008.

[79] T. S. Messerges. Using second-order power analysis to attack DPA resistant software. In C. K. Koç and C. Paar [22], 238–251.

[80] T. S. Messerges, E. A. Dabbish, and R. H. Sloan. Investigations of power analysis attacks on smartcards. In *WOST'99: Proceedings of the USENIX Workshop on Smartcard Technology on USENIX Workshop on Smartcard Technology*, 17–17, Berkeley, CA, USA, 1999. USENIX Association.

[81] V. S. Miller. Use of elliptic curves in cryptography. In Hugh C. Williams, editor, *CRYPTO*, volume 218 of *Lecture Notes in Computer Science*, 417–426. Springer, Berlin Heidelberg, 1985.

[82] P. L. Montgomery. Speeding the pollard and elliptic curve methods for factorizations. *Mathematics of Computation*, 48:243–264, 1987.

[83] S. W. Moore, R. J. Anderson, R. D. Mullins, G. S. Taylor, and J. J. A. Fournier. Balanced self-checking asynchronous logic for smart card applications. *Microprocessors and Microsystems*, 27(9):421–430, 2003.

[84] A. Moradi, M. Khatir, M. Salmasizadeh, and M. T. M. Shalmani. Charge recovery logic as a side channel attack countermeasure. In *ISQED*, 686–691. IEEE, 2009.

[85] S. Morioka and A. Satoh. An optimized s-box circuit architecture for low power AES design. In Kaliski Jr. et al. [51], 172–186.

[86] J. M Moya, J. C. Vallejo, D. Fraga, A. Araujo, D. Villanueva, and J.-M. de Goyeneche. Using reputation systems and non-deterministic routing to secure wireless sensor networks. *Sensors*, 9(5):3958–3980, 2009.

[87] R. Muresan and C. H. Gebotys. Current flattening in software and hardware for security applications. In A. Orailoglu, P. H. Chou, P. Eles, and A. Jantsch, editors. *CODES+ISSS*, 218–223. ACM, 2004.

[88] M. Nemani and F. N. Najm. Towards a high-level power estimation capability [digital ICS]. *IEEE Trans. on CAD of Integrated Circuits and Systems*, 15(6):588–598, 1996.

[89] K. Okeya, H. Kurumatani, and K. Sakurai. Elliptic curves with the Montgomery-form and their cryptographic applications. In H. Imai and Y. Zheng, editors, *Public Key Cryptography*, volume 1751 of *Lecture Notes in Computer Science*, 238–257. Springer, Berlin Heidelberg, 2000.

[90] K. Okeya and K. Sakurai. Power analysis breaks elliptic curve cryptosystems even secure against the timing attack. In B. K. Roy and E. Okamoto, eds, *INDOCRYPT*, volume 1977 of *Lecture Notes in Computer Science*, 178–190. Springer, Berlin Heidelberg, 2000.

[91] S. B. Örs, F. K. Gürkaynak, E. Oswald, and B. Preneel. Power-analysis attack on an ASIC AES implementation. In *ITCC (2)*, 546–552. IEEE Computer Society, 2004.

[92] E. Oswald and S. Mangard. Template attacks on masking - resistance is futile. In M. Abe, editor, *CT-RSA*, volume 4377 of *Lecture Notes in Computer Science*, 243–256. Springer, Berlin Heidelberg, 2007.

[93] E. Oswald, S. Mangard, C. Herbst, and S. Tillich. Practical second-order DPA attacks for masked smart card implementations of block ciphers. In D. Pointcheval, editor, *CT-RSA*, volume 3860 of *Lecture Notes in Computer Science*, 192–207. Springer, Berlin Heidelberg, 2006.

[94] P. Paillier and I. Verbauwhede, editors. *Cryptographic Hardware and Embedded Systems - CHES 2007, 9th International Workshop, Vienna, Austria, September 10-13, 2007, Proceedings*, volume 4727 of *Lecture Notes in Computer Science*. Springer, Berlin Heidelberg, 2007.

[95] T. Popp, M. Kirschbaum, T. Zefferer, and S. Mangard. Evaluation of the masked logic style mdpl on a prototype chip. In Paillier and Verbauwhede [94], 81–94.

[96] T. Popp and S. Mangard. Masked dual-rail pre-charge logic: DPA-resistance without routing constraints. In *Cryptographic Hardware and Embedded Systems - CHES 2005, 7th International Workshop*, 172–186. Springer, Berlin Heidelberg, 2005.

[97] N. Pramstaller, F. K. Gürkaynak, S. Häne, H. Kaeslin, N. Felber, and W. Fichtner. Towards an AES crypto-chip resistant to differential power analysis. In C. L. C. M. Steyaert, editor, *Proceedings of the 30th European Solid-State Circuits Conference*, 307–310. IEEE, 2004.

[98] J.-J. Quisquater and D. Samyde. Eddy current for Magnetic Analysis with Active Sensor. In *Esmart 2002, Nice, France*, 9 2002.

[99] J. R. Rao and B. Sunar, editors. *Cryptographic Hardware and Embedded Systems - CHES 2005, 7th International Workshop, Edinburgh, UK, August 29 - September 1, 2005, Proceedings*, volume 3659 of *Lecture Notes in Computer Science*. Springer, Berlin Heidelberg, 2005.

[100] S. Ravi, P. C. Kocher, R. B. Lee, G. McGraw, and A. Raghunathan. Security as a new dimension in embedded system design. In S. Malik, L. Fix, and A. B. Kahng, eds, *DAC*, 753–760. ACM, 2004.

[101] S. Ravi, A. Raghunathan, and S. T. Chakradhar. Tamper resistance mechanisms for secure, embedded systems. In *Proceedings of the 17th International Conference on VLSI Design*, VLSID 204, 605–611, Mumbai, IN. IEEE Computer Society, 2004.

[102] S. Ravi, A. Raghunathan, N. Potlapally, and M. Sankaradass. System design methodologies for a wireless security processing platform. In *Proceedings of the 39th Conference on Design Automation*, 777–782, New Orleans, LA, USA, 2002. ACM.

[103] P. Schaumont and K. Tiri. Masking and dual-rail logic don't add up. In Paillier and Verbauwhede [94], 95–106.

[104] W. Schindler, K. Lemke, and C. Paar. A stochastic model for differential side channel cryptanalysis. In Rao and Sunar [99], 30–46.

[105] K. Schramm, G. Leander, P. Felke, and C. Paar. A collision-attack on AES: Combining side channel- and differential-attack. In Joye and Quisquater [49], 163–175.

[106] K. Schramm, T. J. Wollinger, and C. Paar. A new class of collision attacks and its application to des. In Thomas Johansson, editor, *FSE*, volume 2887 of *Lecture Notes in Computer Science*, 206–222. Springer, Berlin Heidelberg, 2003.

[107] A. Shamir. Protecting smart cards from passive power analysis with detached power supplies. In C. K. Koç and C. Paar [22], 71–77.

[108] A. Shamir and E. Tromer. Acoustic cryptanalysis: on nosy people and noisy machines. [online].

[109] O. Sibert, P. A. Porras, and R. Lindell. An analysis of the intel 80x86 security architecture and implementations. *IEEE Trans. Software Eng.*, 22(5):283–293, 1996.

[110] S. P. Skorobogatov and R. J. Anderson. Optical fault induction attacks. In Kaliski Jr. et al. [51], 2–12.

[111] D. Sokolov, J. Murphy, A. Bystrov, and A. Yakovlev. Design and analysis of dual-rail circuits for security applications. *IEEE Trans. Comput.*, 54(4):449–460, 2005.

[112] D. Sokolov, J. Murphy, A. V. Bystrov, and A. Yakovlev. Improving the security of dual-rail circuits. In Joye and Quisquater [49], 282–297.

[113] F-X. Standaert, E. Peeters, C. Archambeau, and J.-J. Quisquater. Towards security limits in side-channel attacks. In Goubin and Matsui [42], 30–45.

[114] G. E. Suh, D. E. Clarke, B. Gassend, M. van Dijk, and S. Devadas. Aegis: Architecture for tamper-evident and tamper-resistant processing. In U. Banerjee, K. Gallivan, and A. González, editors, *ICS*, 160–171. ACM, 2003.

[115] D. Suzuki and M. Saeki. Security evaluation of DPA countermeasures using dual-rail pre-charge logic style. In Goubin and Matsui [42], 255–269.

[116] D. Suzuki, M. Saeki, and T. Ichikawa. DPA leakage models for CMOS logic circuits. In Rao and Sunar [99], 366–382.

[117] D. Suzuki, M. Saeki, and T. Ichikawa. Random switching logic: A new countermeasure against DPA and second-order DPA at the logic level. *IEICE Trans. Fundam. Electron. Commun. Comput. Sci.*, E90-A(1):160–168, 2007.

[118] K. Tiri. Side-channel attack pitfalls. In *DAC*, 15–20. IEEE, 2007.

[119] K. Tiri, D. Hwang, A. Hodjat, B. -C. Lai, S. Yang, P. Schaumont, and Ingrid Verbauwhede. A side-channel leakage free coprocessor ic in 0.18μm CMOS for embedded AES-based cryptographic and biometric processing. In Joyner Jr. et al. [52], 222–227.

[120] K. Tiri and P. Schaumont. Changing the odds against masked logic. In Eli Biham and Amr M. Youssef, editors, *Selected Areas in Cryptography*, volume 4356 of *Lecture Notes in Computer Science*, 134–146. Springer, Berlin Heidelberg, 2006.

[121] K. Tiri and I. Verbauwhede. Securing encryption algorithms against DPA at the logic level: Next generation smart card technology. In Walter et al. [130], 125–136.

[122] K. Tiri and I. Verbauwhede. A logic level design methodology for a secure DPA resistant ASIC or FPGA implementation. In *DATE '04: Proceedings of the Conference on Design, Automation and Test in Europe*, page 10246, Washington, DC, USA, 2004. IEEE Computer Society.

[123] K. Tiri and I. Verbauwhede. Place and route for secure standard cell design. In *CARDIS, 2004*, 143–158. Kluwer Academic Publishers, Toulouse, FR, 2004.

[124] K. Tiri and I. Verbauwhede. Simulation models for side-channel information leaks. In Joyner Jr. et al. [52], 228–233.

[125] K. Tiri and I. Verbauwhede. A digital design flow for secure integrated circuits. *IEEE Trans. on CAD of Integrated Circuits and Systems*, 25(7):1197–1208, 2006.

[126] V. Tiwari, S. Malik, and A. Wolfe. Power analysis of embedded software: A first step towards software power minimization. *IEEE Trans. VLSI Syst.*, 2(4):437–445, 1994.

[127] C.-C. Tsai and M. Marek-Sadowska. Boolean functions classification via fixed polarity reed-muller forms. *IEEE Trans. Comput.*, 46(2):173–186, 1997.

[128] Y. Tsunoo, T. Saito, T. Suzaki, M. Shigeri, and H. Miyauchi. Cryptanalysis of DES implemented on computers with cache. In Walter et al. [130], 62–76.

[129] J. Waddle and D. Wagner. Towards efficient second-order power analysis. In Joye and Quisquater [49], 1–15.

[130] C. D. Walter, Ç. K. Koç, and C. Paar, editors. *Cryptographic Hardware and Embedded Systems - CHES 2003, 5th International Workshop, Cologne, Germany, September 8-10, 2003, Proceedings*, volume 2779 of *Lecture Notes in Computer Science*. Springer, Berlin Heidelberg, 2003.

[131] N. H. E. Weste, D. Harris, and A. Banerjee. *CMOS VLSI Design: A Circuits and Systems Perspective*. Pearson/Addison-Wesley, 2005.

[132] M. J. Wiener, editor. *Advances in Cryptology - CRYPTO '99, 19th Annual International Cryptology Conference, Santa Barbara, California, USA, August 15-19, 1999, Proceedings*, volume 1666 of *Lecture Notes in Computer Science*. Springer, Berlin Heidelberg, 1999.

[133] S.-M. Yen, W.-C. Lien, S.-J. Moon, and J. C. Ha. Power analysis by exploiting chosen message and internal collisions — vulnerability of checking mechanism for RSA-decryption. In E. Dawson and S. Vaudenay, editors, *Mycrypt*, Volume 3715 of *Lecture Notes in Computer Science*, 183–195. Springer, Berlin Heidelberg, 2005.

Index